Mathematical
TEX by Example

Mathematical TeX by Example

Arvind Borde

ACADEMIC PRESS, INC.
Harcourt Brace Jovanovich, Publishers

Boston   San Diego   New York
London   Sydney   Tokyo   Toronto

ACADEMIC PRESS, INC.
1250 Sixth Avenue, San Diego, CA 92101–4311

Apart from its covers, the book was typeset in TeX by the author, with
assistance from the production department of Academic Press, Boston.
A few examples were typeset with $\mathcal{AMS}$-TeX, a package of commands
distributed by the American Mathematical Society. The PostScript®
examples were produced with dvips, from Radical Eye Software.
The formatting commands for the book are shown in Chapter 6.

The final copy was produced by the American Mathematical Society.
The PostScript examples were typeset on an Agfa/Compugraphic 9600
imagesetter, all other pages on an Autologic APS Micro-5 phototypesetter.

'TeX' and '$\mathcal{AMS}$-TeX' are trademarks of the American Mathematical Society.
'PostScript' is a trademark of Adobe Systems, Inc.

United Kingdom Edition published by
ACADEMIC PRESS LIMITED
24–28 Oval Road, London NW1 7DX

Library of Congress Cataloging-in-Publication Data

Borde, Arvind, date.
   Mathematical TeX by Example / Arvind Borde.
       p.   cm.
   Includes bibliographical references and index.
   ISBN 0-12-117645-2
   1. TeX (Computer file) 2. Computerized typesetting—Computer
programs. 3. Mathematics printing—Computer programs.   I. Title.
Z253.4.T47B67 1992b
686.2'2544—dc20   92-29048
                                                                         CIP

Printed in the United States of America
92 93 94 95 96 97 MV 9 8 7 6 5 4 3 2 1

To the memory of
Professor M. S. Huzurbazar

Within another ten years or so, I expect that the typical typewriter will be replaced by a television screen attached to a keyboard and to a small computer. It will be easy to make changes to a manuscript, to replace all occurrences of one phrase by another and so on, and to transmit the manuscript either to the television screen, or to a printing device, or to another computer. ... It won't be long before these machines change the traditional methods of manuscript preparation in universities and technical laboratories.

DONALD E. KNUTH, *Mathematical Typography*
Josiah Willard Gibbs lecture, given to the American Mathematical Society (1978)

Contents

Preface

T HIS BOOK is aimed at readers who are already broadly familiar with the basic uses of TeX: it seeks to introduce them to a variety of additional tools and techniques—mainly related to typesetting mathematics. Near the start of the book there is a series of reworkings in TeX of samples from standard mathematics books and journals. Some of these have been done with Plain TeX, others with $\mathcal{AMS}$-TeX, others with still other packages. The spirit here is precisely the one-example-is-worth-a-hundred-explanations attitude of my earlier book, *TeX by Example*. It is hoped that these real-world applications will prove a useful source of typesetting ideas.

The examples have been chosen with an eye towards variety. Because of this, they touch on a number of interesting additional issues: how TeX handles languages other than English, how non-standard typefaces may be used with TeX, how pictures may be included in TeX documents. Some readers may, in fact, find these discussions of additional issues to be the main attraction of the book. These issues are not covered in detail, except for a long discussion of typeface selection in TeX. Complete sets of alternate faces for typesetting mathematics with TeX have recently become available, and a systematic and thorough description seemed warranted. The book ends with a long Glossary/Index that covers all Plain TeX and $\mathcal{AMS}$-TeX commands that are likely to be of interest in typesetting mathematics.

Academic Press has been generously supportive throughout my association with them. I particularly thank David Pallai, my publisher, and Jenifer Swetland, my editor, for their original suggestion that I put together this book, and for their patience as it ballooned beyond the original plan for a slim mathematical companion to *TeX by Example*. I also thank Sue Purdy Pelosi, my production editor, for her suggestions and advice. Many other people have helped as well, some by reading parts of the manuscript, others by making suggestions or answering questions. They include Barbara Beeton, Matthew Choptuik, Rosanne Di Stefano, Victor Eijkhout, Michael Ferguson, Tomas Rokicki, Rainer Schöpf, Jim Simone and Michael Spivak. I thank them all. And, of course, I thank Donald Knuth—not only for making it all possible, but also for making it all so enjoyable.

Arvind Borde
Cambridge, Massachusetts
August 1992

But we may conclude, I think, that its matter will limit us somewhat; a work on differential calculus, ... a statesman's speeches, ... a treatise on manures, such books, though they may be handsomely and well printed, would scarcely receive ornament with the same exuberance as a volume of lyrical poems, ...

WILLIAM MORRIS, *The Ideal Book*
A lecture (1893)

1

Introduction

ᴅᴏɴᴀʟᴅ Kɴᴜᴛʜ composed TeX not just for "the creation of beautiful books," but "especially for books that contain a lot of mathematics." He has succeeded remarkably well: the briefest of reflections on the scissors-and-paste manipulations and the inking-in by hand of symbols that had to be undertaken in the old pre-TeX days, or on the clumsily composed formulas that emerged from other systems (and still emerge, even from some contemporary ones), will confirm the truth of this. Without any exaggeration at all, it can be asserted that TeX has wrought a quiet revolution in the way in which scientists and mathematicians are able to present their work, and even in how they communicate with each other.

But progress has its price. The very power of TeX leads its users to expect it to be all programs to all people, and when it isn't, there are grumblings that it is deficient in this way or that. Some of these grumblings are widely aired, and some touch—either directly or peripherally—on the typesetting of mathematics. Addressing such complaints is a secondary aim of this book. A discussion of what are often seen as weaknesses in TeX is particularly important at the present time. Many of the early contributors to TeX have moved on to other work, and even Donald Knuth has announced this as his intention (ᴋɴᴜᴛʜ [1990b]). There is much active debate, therefore, on how best to build on the successes of TeX and on how to maintain various TeX systems in the future. Some of these systems have institutional support—$\mathcal{AMS}$-TeX, for instance. Others are likely to continue to be supported by their creators, or are already officially in other hands—LaTeX, for example, has been handed over by Leslie Lamport to an international committee (which has undertaken a major enhancement to that system).

Since developments are occurring at a rapid clip, it is impossible in a book of this kind to be fully up-to-date. Current information can usually be obtained from the TeX Users Group, or by posting questions on electronic bulletin boards. The inside back cover gives more information about these resources.

1

1.1 TEX Today

TEX is widely used. It not only serves as the basic typesetting mechanism for a number of mathematical and scientific journals, it has also been used to typeset a vast range of other material: everything from chess to Chinese. Designed from the start to be portable, it now also serves as the almost universal language of discourse when technical messages are exchanged by electronic mail. Still, criticisms are often voiced. Here is a list of prominent contemporary complaints, each followed by a few comments:

- *TEX is too powerful.* This is heard most commonly from editors and publishers. TEX has given their authors too much power. They *will* go ahead and do their own, often incompatible, things. They *will* define lots and lots of private new commands, perhaps even redefine some basic commands, then forget to tell anybody (and forget to supply files that contain the new definitions). They *will* fiddle with things like fine spacing, instead of confining themselves to the content of their documents and leaving the detailed adjustments to the professionals.

There is truth to all of these. Authors don't distinguish between privately produced reports like preprints—where they often have to adjust spacing, and so forth, themselves—and articles and books submitted for professional publication. Publishers invariably have their own style guidelines, and it is important (if the efficiencies that TEX promises are indeed to be realized) that authors not introduce commands of their own that might interfere with matters of house style. It is now becoming common—but is perhaps not yet done to a great enough extent—for publishing houses to supply instructions to their authors about what they can and cannot do, along with a style file (see § 1.7) that the authors are required to use. Following these guidelines may make for some tedium for authors, but it would be a lot more tedious for them in the long run if, in a backlash against the indiscriminate use of private commands, publishers stop accepting submissions in TEX altogether.

One of the problems with private commands is the possibility of clashes with others of the same name. There are mechanisms available to reduce the incidence of such clashes; for example, $\mathcal{AMS}$-TEX provides a command called \define (to be used in place of \def) that first checks to see if a command with the chosen name already exists.

There is an extreme point of view that would ban any attempt by users to define their own formats, directing them always to some preexisting and professionally designed package. For documents designed for actual publication, this is a sound course of action. But TEX has a multiplicity of uses and, for at least one very important one, this course isn't possible: the use of TEX as a language of electronic communication. TEX has made it possible for users to transfer documents back and forth, and to print hard copy at any installation at all, confident that it will be identical in appearance in all essential aspects to copies printed elsewhere. What is needed for such uses is the continued availability and use of small TEX packages, ones that travel light.

- *TEX is not powerful enough.* This complaint usually comes from authors, and it is more and more commonly heard. After the initial excitement of seeing TEX's beautifully formed formulas wears off (and the fact that TEX reliably produces such output every time becomes familiar and unremarkable), users sometimes start making comparisons with commercial word processors. Compared to the flashy effects that are often easy to achieve with such programs, the quiet pleasures of TEX—the perfectly placed superscript, the precise spacing around a binary relation—can seem unsatisfying, and the program can begin to appear limiting.

The perception about power is mostly false, though it can sometimes seem indisputably true. In principle, any of the showy effects of a commercial word processor or desktop publishing program can not only be imitated, but far surpassed by TEX and its companion program METAFONT. In practice, however, a lot of the necessary programming remains to be done, or when done, remains to be widely disseminated. Lists of companies and agencies that provide extensions to basic TEX can be obtained from several sources (see the inside back cover).

- *TEX offers only a few typefaces; it doesn't draw pictures.* These are special cases of the *not powerful enough* objection, but are so frequently voiced that they warrant explicit expression. The complaint about pictures is a long-standing one; the complaint about typefaces—and the sameness of appearance of all documents produced with TEX—is more recent.

Though both complaints are true of standard TEX distributions, there are solutions easily within everybody's grasp. For example, many different sets of typefaces may be used with TEX. Examples 7–9 in this book, for instance, have been done with non-standard typefaces and—for those interested in it—Chapter 5 tells the full story behind typeface selection in TEX. There are several sources of new typefaces, public domain and commercial. As for pictures, TEX has a simple mechanism whereby it can import graphics from other sources and, in fact, the program itself can be used to draw simple pictures through packages like PICTEX. Examples 10–13 display several pictures.

- *TEX is difficult to use.* People with mouse-driven word processors might encounter problems by clicking on the wrong thing and making unintended selections—problems usually easily remedied; mean-

while, TₑX users might be staring blankly at
```
Overfull \hbox
```
or even more incomprehendingly at something like
```
! Missing { inserted.
<to be read again>
                         \fam
\rm ->\fam
         \z@ \tenrm
```
on the screen.

There is something to this complaint. For simple applications with simple layouts—preprints, technical reports, even simple books—using TₑX is simplicity itself. For more complicated tasks, standard TₑX packages are difficult to use, and they can quickly lead into a morass of incomprehensible error messages or to wildly unexpected results in the output. Again, the solution is to use a higher-level package of TₑX commands, one that automatically provides the needed features—and several such packages now exist. Most features needed for typesetting mathematics are, for example, provided by $\mathcal{AMS}$-TₑX, $\text{L}\mathcal{AMS}$-TₑX or $\mathcal{AMS}$-LATₑX, and users are urged to investigate them. Higher-level commands may also be directly defined, but those who wish to do so must be prepared to invest a certain amount of initial time and effort. Chapter 6 carries the definitions of several such commands that were used in the making of this book; they can perhaps be used as models.

- *TₑX doesn't distinguish between the logical structure of a document and its visual structure.*

This is partially true (even though the complaint itself will seem incomprehensible to those whose uses of TₑX have so far kept them insulated from issues of design philosophy). The question is discussed in § 1.7.

- *TₑX doesn't cost anything.* This isn't a complaint, exactly, but a source of suspicion among commercial publishers and other business organizations. Used for so long to having to pay good money for nothing, they find it hard to believe that it is possible to pay no money and get something. The makers of computer programs are, after all, allowed some of the quickest conversions of pure thought to pure cash known in human history, and it is hard for a lot of people to accept that a good program—let alone a great one—can be in the public domain.

There isn't a real answer to this, except to say, "Try it and see." But there is also a rational component to this otherwise irrational fear of the free: Who will provide support in case there are problems? Who will answer questions? The solution here is to shop around among vendors of TₑX software—and many exist—to find one with satisfactory support policies. TEX also comes bundled with some advanced small computer systems (usually those in the workstation class), and there is sometimes support from the group that supplied the TEX part of the bundle.

The wary relationship between TEX and its potential users outside the academic world is compounded by a further factor. TEX attracts a large number of highly enthusiastic adherents who have used its programmable aspects over the years to construct a fascinating variety of additions—some of them quite brilliant—often purely for the fun of it. But this very enthusiasm can seem like play from the outside—especially to those for whom somberness is synonymous with seriousness of purpose. Contributors to TEX are usually serious about what they do, but rarely somber.

1.2 The Flavors of TEX

At the core of TEX lies an immensely powerful, highly programmable computer system—a computer language, in some senses—specially directed to the needs of typesetting. Strictly speaking, it is just this core that constitutes TEX; the name, however, is almost always used more loosely to include an additional package (as discussed below). The commands that TEX (the core) comes with— called primitive commands—can be used to create practically any format at all. They are also practically impossible to use directly, except by people with a large amount of time, a large degree of skill and a large sense of humor (since working with the core of TEX can sometimes be like working with a very intelligent but very willful child).

It was never the intention, however, that all users of TEX get down on their hands and knees, logical wrench in hand, and create smoothly running typesetting machines. Rather, it was intended that professional format design- ers, building on the core that TEX provides, construct higher-level structures to meet the needs of users.

There are two widely used higher-level structures. The first is Plain TEX— though most of its users do not know it by that name. The structure is so widely used, and the functions that it performs are mostly so basic, that people usually refer to Plain TEX simply as TEX. In fact, the word "Plain" is conventionally not even capitalized. This book will break with that convention, since it is (or so it seems to the author) important that users realize the relationship between flavors of TEX, a relationship that is obscured when "plain TEX" is confused with a description of TEX rather than being seen as a name.

The other widely used structure, LATEX, is also built (by definition) out of the primitive commands of TEX. Thus, the LATEX-versus-TEX debates that sometimes rage are meaningless: the participants are really comparing LATEX

**The Flavors
of TEX**

with Plain TEX. LATEX does offer considerably more to users, but it also imposes a certain rigidity of style. Among other offerings, it allows a choice—actually, it requires a choice—of document styles, and then provides a great deal of fully automated typesetting to go with that style; it automatically generates, on request, tables of contents; and it provides rudimentary graphical tools and a wider range of typefaces.

Both of these structures—geared as they originally were towards meeting the needs of the authors of technical reports and books, with their usually straightforward layouts—have deficiencies. These deficiencies have become more noticed and more commented on, as users have become more demanding. But some remedies are on hand. Version 3.0 of LATEX, for example, will provide more flexible styles. And, though Plain TEX superstructures are somewhat limited in their range of styles, they at least frequently meet the special needs of special groups of users. $\mathcal{A}_{\mathcal{M}}\mathcal{S}$-TEX, distributed by the American Mathematical Society, is the most well known. Other packages include $\text{L}\mathcal{A}_{\mathcal{M}}\mathcal{S}$-TEX (discussed in some places in this book), TEXsis (a format developed by Eric Myers and Frank Paige) and `eplain` (an extension to Plain TEX developed by Karl Berry). They are all available at TEX archives. There is also an interesting recent 'meta-format' called `lollipop` (EIJKHOUT and LENSTRA [1991]).

Plain TEX, LATEX, and $\mathcal{A}_{\mathcal{M}}\mathcal{S}$-TEX are the most widely used of TEX superstructures. Here is a diagram that roughly illustrates their relationships to each other and to TEX:

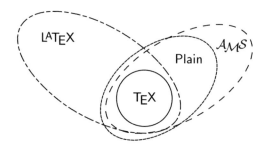

The diagram isn't drawn to any particular scale, but it is meant to make clear that there are several commands (beyond the primitive ones) that are common to all three packages, and that LATEX otherwise goes off in a different direction from the other two. There are also a couple of commands that LATEX and Plain TEX have in common by themselves, and a few unique to Plain TEX. Other than that, Plain TEX is functionally pretty much a subset of $\mathcal{A}_{\mathcal{M}}\mathcal{S}$-TEX.

In this book, "TEX" will usually mean the area common to all these major packages; commands and attitudes particular to a particular package will be labelled as such. It will also be explicitly stated when the name is used in the strict sense, referring only to the primitive commands at the core.

Part of the reason why there is confusion in the minds of users about Plain TeX is that users often never see it directly. There is a file of commands called `plain.tex`, but it is rarely introduced explicitly with an `\input` statement, since that is an inefficient method for a large package. A more efficient method is to create a *format file* that contains pre-processed versions of the commands in `plain.tex`; this may be done with a variant of TeX called IniTeX. The new file has the extension `.fmt` and it is automatically included when the command to run TeX is given, saving users from having to include it explicitly themselves.

IniTeX can be used to create format files for other packages of commands as well. A particular format, say `pformat.fmt`, can be used by saying something like

    tex &pformat

in place of the usual `tex` command (or by defining a new system command that contains `tex &pformat` within it).

1.3 What Is in this Book?

The main focus of the book is the typesetting of mathematics, although several peripheral issues are discussed as well along the way (as in this chapter).

Chapter 2 is the soul of the book: it contains a series of examples illustrating the uses of various flavors of TeX. Left-hand pages contain TeX input, with occasional notes given below, and right-hand pages contain the corresponding output. The examples consist of standard pieces of mathematics taken from a variety of books and journals, going back to the beginning of the century. It is hoped that these applications of TeX to "real life" will be found useful.

This part of the book is similar in style to the corresponding part of *TeX by Example* (BORDE [1992]). There is, however, an important difference. A deliberate attempt was made in that book to keep the examples as simple as possible. This meant that the input was also simple, allowing users to pretty much copy what they needed, just as it appeared there. Output pages here are more complex than those in that book, and the input is also more complex. Commands that appear in the input boxes are often not standard ones at all (see §1.4). In order to use them, their definitions must also be copied. These definitions occur either in the input itself, close to where the command is used, or in one of the formatting files that were put together for the making of this book. The files are reproduced in Chapter 6.

Of course, if the example features a special TeX package like, say, PiCTeX or LAMS-TeX, there is a good chance that a command in the input is not a standard command (it can be looked up in the Glossary to see if it is). In this case, users will have to get hold of the package in order to use the command.

Most of the packages used in this book are in the public domain: information about them is on the inside back cover.

The examples intentionally make many different points. They all show pieces of mathematics (in a few cases, diagrams), but some demonstrate additional effects as well: document formatting, typeface switching, and the like. It is, in fact, under headings corresponding to some of these additional effects that the examples have been grouped, since it is a given that they all illustrate mathematics.

A piece of mathematics selected for display sometimes bears very little connection to the topic of its source: for instance, Example 2 is from a book called *Semi-Riemannian Geometry*, but the pieces of mathematics displayed there show partitioned matrices. This may make the classification chosen for the examples—by source, under broad categories that correspond to additional effects—seem perverse, but the apparently more logical grouping under typographical topic proved impossible. Real-world pieces of mathematics often simultaneously contain so many different typographical features—spacing, the positioning of equation numbers, an unusual alignment, and so on—that it was difficult to decide where to put some of them. The final decision on classification was a compromise, but one made more palatable by including entries in the Glossary for all the topics that might have been used, in a simpler world, to group the examples. The entries point to places in the book where illustrations of the topics are given, and they often briefly discuss the topics on the spot.

No effort was made to find "interesting mathematics" for the examples: the reasons for or against inclusion were purely typographical.

Chapter 3 gives a quick summary of the principal features of $\mathcal{AMS}$-TEX. The chapter provides only an overview to go along with the uses of $\mathcal{AMS}$-TEX in Chapter 2. Detailed information on $\mathcal{AMS}$-TEX may be found in Michael Spivak's excellent manual, *The Joy of TEX* (SPIVAK [1982, 1990]).

Chapter 4 gives a very brief summary of $\mathcal{LAMS}$-TEX and $\mathcal{AMS}$-LATEX. The intention here is merely to let users know what is out there, not to do the packages justice.

Chapter 5 gives a detailed and systematic account of how TEX makes typeface decisions and how users can impose their own choices. TEX users have been timid in the past about exploring typeface lifts for their documents, but that now appears to be changing as more and more of them realize that it is indeed possible to enjoy the full power of TEX without also having to adopt its regulation appearance. An attempt has been made in Chapter 5 to tell the full story, starting at the beginning and omitting nothing. On the one hand, the discussion has been designed to be accessible even to somebody who has used TEX just once or twice and knows that {\bf bold type} will give **bold type**, {\it italic type} will give *italic type*, {\sl slanted type} will give *slanted type*, and {\tt typewriter type} will give `typewriter type`, but nothing else. On the other hand, no details have been left out, so this discussion should also serve as a useful reference for experienced users.

The material in Chapter 5 should allow users to make practically arbitrary typeface changes—as long as the typefaces they want are indeed available on their computer systems.

Chapter 6 reproduces the definitions of most of the new commands that are introduced in this book.

These chapters are followed by a Bibliography. References in the book (e.g., ZAPF [1987]) are to sources listed here.

The book ends with a long Glossary/Index that explicitly lists all the commands of Plain TeX and $\mathcal{A}\mathcal{M}\mathcal{S}$-TeX that are useful in typesetting mathematics. It also lists and discusses (from the typesetting point of view) a large number of mathematical topics, e.g., continued fractions or integrals.

1.4 Conventions

In the input for Examples 1 to 9, commands whose names are entirely in lower case are Plain TeX commands. They may be used freely. Examples 10 to 18 cover specialized packages, and the commands used there are often a mix of ones from the package and from Plain TeX.

Commands that contain capital letters in their names are usually special commands defined for this book. The definitions are all given in Chapter 6, and the definitions must first be copied if the commands are to be used. Occasionally, an $\mathcal{A}\mathcal{M}\mathcal{S}$-TeX or a L$\mathcal{A}\mathcal{M}\mathcal{S}$-TeX command will also contain capital letters. It will be clearly indicated when that is the case.

Terms that appear in sans serif type in the text are listed in the Glossary. In order to avoid excessive typographical distraction, such type has not been used promiscuously, only when it may indeed be worthwhile to look up more information.

The next few lines are aimed at experts; normal readers may safely skip to the next paragraph. Most of TeX's specialized expressions will be deliberately avoided throughout this book in the interest of making it accessible even to casual users who—like it or not—will want to browse. The expression "control sequence" will not be used at all—in most cases the word "command" will be used instead—and the word token will be used only when absolutely necessary.

A command that is reproduced in the middle of text will usually just be reproduced straight, as it would appear in the input: \hfil, \vfil, etc. If, however, there appears to be the possibility of misunderstanding what is part of the command and what is surrounding text, single quotes will be used around the command: for example, '\hskip 1in', or '\.'.

1.5 Assumptions

This is not an introductory book. It will be assumed that readers are broadly familiar with most of the following commands:

```
\it, \bf, \rm, \tt, \sl
\centerline, \rightline, \leftline, \line
\hskip, \quad, \qquad
\baselineskip, \vskip, \smallskip, \medskip, \bigskip
\hfil, \hfill, \vfil, \vfill, \break, \eject
\item, \itemitem, \noindent, \parindent, \parskip
\hsize, \vsize, \hoffset, \voffset
\halign, \settabs
\hbox, \vbox
\def, \let, \end, \bye
```

The commands are all briefly explained in the Glossary. Some of them may be discussed at greater length in places, but readers should view such occurrences as pieces of luck, not as a right. If they need longer explanations, they will have to turn to an introductory book.

It helps in understanding the input for the examples if readers are familiar with the uses of various mathematics spacing commands. Here is a summary:

$\!$ gives a negative thin space; for example, $[\!]$ gives [].

$\,$ gives a thin space; for example, $[\,]$ gives [].

$\>$ gives a medium space; for example, $[\>]$ gives [].

$\;$ gives a thick space; for example, $[\;]$ gives [].

It also helps to be aware that standard text spacing commands can continue to be used within formulas: $[\ ]$—i.e., one input space after the '\'—gives [ ], $[\quad]$ gives [   ] and $[\qquad]$ gives [   ]. A final non-obvious command is the use of '~' as a tie: A~B will give a single space between 'A' and 'B', and a line break will not be allowed there. Such ties are typically used to block awkward line breaks when one says things like Theorem~1.

Another piece of background knowledge that will be mostly taken for granted is the role played by the "command characters":

$$\backslash \quad \{ \quad \} \quad \$ \quad \& \quad \# \quad \hat{} \quad \_ \quad \sim \quad \%$$

These characters are usually interpreted specially by TeX, and it helps to know what their uses are; for example, { and } are used to **group** portions of input text (often to confine the effects of certain commands), $ signals the start and the end of a formula and % is used to make comments.

1.6 Typesetting Mathematics

The typesetting of mathematics has traditionally placed extra burdens on compositors and printers. Indeed, in some cases, mathematical notation has been driven by what could or could not easily be printed. It is possible (but is as yet an unrealized bonus) that the enormous flexibility of a system like TEX will free mathematicians from practical constraints and will allow them to develop new notations that even better express their ideas.[†]

Using TEX to typeset mathematics comes with some special problems of its own. The input is a one-dimensional sequence of typesetting instructions that gets translated (superbly) into the essentially two-dimensional structure of a mathematical formula. But this makes input and output quite unlike each other, and errors can be hard to track down. The examples in this book will provide illustrations of how input can be kept fairly simple and fairly readable. They will also supply hints on how the appearance of the output can be improved. The chapters on typesetting mathematics in *The TEXbook* (KNUTH [1984, 1986]), as well as many parts of *The Joy of TEX* (SPIVAK [1982, 1990]), also provide a variety of hints and tips.

1.7 Logical and Visual Design

Commands like \beginsection (Plain TEX), \chapter (LATEX), or \abstract (AMS-TEX) are clearly different in spirit from those like \hfil or \bigskip. The commands in the first group say something about the logical organization of the document; those in the second are intended only to have some specific visual effect.

It has been argued—within the TEX community most forcefully by Leslie Lamport (see LAMPORT [1987])—that a document should contain only logical commands, not visual ones. The point of view also underlies a recent interest in the notion of *generalized markup*. "Markup" is an old typesetting term: it refers to marking up a manuscript with instructions for the printer. The instructions will indicate what typefaces are to be used at various places, where extra space should be left, etc. "Generalized markup" refers to marking up a document, but with indications of its logical structure: Which of its elements are titles? Where do paragraphs begin and end? Where do equations begin and end? The concept has received a boost in recent years with the increasingly broad acceptance of a single standard for generalized markup, called the Standard Generalized Markup Language (SGML, for short).

[†] This hope was expressed to the author by Robert Melter.

**Logical
and Visual
Design**

Although the approach sometimes attracts zealots, there are significant advantages to using generalized markup, or logical (as opposed to visual) formatting. Not only does this approach lay bare the logical structure of a document, it actually makes the final, physical formatting a lot easier. Consider, for example, a simple document divided into sections, each with a title. Though it may seem a roundabout approach, things are made a lot easier, eventually, if every section is introduced by a made-up command called, say, \NewSection. Thus a section called "Introduction" would begin, for example, with something like

\NewSection{Introduction}

The computer file containing the document would then consist of paragraphs separated by blank lines and grouped into sections by the use of the command \NewSection. Once the file is ready, and the content of the document taken care of, the author—or, preferably, a document designer—can think about its appearance. If, say, it is decided that all section titles are to appear flush left in boldface, with a \bigskip worth of space above them and a \medskip below, \NewSection can be defined as

\def\NewSection #1{\bigskip \leftline{\bf#1} \medskip}

The advantage of this approach should be clear: global changes are easy to make (for example, the titles can all be centered just by changing \leftline to \centerline in the definition), refinements can be added at will (indentation can be suppressed in the paragraph that follows the title, page breaks can be encouraged above the title and prohibited below it, and so forth), and the final appearance of the document has a consistency that it usually lacks if explicit visual formatting is indiscriminately injected throughout its body.

LaTeX tries to enforce this approach, with mixed success. The attitude also underlies several $\mathcal{AMS}$-TEX commands: abstracts, the addresses of authors, etc., are tagged by suitably named commands. When producing preprints the commands are interpreted in one way (the definitions are all stored in a style file that is used as input at the top of the document), and when the document is printed in a book or a journal, often in another (a different style file is used). Sticking to commands that come with style files and avoiding, as far as possible, explicit formatting commands of one's own can go a long way towards resolving the first complaint of § 1.1.

Even with the best of intentions, a TEX document will contain a mixture of logical and visual commands, as the input shown in the examples in this book reveals. In particular, it is hard to properly typeset mathematics without adjusting things occasionally by hand. An effort to at least minimize explicitly visual formatting does, however, usually pay off.

Introduction §1.8 | 13

The @tt@ck
of the Killer
@ Signs

1.8 The @tt@ck of the Killer @ Signs

Anybody who has tried to read a large file of TeX commands will have encountered puzzling commands like \z@ or \m@th or \sixt@@n. Sometimes such commands pop up even in error messages. What do they mean? What possible end can they serve?

In all widely used flavors of TeX, command names are either permitted to consist of only a single character, in which case any character at all may be used, or they are required to consist only of letters of the alphabet. Thus, \%, \- and \7 are all acceptable as single-character command names, and \name, \Command and so forth as names with many characters. But, say, \Command_name and \Name-of-command are normally not accepted. (This glosses over some subtleties, but it is broadly true.)

As far as TeX is concerned, a "letter" is any character that is categorized as such. Character categories are discussed in §5.2; readers who need to know more about the notion may want to look there first (since the discussion given below is a little sketchy).

There are 16 categories of characters in TeX, numbered from 0 to 15; category number 11 corresponds to letters. Normally, just the 52 letters A, ..., z are placed in this category, but in principle any character at all may be put there. Categories are assigned through a command called \catcode.

When large packages of TeX commands are written, it usually happens that some commands are used purely internally. In order to prevent users from inadvertently redefining them, the following device is used: a non-letter character is chosen and temporarily assigned to category 11. This character is used in the names of internal commands, then replaced in its original category at the end. The device makes it impossible for regular users to accidentally alter the inner workings of the package.

The most common choice for such a safe-guard character is @. It normally belongs to category 12 (called "other", for characters that don't fit anywhere else). The Plain TeX file, for example, says \catcode'\@=11 towards the start, uses @ in the names of several commands, then replaces @ in its original category by saying \catcode'\@=12 at the end. LaTeX goes a step further and defines the commands \makeatletter and \makeatother; the first places @ in category 11 and the second replaces it in category 12.

Other characters are sometimes used for the same purpose. PiCTeX, for instance, uses !. Also, since the device, as it is normally employed, makes commands s@mewh@t unre@d@ble, there are moves to urge the writers of TeX packages to name their private commands thus: \@sixteen, \@math, etc. This will have the same protective property as before, but it will make the command names easier to read.

```
 1  \input OutForm.tex       % Input the 'output format' file.
 2  \input BookForm.tex      % Input the 'output format' file.
 3
 4  % Formatting commands for the full right-hand pages:
 5  \BookTitle={Transactions of the American Mathematical Society}
 6  \RealPno=15
 7
 8  \Title 2/Examples
 9
10  \ExampleType={Plain \TeX}%
11  \NewExample
12  Eight decades of mathematics from the {\it AMS Transactions\/} (1900--1977)
13  are shown here, redone with \TeX. The contents of the sample pages
14  are mild (typographically speaking), but they are useful
15  nevertheless as reminders of the basic ways in which standard mathematical
16  expressions can be typeset with \TeX.
17        ....................................................................
18  For the most part it was possible to get \TeX\ to imitate the original sources
19  reasonably closely. Still, some liberties were taken: a fixed page size was
20  used for all the examples, despite slight variations in the originals; no
21  attempt was made to match all the typeface sizes exactly; and in the case of
22  Example~1i, the original has so spindly an appearance (according to Knuth,
23  the {\sl AMS}, finding the costs of traditional typesetting too high, had
24  resorted for a while to a ''form of typewriter composition'') that the
25  attempt to match it was abandoned.
26  \EndPage
27
28  % Formatting commands for the 'output' part of the right hand pages:
29  \SmallFootnotes      % Sets small type for footnotes.
30  \hsize 4.8125in      % Sets the horizontal page size.
31  \vsize 7.3125in      % Sets the vertical page size.
32  \parindent15pt       % Sets the paragraph indentation.
33  \def\:{\thinspace } % Used at a few points to save typing.
34  % The commands '\Lcase', '\Eightrm' and '\Smallcaps' that are used below are
35  % not standard; the definitions are in Chapter 6.
36  \def\Header #1#2{\headline={\ifodd\pageno
37      \rlap{\Eightrm #1}\hfil{\Smallcaps \Lcase{#2}}\hfil\llap{\folio} \else
38      \rlap{\folio}\hfil{\Smallcaps \Lcase{#2}}\hfil\llap{\Eightrm #1} \fi}}
39
```

NOTES

1–2: Two files are used as input at the start of this file of examples. The file `OutForm.tex` is needed for all the examples: it describes the overall format of the full output (right-hand) pages. The second file, `BookForm.tex`, is used for the internal format of those examples that imitate the page layout of books and journals. These files are set up to allow the page-within-a-page structure of the right-hand pages to succeed: standard Plain TEX commands like \hsize, \vsize, \headline, \pageno, and so forth, continue to have their customary effects on the material within the examples, but do not affect the overall page format. New commands, like \BookTitle, \RealPno, etc., made available through the `OutForm` file, allow control over the full page. All the non-standard commands are partially capitalized to allow them to be kept separate from the standard ones. Their functions should be obvious from their names; the definitions are given in Chapter 6.

5, 6, 8, 10, 11: These are all non-standard commands; \RealPno, for instance, sets the real page number for the right hand pages (the page number from the source is set with \pageno).

26: An abbreviation for \hfil\break.

29: A non-standard command that forces footnotes to appear in 8-point type.

30–32: These set the page for the actual examples.

36–38: A prototype command that allows the headlines on right- and left-hand pages to differ. Further examples are given in Chapter 6 and \Lcase is also defined there.

2

Examples

Plain TₑX

———————— **Example 1** ————————

From *Transactions of the American Mathematical Society*

Eight decades of mathematics from the *AMS Transactions* (1900–1977) are shown here, redone with TₑX. The contents of the sample pages are mild (typographically speaking), but they are useful nevertheless as reminders of the basic ways in which standard mathematical expressions can be typeset with TₑX.

The pages shown have an historical significance: it was with a discussion of samples from precisely these pages that Donald Knuth began his 1978 Gibbs lecture on Mathematical Typography to the American Mathematical Society (KNUTH [1979a]), a lecture in which he discussed a new computer typesetting system that he was developing, at that time variously called Tau Epsilon Chi, or "the uppercase form of $\tau\epsilon\chi$," TₑX.

Readers with an interest in typesetting are urged to look at Knuth's lecture and at the original pages on which this example is based, not only to observe the changes in typesetting styles that have occurred over this century, but also to see how TₑX's output compares with these samples from the past.

For the most part it was possible to get TₑX to imitate the original sources reasonably closely. Still, some liberties were taken: a fixed page size was used for all the examples, despite slight variations in the originals; no attempt was made to match all the typeface sizes exactly; and in the case of Example 1i, the original has so spindly an appearance (according to Knuth, the *AMS*, finding the costs of traditional typesetting too high, had resorted for a while to a "form of typewriter composition") that the attempt to match it was abandoned.

The sample pages in this example have been produced with the permission of the American Mathematical Society, the copyright holder.

```
1   \Example={1a}     \Volume 1, 2 (1900) % These place stuff in the right margin.
2   \Header{[January}{H. S. White: Conics and Curves} % Defined on page 14.
3   \pageno=2
4
5   \noindent If this is equated to zero it denotes in variables $x_1$, $x_2$,
6   $x_3$, the poloconic of the line
7   $$ u_1x_1 + u_2x_2 + u_3x_3 = 0. $$
8   Or if $u_1$, $u_2$, $u_3$ be taken as variables it is the tangential
9   equation of the conic polar of a point ($x$). If now the variables ($u$) are
10  replaced by symbols of any ternary $n$-ic $A_x^n$, we shall say that the
11  equation
12  $$ (aa'A)^2a_xa'_xA_x^{n-2} = 0 \leqno(1) $$
13  denotes the poloconic, polocubic, or polo-n-ic of the curve $A_x^n=0$ with
14  respect to the cubic $a_x^3=0$. {\Smallcaps Hilbert} proposes to determine
15  those conics, if any, which coincide with their poloconics. This requires
16  the determination of a value $\lambda$ such that
17  $$ (aa'A)^2a_xa'_x\equiv \lambda\!\cdot\!A_x^2. \leqno(2) $$
18  It is found that the six values of $\lambda$ resulting from elimination
19  of the unknown $A$'s coincide by threes. Hence to the two values
20  $$ \lambda = \pm\sqrt{\textstyle{1\over 6}S} = \pm\sqrt{{\textstyle{1\over 6}}
21      (aa'a'')(aa'a''')(aa''a''')(a'a''a''')} $$
22  there correspond two quadratic forms each containing linearly three arbitrary
23  parameters. So much {\Smallcaps Hilbert} states. In order to recognize these
24  autopoloconics as known systems it will be convenient to use a canonical form
25  of the fundamental cubic, due to {\Smallcaps Hesse}.\footnote{$^\ast$}
26  {C\:r\:e\:l\:l\:e\:'\:s\:\ J\:o\:u\:r\:n\:a\:l, vol.\ 28, pp. 68--96, 1844.}
27
28  Referred to an inflexional triangle, the equation of the cubic takes the form
29  $$ a_x^3 = x_1^3 + x_2^3 + x_3^3 + 6mx_1x_2x_3 = 0. \leqno(3) $$
30  All conic polars accordingly have the form:
31  $$ a_ya_x^2 = (y_1x_1^2 + y_2x_2^2 + y_3x_3^2) + 2m(y_1x_2x_3 + y_2x_3x_1
32      + y_3x_1x_2) = 0. \leqno(4) $$
33  With respect to this triangle, the quantity $m$ may be termed
34  indifferently the parameter of the cubic or the parameter of any one of its
35  conic polars. The defining equation of the autopoloconics may be written with
36  the aid of a bordered Hessian, as follows:
37  $$ 2\left\vert\matrix{\hfill x_1 & \hfill mx_3 & \hfill mx_2 & A_1 \cr
38                        \noalign{\vskip2pt}
39                        \hfill mx_3 & \hfill x_2 & \hfill mx_1 & A_2 \cr
40                        \noalign{\vskip2pt}
41                        \hfill mx_2 & \hfill mx_1 & \hfill x_3 & A_3 \cr
42                        \noalign{\vskip2pt}
43                        \hfill A_1 & \hfill A_2 & \hfill A_3 & 0 \hfill \cr}
44      \right\vert \equiv \lambda A_x^2. \leqno(5) $$
45  Expanded, this identity gives three pairs of equations of conditions like the
46  following,
47  \EndPage
```

NOTES

This is a fairly straightforward page of mixed text and mathematics.

1: These commands store information given to them. They are made to divulge the information in the file `OutForm.tex` when the right margin of the full page is being put together.

2: This shows a use of the `\Header` command defined on the previous page.

14, 23, 25: `\Smallcaps` is non-standard; it gives small capitals.

26: `\:` is an abbreviation; see the previous page.

38, etc.: This is one way to get more space between rows when using the `\matrix` command.

Source:

*Transactions
of the
American
Mathematical
Society*

1, 2 (1900)

If this is equated to zero it denotes in variables x_1, x_2, x_3, the poloconic of the line

$$u_1 x_1 + u_2 x_2 + u_3 x_3 = 0.$$

Or if u_1, u_2, u_3 be taken as variables it is the tangential equation of the conic polar of a point (x). If now the variables (u) are replaced by symbols of any ternary n-ic A_x^n, we shall say that the equation

(1)
$$(aa'A)^2 a_x a_x' A_x^{n-2} = 0$$

denotes the poloconic, polocubic, or polo-n-ic of the curve $A_x^n = 0$ with respect to the cubic $a_x^3 = 0$. HILBERT proposes to determine those conics, if any, which coincide with their poloconics. This requires the determination of a value λ such that

(2)
$$(aa'A)^2 a_x a_x' \equiv \lambda \cdot A_x^2.$$

It is found that the six values of λ resulting from elimination of the unknown A's coincide by threes. Hence to the two values

$$\lambda = \pm\sqrt{\tfrac{1}{6}S} = \pm\sqrt{\tfrac{1}{6}(aa'a'')(aa'a''')(aa''a''')(a'a''a''')}$$

there correspond two quadratic forms each containing linearly three arbitrary parameters. So much HILBERT states. In order to recognize these autopoloconics as known systems it will be convenient to use a canonical form of the fundamental cubic, due to HESSE.[*]

Referred to an inflexional triangle, the equation of the cubic takes the form

(3)
$$a_x^3 = x_1^3 + x_2^3 + x_3^3 + 6m x_1 x_2 x_3 = 0.$$

All conic polars accordingly have the form:

(4)　$$a_y a_x^2 = (y_1 x_1^2 + y_2 x_2^2 + y_3 x_3^2) + 2m(y_1 x_2 x_3 + y_2 x_3 x_1 + y_3 x_1 x_2) = 0.$$

With respect to this triangle, the quantity m may be termed indifferently the parameter of the cubic or the parameter of any one of its conic polars. The defining equation of the autopoloconics may be written with the aid of a bordered Hessian, as follows:

(5)
$$2\begin{vmatrix} x_1 & mx_3 & mx_2 & A_1 \\ mx_3 & x_2 & mx_1 & A_2 \\ mx_2 & mx_1 & x_3 & A_3 \\ A_1 & A_2 & A_3 & 0 \end{vmatrix} \equiv \lambda A_x^2.$$

Expanded, this identity gives three pairs of equations of conditions like the following,

[*] Crelle's Journal, vol. 28, pp. 68–96, 1844.

```
1   \Example={1b}  \Volume 13, 135 (1912) % To place stuff in the right margin.
2   \Header{1912}}{Weierstrass' Preparation Theorem} % Defined on page 14.
3   \pageno=135
4
5   \noindent ditions on the coefficients of the $F_\alpha$ essential to the proof
6   of the theorem are that the series $F_\alpha$ converge, and that the resultant
7   $R$ of the polynomials $f_\alpha$ shall be different from zero.
8
9   {\Smallcaps Kistler}\footnote{$^\ast$}{{\it \"Uber Funktionen von mehreren
10  komplexen Ver\"anderlichen}, Dissertation, G\"ottingen (1905).} has considered
11  the solutions of any set of equations
12  $$\displaylines{\hfill F_\alpha(x_1, x_2, \cdots, x_m) = 0 \hfill
13  \llap{$(\alpha=1,2,\cdots,n)$}} $$
14  in which the functions $F$ are analytic. By an extension of Kronecker's
15  algorithm for the case when the functions $F$ are polynomials, he shows that
16  the solutions of such a system can be classified into
17  ``\:Mannigfaltigkeiten\:'' of dimensions greater than or equal to $m-n$.
18  Furthermore each ``\:Mannigfaltigkeit\:'' separates into a number of
19  ``\:analytische Gebilde\:'' for which the variables $x_1$, $x_2,\cdots,x_m$
20  are expressible as analytic functions of $k$ parameters $u$, where $k$ is the
21  dimension of the Mannigfaltigkeit. The theorem of the present paper has a
22  relationship to these results which is similar to Kronecker's algorithm in
23  algebra.
24
25  In the first volume of his {\it M\'ecanique C\'eleste\/} (page~72)
26  {\Smallcaps Poincar\'e} makes the statement that the solutions of a system of
27  equations (3), at least so far as the solutions in the neighborhood of the
28  origin are concerned, is equivalent to the solution of a system
29  $$\displaylines{\rlap{(5)} \hfill \varphi_\alpha(x_1, x_2, \cdots, x_m :
30  y_1, y_2, \cdots, y_n) = 0 \hfill \llap{$(\alpha=1,2,\cdots,n),$}} $$
31  in which the $\varphi_\alpha$ are polynomials in $y_1$, $y_2,\cdots,y_n$. The
32  proof of this theorem has been considered recently by {\Smallcaps W. D.
33  MacMillan}. By applying the usual algebraic elimination theory to the
34  polynomials $\varphi_\alpha$, a polynomial $\Phi (x_1,x_2,\cdots,x_m:y_n)$
35  would be found for which the lowest term in $\Phi (0,0,\cdots,0:y_n)$ would
36  be, say, of degree $\rho$. By applying the preparation theorem to
37  $\Phi (x_1,x_2,\cdots,x_m:y_n)$, therefore, a polynomial $\Psi$ of type (1)
38  of degree $\rho$ would appear. This is not in general the polynomial whose
39  existence is sought in this paper, as may readily be shown by means of an
40  example. The polynomials $\varphi_\alpha$ may have roots for which $y_n$ is
41  near to zero, while the values of $y_1$, $y_2\cdots,y_{n-1}$ are not small.
42  Such values of $y_n$ must appear as roots of the polynomial $\Psi$, and its
43  degree $\rho$ is therefore in general too great.\footnote{\dag}{The
44  polynomials
45  $$ y^2 + z^2 - 2y - 2z + x = 0,\quad 2y^2 - z^2 - 4y + 2z = 0 $$
46  have only one solution for which $y$ and $z$ are power series in $x$
47  vanishing with $x$. The resultant of these two equations after the
48  elimination of $y$ is, however,
49  $$ 9z^4 - 36z^3 + 12(x+3)z^2 - 24xz + 4x^2. $$
50  If the preparation theorem is applied to this, a polynomial in $z$ of
51  degree~2 is found, one of the roots of which must therefore correspond to a
52  solution $(y,z)$ for which $y$ is not in the neighborhood of the origin. The
53  same is true of the polynomial in $y$ found by eliminating $z$.} \EndPage
```

NOTES

17, 18, 19: \: gives a small space (see page 14); it is used here to imitate the style in the source.
29–30: \rlap and \llap are right- and left-overlap commands. They allow the main equation to be centered horizontally, regardless of the width of the material at the ends of the line.

| 19

Example 1b

Source:
*Transactions
of the
American
Mathematical
Society*
13, 135 (1912)

ditions on the coefficients of the F_α essential to the proof of the theorem are that the series F_α converge, and that the resultant R of the polynomials f_α shall be different from zero.

KISTLER* has considered the solutions of any set of equations

$$F_\alpha(x_1, x_2, \cdots, x_m) = 0 \qquad (\alpha = 1, 2, \cdots, n)$$

in which the functions F are analytic. By an extension of Kronecker's algorithm for the case when the functions F are polynomials, he shows that the solutions of such a system can be classified into "Mannigfaltigkeiten" of dimensions greater than or equal to $m - n$. Furthermore each "Mannigfaltigkeit" separates into a number of "analytische Gebilde" for which the variables $x_1, x_2, \cdots, x_m$ are expressible as analytic functions of k parameters u, where k is the dimension of the Mannigfaltigkeit. The theorem of the present paper has a relationship to these results which is similar to Kronecker's algorithm in algebra.

In the first volume of his *Mécanique Céleste* (page 72) POINCARÉ makes the statement that the solutions of a system of equations (3), at least so far as the solutions in the neighborhood of the origin are concerned, is equivalent to the solution of a system

$$(5) \qquad \varphi_\alpha(x_1, x_2, \cdots, x_m : y_1, y_2, \cdots, y_n) = 0 \qquad (\alpha = 1, 2, \cdots, n),$$

in which the φ_α are polynomials in $y_1, y_2, \cdots, y_n$. The proof of this theorem has been considered recently by W. D. MACMILLAN. By applying the usual algebraic elimination theory to the polynomials φ_α, a polynomial $\Phi(x_1, x_2, \cdots, x_m : y_n)$ would be found for which the lowest term in $\Phi(0, 0, \cdots, 0 : y_n)$ would be, say, of degree ρ. By applying the preparation theorem to $\Phi(x_1, x_2, \cdots, x_m : y_n)$, therefore, a polynomial Ψ of type (1) of degree ρ would appear. This is not in general the polynomial whose existence is sought in this paper, as may readily be shown by means of an example. The polynomials φ_α may have roots for which y_n is near to zero, while the values of $y_1, y_2 \cdots, y_{n-1}$ are not small. Such values of y_n must appear as roots of the polynomial Ψ, and its degree ρ is therefore in general too great.†

---

* *Über Funktionen von mehreren komplexen Veränderlichen*, Dissertation, Göttingen (1905).

† The polynomials

$$y^2 + z^2 - 2y - 2z + x = 0, \quad 2y^2 - z^2 - 4y + 2z = 0$$

have only one solution for which y and z are power series in x vanishing with x. The resultant of these two equations after the elimination of y is, however,

$$9z^4 - 36z^3 + 12(x+3)z^2 - 24xz + 4x^2.$$

If the preparation theorem is applied to this, a polynomial in z of degree 2 is found, one of the roots of which must therefore correspond to a solution (y, z) for which y is not in the neighborhood of the origin. The same is true of the polynomial in y found by eliminating z.

```
 1 \Example={1c}   \Volume 23, 216 (1922) % To place stuff in the right margin.
 2 \Header{[March]}{Karl Schmidt} % Defined on page 14.
 3 \pageno=216
 4 \def\B #1{\bar{\vphantom{b}#1}}
 5 \def\AB {a\B b + \B ab} % To save typing a frequently occurring expr.
 6 \def\<{\mathrel{<\kern-.2pt\llap{$-\,$}}}
 7
 8 \noindent where $v$ may take any value between $Z$ and $U$. By this
 9 representation every effective value of $x$ is split into two parts, namely
10 $ab$ and $v(\AB)$. It is the $ab$ part which is ineffective, i.e., it
11 contributes nothing to the value of $f(x)$; or
12 $$ f[ab + v(\AB)] = f[v(\AB)]. $$
13 For
14 $$\eqalign{
15 f[ab+v(\AB)] = a[ab+v(\AB)] + b[(\B a+\B b)(\B v + ab + \B a\B b)] \cr
16 = ab + va\B b + \B ab\B v = a(b+v) + \B ab\B v = b(a+\B v) + av,\Hfil \cr}
17 \leqno(1) $$
18 and
19 $$\eqalign{
20 f[v(\AB)] = va\B b + b(\B v+ab+\B a\B b) = va\B b + b\B v + ab \cr
21 = a(b+v) + b\B v = b(a+\B v) + av.\Hfil \cr} \leqno(2) $$
22
23 I call this ineffective part of $x_e$ ``{\it innocuous\/}'' to indicate that
24 it does not invalidate the fundamental proposition
25 $$ [f(x'_e \ne f(x''_e)] = (x'_e \ne x''_e) $$
26 which was proved above (P. 4) for effective values of $x$. The reason why
27 this ineffective part of $x_e$ is innocuous is clear: it, {\it as a whole}, is
28 part of every $x_e$, so that the variation of $x_e$ does not take place in it
29 at all.
30
31 {\bf D. 3.} But this consideration leads to the {\it definition of the
32 wholly-effective range of x}. By this I mean the collection of values which
33 lies between $Z$ and $\AB$, i.e.,
34 $$ Z \< x_\omega \< \AB, $$
35 where $x_\omega$ designates the wholly-effective values of $x$. It is obvious
36 that the wholly-effective range of $x$ is obtained by omitting the ineffective
37 part of $ab$ from the effective range.
38
39 We can write $x_\omega$ in the form:
40 $$ x_\omega = v(\AB) $$
41 where
42 $$ Z \< v \< U. $$
43
44 If we choose the regional interpretation  of our Boolean entities, which in
45 the present paper is the preferred interpretation, then the wholly-effective
46 range is, in general, represented by a disconnected region. A value in this
47 range can be expressed thus:
48 $$ (x_\omega = x_1 + x_2)(x_1 \< ab)(x_2 \< \B ab). $$
49 \EndPage
```

NOTES

4, 5: Abbreviations that are found convenient here.
6: Construction of a needed symbol by superposing '<' and '−'. The −.2pt spacing adjustment was arrived at after doing trial runs. \mathrel tells TeX that the new symbol will be used as a relation: the program will now use suitable spacing around it. The symbol will work without that statement, but the spacing will be terrible.

16, 21: \Hfil is defined in BookForm.tex; it is like \hfil, but it works inside \eqalign (whereas \hfil does not). It was needed in order to center the second line of the alignments. (\displaylines wasn't used here because the equation number had to be attached to the whole display.)

| | 21

Example 1c

Source:

Transactions of the American Mathematical Society

23, 216 (1922)

where v may take any value between Z and U. By this representation every effective value of x is split into two parts, namely ab and $v(a\bar{b} + \bar{a}b)$. It is the ab part which is ineffective, i.e., it contributes nothing to the value of $f(x)$; or

$$f[ab + v(a\bar{b} + \bar{a}b)] = f[v(a\bar{b} + \bar{a}b)].$$

For

(1)
$$\begin{aligned} f[ab + v(a\bar{b} + \bar{a}b)] &= a[ab + v(a\bar{b} + \bar{a}b)] + b[(\bar{a} + \bar{b})(\bar{v} + ab + \bar{a}\bar{b})] \\ &= ab + va\bar{b} + \bar{a}b\bar{v} = a(b + v) + \bar{a}b\bar{v} = b(a + \bar{v}) + av, \end{aligned}$$

and

(2)
$$\begin{aligned} f[v(a\bar{b} + \bar{a}b)] &= va\bar{b} + b(\bar{v} + ab + \bar{a}\bar{b}) = va\bar{b} + b\bar{v} + ab \\ &= a(b + v) + b\bar{v} = b(a + \bar{v}) + av. \end{aligned}$$

I call this ineffective part of x_e "*innocuous*" to indicate that it does not invalidate the fundamental proposition

$$[f(x'_e) \neq f(x''_e)] = (x'_e \neq x''_e)$$

which was proved above (P. 4) for effective values of x. The reason why this ineffective part of x_e is innocuous is clear: it, *as a whole*, is part of every x_e, so that the variation of x_e does not take place in it at all.

D. 3. But this consideration leads to the *definition of the wholly-effective range of x*. By this I mean the collection of values which lies between Z and $a\bar{b} + \bar{a}b$, i.e.,

$$Z < x_\omega < a\bar{b} + \bar{a}b,$$

where x_ω designates the wholly-effective values of x. It is obvious that the wholly-effective range of x is obtained by omitting the ineffective part of ab from the effective range.

We can write x_ω in the form:

$$x_\omega = v(a\bar{b} + ab)$$

where

$$Z < v < U.$$

If we choose the regional interpretation of our Boolean entities, which in the present paper is the preferred interpretation, then the wholly-effective range is, in general, represented by a disconnected region. A value in this range can be expressed thus:

$$(x_\omega = x_1 + x_2)(x_1 < ab)(x_2 < \bar{a}b).$$

```
 1  \Example={1d}   \Volume 25, 10 (1923) % To place stuff in the right margin.
 2  \Header{[January]}{F. R. Sharpe and Virgil Snyder} % Defined on page 14.
 3  \pageno=10
 4  \def\Pt(#1#2#3#4){(#1,~#2,~#3,~#4)}
 5
 6  \noindent the point \Pt(0001) in ($y$) has for image in ($X$) the point
 7  having the same co\"ordinates. Conversely, therefore, the point \Pt(0001) in
 8  ($X$) has one image at \Pt(0001) in ($y$); the other is the entire plane
 9  $y_4=0$, since any plane of ($X$) through this point has for image a sextic
10  surface having the fixed component $y_4=0$. The residual quintic surface
11  meets $y_4=0$ in a composite quintic consisting of three lines of $\gamma_6$
12  and a variable conic. Hence the point \Pt(0001) is 5-fold on $s'_9$.
13
14  11. {\bf Lines of the composite curve $\gamma_6$.} The image of any plane of
15  the pencil through the line $X_1=0$, $X_2=0$ has the plane $y_1+y_2=0$ for
16  fixed component in addition to the planes, images of the vertices on the line.
17  Hence the plane is the image of the line. The sextic surface, image of a
18  general plane in ($X$) meets the plane $y_1+y_2=0$ in a cubic curve apart from
19  fixed fundamental elements, hence the curve is the image of a point on the
20  line $X_1=0$, $X_2=0$ and the line is triple on every $s'_9$ of the system.
21
22  {\it The point\/} \Pt(1111). The images of planes through the point \Pt(1111)
23  have in common in addition to fundamental elements common to all the sextic
24  surfaces of the web, the three lines of the type $y_1-y_2=0$, $y_3-y_4=0$;
25  hence the image of the point is these three lines.
26
27  The surfaces $s'_9$ all have the vertices of the composite $\gamma_6$ for
28  five-fold points, the edges three-fold. Moreover, they also have the lines
29  of the type $X_1-X_2=0$, $X_3-X_4=0$ double. The image in ($y$) of an $s'_9$
30  of the system is of order~54. It consists of $s_1$, its conjugate $s_3$ in I,
31  and of the following fundamental elements: the four planes $y_i=0$, each
32  counted five times, the six planes $y_i+y_k=0$, each counted three times,
33  and the three quadrics of the type $y_1y_2-y_3y_4=0$, each counted twice.
34
35  We have seen that any point on the line $y_1+y_2=0$, $y_3+y_4=0$ has for
36  image in ($X$) the whole line $X_1+X_2=0$, $X_3+X_4=0$. Since any plane in
37  ($y$) meets the line in one point, its image $s'_9$ contains the whole line.
38  Hence the system $s'_9$ has also the three lines of this type for basis
39  elements.
40
41  12. {\bf Algebraic procedure.} The plane containing the points ($y$), ($y'$),
42  and the vertex \Pt(1000) has the equation
43  $$ p_{34}x_2 + p_{42}x_3 + p_{23}x_4 = 0. $$
44  Since ($y$) and ($y'$) both satisfy this equation we may write
45  $$ p_{34}y_2 + p_{42}y_3 + p_{23}y_4 = 0 $$
46  and
47  $$ {p_{34}\over y_2} + {p_{42}\over y_3} + {p_{23}\over y_4} = 0; $$
48  hence
49  $$ {y_4\over y_3} + {y_3\over y_4} + {p_{23}^2 + p_{24}^2 - p_{34}^2\over
50  p_{23}p_{34}} = 0. $$
51  \EndPage
```

23

Example 1d

Source:

Transactions of the American Mathematical Society
25, 10 (1923)

the point $(0, 0, 0, 1)$ in (y) has for image in (X) the point having the same coördinates. Conversely, therefore, the point $(0, 0, 0, 1)$ in (X) has one image at $(0, 0, 0, 1)$ in (y); the other is the entire plane $y_4 = 0$, since any plane of (X) through this point has for image a sextic surface having the fixed component $y_4 = 0$. The residual quintic surface meets $y_4 = 0$ in a composite quintic consisting of three lines of γ_6 and a variable conic. Hence the point $(0, 0, 0, 1)$ is 5-fold on s_9'.

11. **Lines of the composite curve γ_6.** The image of any plane of the pencil through the line $X_1 = 0$, $X_2 = 0$ has the plane $y_1 + y_2 = 0$ for fixed component in addition to the planes, images of the vertices on the line. Hence the plane is the image of the line. The sextic surface, image of a general plane in (X) meets the plane $y_1 + y_2 = 0$ in a cubic curve apart from fixed fundamental elements, hence the curve is the image of a point on the line $X_1 = 0$, $X_2 = 0$ and the line is triple on every s_9' of the system.

The point $(1, 1, 1, 1)$. The images of planes through the point $(1, 1, 1, 1)$ have in common in addition to fundamental elements common to all the sextic surfaces of the web, the three lines of the type $y_1 - y_2 = 0$, $y_3 - y_4 = 0$; hence the image of the point is these three lines.

The surfaces s_9' all have the vertices of the composite γ_6 for five-fold points, the edges three-fold. Moreover, they also have the lines of the type $X_1 - X_2 = 0$, $X_3 - X_4 = 0$ double. The image in (y) of an s_9' of the system is of order 54. It consists of s_1, its conjugate s_3 in I, and of the following fundamental elements: the four planes $y_i = 0$, each counted five times, the six planes $y_i + y_k = 0$, each counted three times, and the three quadrics of the type $y_1 y_2 - y_3 y_4 = 0$, each counted twice.

We have seen that any point on the line $y_1 + y_2 = 0$, $y_3 + y_4 = 0$ has for image in (X) the whole line $X_1 + X_2 = 0$, $X_3 + X_4 = 0$. Since any plane in (y) meets the line in one point, its image s_9' contains the whole line. Hence the system s_9' has also the three lines of this type for basis elements.

12. **Algebraic procedure.** The plane containing the points (y), (y'), and the vertex $(1, 0, 0, 0)$ has the equation

$$p_{34} x_2 + p_{42} x_3 + p_{23} x_4 = 0.$$

Since (y) and (y') both satisfy this equation we may write

$$p_{34} y_2 + p_{42} y_3 + p_{23} y_4 = 0$$

and

$$\frac{p_{34}}{y_2} + \frac{p_{42}}{y_3} + \frac{p_{23}}{y_4} = 0;$$

hence

$$\frac{y_4}{y_3} + \frac{y_3}{y_4} + \frac{p_{23}^2 + p_{24}^2 - p_{34}^2}{p_{23} p_{34}} = 0.$$

```
1  \Example={1e}   \Volume 28, 207 (1926)
2  \headline{\hfil}  \footline{\hfil\Eightrm 207\hfil}
3  \topglue .75in
4
5  \centerline{\bf NEW DIVISION ALGEBRAS\footnote{$^\ast$}{Presented to the
6  Society, October 31, 1925; received by the editors in October, 1925.}}
7  \medskip
8  \centerline{\Sixrm BY}
9  \smallskip
10 \centerline{\Smallcaps l.\ e.\ dickson}
11 \medskip
12 1. {\bf Introduction.} The chief outstanding problem in the theory of linear
13 algebras (or hypercomplex numbers) is the determination of all division
14 algebras. We shall add here very greatly to the present meager knowledge of
15 them, since we shall show how to construct one or more types of division
16 algebras of order $n^2$ corresponding to every solvable group of order $n$.
17
18 While it was long thought that the theory of continuous groups furnishes an
19 important tool for the study of linear algebras, the reverse position is now
20 taken. But this memoir shows how vital a r\^ole the theory of finite groups
21 plays in the theory of division algebras.
22
23 Fields and the algebra of real quaternions were the only known division
24 algebras until the writer's discovery in 1905 of a division algebra $D$,
25 over a field $F$, whose $n^2$ basal units are $i^aj^b$ ($a,b=0,1,\cdots,n-1)$,
26 where $i$ is a root of an irreducible cyclic equation of degree $n$ for $F$.
27 Recently, Cecioni\footnote{\dag}{R\:e\:n\:d\:i\:c\:o\:n\:t\:i\:\ d\:e\:l\:\
28 C\:i\:r\:c\:o\:l\:o\:\ M\:a\:t\:e\:m\:a\:t\:i\:c\:o\:\ d\:i\:\
29 P\:a\:l\:e\:r\:m\:o, vol.\ 47 (1923), pp.\ 209--54.} gave a further division
30 algebra of order~16.
31
32 It is here shown that the algebras $D$ form only the first of an infinitude
33 of systems of division algebras. The next system is composed of algebras
34 $\Gamma$ of order $p^2q^2$ over $F$ with the basal units $i^aj^bk^c$
35 ($a<pq$, $b<q$, $c<p$). We start with an irreducible equation of degree $pq$,
36 three of whose roots are $i$, and the rational functions $\theta(i)$ and
37 $\psi(i)$ with coefficients in $F$, such that the $q$th iterative
38 $\theta^q(i)$ of $\theta(i)$ is $i$, and likewise $\psi^p(i)=i$, while all
39 the roots are given by
40 $$ \theta^k[\psi^r(i)]=\psi^r[\theta^k(i)] \qquad
41 (k=0,1,\cdots,q-1; r=0,1,\cdots,p-1). $$
42 The complete multiplication table of the units follows by means of the
43 associative law from
44 $$ i^q=g, \qquad k^p=\gamma, \qquad kj=\alpha jk, \qquad ji=\theta(i)j,
45 \qquad ki=\psi(i)k, $$
46 where $g$, $\gamma$ and $\alpha$ are in the field $F(i)$. The conditions for
47 associativity all reduce~to
48 $$\matrix{g=g(\theta),\quad\hfill & \hfill
49 \alpha\alpha(\theta)\alpha(\theta^2)\cdots\alpha(\theta^{q-1})g=g(\psi), \cr
50 \gamma=\gamma(\psi),\quad\hfill & \hfill
51 \alpha\alpha(\psi)\alpha(\psi^2)\cdots\alpha(\psi^{p-1})\gamma(\theta)=\gamma.
52 } $$   \EndPage
```

NOTES

2, 8, 10: \Eightrm, \Sixrm and \Smallcaps are all non-standard typeface commands; they give 8- and 6-point roman and small capitals, respectively.

3: One way to leave space at the top of a page; the command was introduced in Version 3.0 of TeX.

Source:

*Transactions
of the
American
Mathematical
Society*

28, 207 (1926)

NEW DIVISION ALGEBRAS*

BY

L. E. DICKSON

1. **Introduction.** The chief outstanding problem in the theory of linear algebras (or hypercomplex numbers) is the determination of all division algebras. We shall add here very greatly to the present meager knowledge of them, since we shall show how to construct one or more types of division algebras of order n^2 corresponding to every solvable group of order n.

While it was long thought that the theory of continuous groups furnishes an important tool for the study of linear algebras, the reverse position is now taken. But this memoir shows how vital a rôle the theory of finite groups plays in the theory of division algebras.

Fields and the algebra of real quaternions were the only known division algebras until the writer's discovery in 1905 of a division algebra D, over a field F, whose n^2 basal units are $i^a j^b$ $(a, b = 0, 1, \cdots, n-1)$, where i is a root of an irreducible cyclic equation of degree n for F. Recently, Cecioni† gave a further division algebra of order 16.

It is here shown that the algebras D form only the first of an infinitude of systems of division algebras. The next system is composed of algebras Γ of order $p^2 q^2$ over F with the basal units $i^a j^b k^c$ $(a < pq, b < q, c < p)$. We start with an irreducible equation of degree pq, three of whose roots are i, and the rational functions $\theta(i)$ and $\psi(i)$ with coefficients in F, such that the qth iterative $\theta^q(i)$ of $\theta(i)$ is i, and likewise $\psi^p(i) = i$, while all the roots are given by

$$\theta^k[\psi^r(i)] = \psi^r[\theta^k(i)] \qquad (k = 0, 1, \cdots, q-1; r = 0, 1, \cdots, p-1).$$

The complete multiplication table of the units follows by means of the associative law from

$$i^q = g, \qquad k^p = \gamma, \qquad kj = \alpha jk, \qquad ji = \theta(i)j, \qquad ki = \psi(i)k,$$

where g, γ and α are in the field $F(i)$. The conditions for associativity all reduce to

$$g = g(\theta), \qquad \alpha \alpha(\theta) \alpha(\theta^2) \cdots \alpha(\theta^{q-1}) g = g(\psi),$$
$$\gamma = \gamma(\psi), \qquad \alpha \alpha(\psi) \alpha(\psi^2) \cdots \alpha(\psi^{p-1}) \gamma(\theta) = \gamma.$$

---

* Presented to the Society, October 31, 1925; received by the editors in October, 1925.

† Rendiconti del Circolo Matematico di Palermo, vol. 47 (1923), pp. 209–54.

```
1   \Example={1f}  \Volume 105, 340 (1962) % To place stuff in the right margin.
2   \pageno=340
3   \footline{\hfil}   % To wipe out traces of the previous example.
4   \Header{[November]}{Jozef Siciak} % Defined on page 14.
5   \def\Thm #1. #2\par{\smallskip{\Smallcaps #1.\ }{\it #2\par}} % See page 38.
6
7   \line{\rlap{(1)}\hfil $ z = e^{i\theta}z^0 \equiv (e^{i\theta}z_1^0,\ldots,
8   e^{i\theta}z_n^0), \qquad 0\leq\theta\leq2\pi, $ \hfil}
9   \noindent belong to $E$.
10
11  The set $E\subset C^n$ is called a Reinhardt circular set if along with the
12  point $z^0=(z_1^0,\ldots,z^0)\in E$ also the set
13  $$\{\,z\bigm\vert \vert z_k\vert = \vert z_k^0\vert,\quad k=1,2,\ldots,n\,\}$$
14  belongs to $E$.
15
16  Let $E$ be a bounded closed subset of $C^n$, unisolvent with respect to
17  homogeneous polynomials. The function $b(z)$ being defined and lower
18  semicontinuous on $E$, let
19  $$ h^{(\nu)} = \{h_1^{(\nu)},\ldots,h_\nu^{(\nu)}\}, \qquad
20  \nu_0=C_{\nu+n-1,n-1} \leqno(2) $$
21  be the $\nu$th extremal system of $E$ defined by (5.5) and (5.6). If
22  $p^{(\nu)}=\{p_1,\ldots,p_{\nu_0}\}$ is an arbitrary unisolvent system of
23  points of $E$, then the functions
24  $$ \psi^{(i)}(z,p^{(\nu)},b) = T^{(i)}(z,p^{(\nu)})e^{\nu b(p_i)}, \qquad
25  i=1,2,\ldots,\nu_0, \leqno(3) $$
26  $T^{(i)}(z,p^{(\nu)})$ denoting the polynomial (2.12), are homogeneous
27  polynomials of degree~$\nu$. Define the extremal functions $\psi_\nu^{(i)}
28  (z,E,b)$, $i=1,2,3,4$, corresponding to $E$ and~$b$ by the formulas
29  $$\leqalignno{\psi_\nu^{(1)}(z,E,b) & = \max_{(i)}\bigl\vert\psi^{(i)}
30  (z,h^{(\nu)},b)\bigr\vert,\qquad z\in C^n, &(4)\cr
31  \psi_\nu^{(2)}(z,E,b) & = {\textstyle\sum\limits_{i=1}^{\nu_0}}
32  \bigl\vert\psi^{(i)}(z,h^{(\nu)},b)\bigr\vert,\qquad z\in C^n, &(5)\cr
33  \psi_\nu^{(3)}(z,E,b) & = \inf_{p^{(\nu)}\subset E}\left\{\max_{(i)}
34  \bigl\vert\psi^{(i)}(z,p^{(\nu)},b)\bigr\vert\right\},\qquad z\in C^n, &(6)\cr
35  \psi_\nu^{(4)}(z,E,b) & = \inf_{p^{(\nu)}\subset E}
36  \left\{{\textstyle\sum\limits_{i=1}^{\nu_0}}
37  \bigl\vert\psi^{(i)}(z,p^{(\nu)},b)\bigr\vert\right\},\qquad z\in C^n. &(7)\cr
38  } $$
39  By reasoning quite analogous to the reasoning of \S6 we may prove (see
40  also~[19])
41
42  \Thm Theorem 1. At any point $z\in C^n$ the sequences
43  $\{(\psi_\nu^{(i)}(z,E,b)^{1/\nu}\},\; i=1,2,3,4$, are convergent to the same
44  limit $\psi(z,E,b)$,
45  $$ \psi(z,E,b) = \lim_{\nu\to\infty}(\psi^{(i)}(z,E,b))^{1/\nu}, \qquad
46  z\in C^n, \quad i=1,2,3,4. \leqno(8) $$ \par
47  \noindent Lemma 1 of \S6 now takes the form of
48
49  \Thm Lemma 1. If $Q_\nu(z)$ denotes an arbitrary homogeneous polynomial of
50  degree $\nu$ such that
51  $$ \bigl\vert Q_\nu(z)\bigr\vert \leq M\exp [\nu b(z)], \qquad z\in E,
52  \quad M={\rm const}, $$  then
53  $$ \bigl\vert Q_\nu(z)\bigr\vert \leq M\psi^\nu(z,E,b), \qquad z\in C^n. $$
54
55  One obtains easily also the following properties of $\psi$. \par
56  $1^\circ$ Function $\psi(z,E,b)$ is given by
57  $$ \psi(z,E,b) = \lim_{\nu\to\infty} \Bigl\{\sup_{Q_\nu\in A_\nu}
58  \bigl\vert Q_\nu(z)\bigr\vert^{1/\nu}\Bigr\}, \qquad z\in C^n,$$ \EndPage
```

Source:
*Transactions
of the
American
Mathematical
Society*
105, 340 (1962)

(1) $$z = e^{i\theta}z^0 \equiv (e^{i\theta}z_1^0, \ldots, e^{i\theta}z_n^0), \qquad 0 \le \theta \le 2\pi,$$

belong to E.

The set $E \subset C^n$ is called a Reinhardt circular set if along with the point $z^0 = (z_1^0, \ldots, z^0) \in E$ also the set

$$\{ z \mid |z_k| = |z_k^0|, \quad k = 1, 2, \ldots, n \}$$

belongs to E.

Let E be a bounded closed subset of C^n, unisolvent with respect to homogeneous polynomials. The function $b(z)$ being defined and lower semicontinuous on E, let

(2) $$h^{(\nu)} = \{h_1^{(\nu)}, \ldots, h_\nu^{(\nu)}\}, \qquad \nu_0 = C_{\nu+n-1,n-1}$$

be the νth extremal system of E defined by (5.5) and (5.6). If $p^{(\nu)} = \{p_1, \ldots, p_{\nu_0}\}$ is an arbitrary unisolvent system of points of E, then the functions

(3) $$\psi^{(i)}(z, p^{(\nu)}, b) = T^{(i)}(z, p^{(\nu)})e^{\nu b(p_i)}, \qquad i = 1, 2, \ldots, \nu_0,$$

$T^{(i)}(z, p^{(\nu)})$ denoting the polynomial (2.12), are homogeneous polynomials of degree ν. Define the extremal functions $\psi_\nu^{(i)}(z, E, b)$, $i = 1, 2, 3, 4$, corresponding to E and b by the formulas

(4) $$\psi_\nu^{(1)}(z, E, b) = \max_{(i)}|\psi^{(i)}(z, h^{(\nu)}, b)|, \qquad z \in C^n,$$

(5) $$\psi_\nu^{(2)}(z, E, b) = \sum_{i=1}^{\nu_0}|\psi^{(i)}(z, h^{(\nu)}, b)|, \qquad z \in C^n,$$

(6) $$\psi_\nu^{(3)}(z, E, b) = \inf_{p^{(\nu)} \subset E}\left\{\max_{(i)}|\psi^{(i)}(z, p^{(\nu)}, b)|\right\}, \qquad z \in C^n,$$

(7) $$\psi_\nu^{(4)}(z, E, b) = \inf_{p^{(\nu)} \subset E}\left\{\sum_{i=1}^{\nu_0}|\psi^{(i)}(z, p^{(\nu)}, b)|\right\}, \qquad z \in C^n.$$

By reasoning quite analogous to the reasoning of §6 we may prove (see also [19])

THEOREM 1. *At any point* $z \in C^n$ *the sequences* $\{(\psi_\nu^{(i)}(z, E, b)^{1/\nu}\}$, $i = 1, 2, 3, 4$, *are convergent to the same limit* $\psi(z, E, b)$,

(8) $$\psi(z, E, b) = \lim_{\nu \to \infty}(\psi^{(i)}(z, E, b))^{1/\nu}, \qquad z \in C^n, \quad i = 1, 2, 3, 4.$$

Lemma 1 of §6 now takes the form of

LEMMA 1. *If* $Q_\nu(z)$ *denotes an arbitrary homogeneous polynomial of degree* ν *such that*

$$|Q_\nu(z)| \le M \exp[\nu b(z)], \qquad z \in E, \quad M = \text{const},$$

then

$$|Q_\nu(z)| \le M\psi^\nu(z, E, b), \qquad z \in C^n.$$

One obtains easily also the following properties of ψ.

1° Function $\psi(z, E, b)$ is given by

$$\psi(z, E, b) = \lim_{\nu \to \infty}\left\{\sup_{Q_\nu \in A_\nu}|Q_\nu(z)|^{1/\nu}\right\}, \qquad z \in C^n,$$

```
1  \Example={1g}   \Volume 114, 216 (1965) % To place stuff in the right margin.
2  \Header{[January]}{S. A. Amitsur} % Defined on page 14.
3  \pageno=216
4  \def\Thm #1. #2\par{\smallskip{\Smallcaps #1.\ }{\it #2\par}} % See page 38.
5  \def\Prf #1.{\smallskip{\bf #1.}} % See page 38.
6  \noindent
7  which proves that $\lambda$ is in the center of $D$, i.e., $\lambda\in C$.
8  It is evident from the definition that $s=\lambda r$, and~(a) is proved.
9
10 To prove~(b) it suffices to show that if $\sum_{i=1}^{k}r_i\alpha_i=0$ then
11 all $r_i=0$. If this is not the case, let $k$ (the number of elements of the
12 last sum) be minimal; then for all $x\in R$:
13 $$ 0 = r_kx\Bigl(\sum r_i\alpha_i\Bigr) - \Bigl(\sum r_i\alpha_i\Bigr)xr_k
14 = \sum_{i=1}^{k-1}(r_kxr_i - r_ixr_k)\alpha_i. $$
15 This element is of lower length. It follows therefore that $r_kxr_i-r_ixr_k
16 =0$ for $i=1,\ldots k$. Hence, (a) yields that $r_i=\lambda_ir_k$, $\lambda_i
17 \in C$. Thus, $\sum r_i\alpha_i=r_k\sum\lambda_i\alpha_i$. Now
18 $r_k\ne 0$, by the minimality of $k$, and $\sum\lambda_i\alpha_i\in F$ which
19 is a field from which we deduce that $\sum\lambda_i\alpha_i=0$. But the
20 $\alpha_i$ are a $C$-base, hence all $\lambda_i=0$ which is impossible
21 since in particular $\lambda_k=1$.
22
23 \Thm Theorem 7. Let $R$ be a dense ring of linear transformations
24 of $V_D$ and let $F$ be a maximal commutative subfield $D$. If $R_F$ contains
25 a linear transformation of finite rank over $F$, then $R$ contains also a
26 linear transformation of finite rank over $D$, and $(D\colon C)<\infty$.
27
28 \Prf Proof. It follows by Lemma~5 that $R_F$ is a dense ring of linear
29 transformations. Let $T\in R_F$ such that $(TV\colon F)<\infty$ and let
30 $T=\sum_{i=1}^kr_i\alpha_i$ with $r_i\in R$ and $\{\alpha_i\}$ a $C$-base of
31 $F$. Among all $T$ with this property we choose $T$ with $k$ minimal. We note
32 that for $x\in R$, $(r_kxT-Txr_k)V\subset(r_kx)TV+TV$. $TV$ is of finite
33 dimension and so is $r_kxTV$ since the latter is an $F$-homomorphic image of
34 $TV$, where the homomorphism is obtained by the mapping $Tv\to r_kxTv$.
35 Consequently, $r_kxT-Txr_k=\sum(r_kxr_i-r_ixr_k)\alpha_i$ is of lower length,
36 hence $r_kxr_i=r_ixr_k$ for all $x\in R$. It follows from (a) of the
37 preceding lemma that $r_i=r_k\lambda_i$ with $\lambda_i\in C$ and, hence,
38 $T=r_k\sum\lambda_i\alpha_i=r_k\alpha$ with $\alpha\in F$. Since $T\ne 0$, we
39 have also $r_k\ne 0$ and $\alpha\ne 0$.
40
41 Consequently, $TV=r_k\alpha V=r_kV$ as $\alpha^{-1}$ exists in $F$. Now $r_kV$
42 is as well a $D$-space since $r_kR$ commutes with the elements of $D$. Hence
43 $\infty>(r_kV\colon F)=(r_kV\colon D)(D\colon F)$ which yields that both
44 $(r_kV\colon D)<\infty$ and $(D\colon F)<\infty$. The finiteness of the first
45 proves the first part of the theorem and the finiteness of $(D\colon F)$
46 yields (e.g. [3, Chapter~VII, Theorem~9.1, p.~175]) that $(D\colon C)<\infty$
47 and since $F$ is maximal we also have $(D\colon C)=(D\colon F)^2$.
48
49 A bound for $(D\colon C)$ can be obtained as follows: \par
50 Let $T=\sum_{i=1}^k r_i\alpha_i$ be such that $(TV\colon F)=m$, then the
51 preceding proof shows that either $TV=r_kV$ or there exists
52 $T'=\sum r'_i\alpha_i$ of lower length and $(T'V\colon F)\leq 2m$. Continuing
53 this way we get an $r\in R$ such that $(rV\colon F)\leq 2^{k-1}m$. Hence from
54 the relation $(rV\colon F)=(rV\colon D)(D\colon F)$, and
55 $(D\colon C)=(D\colon F)^2$ it follows that:
56 \Thm Corollary 8. If $T=\sum_1^k r_i\alpha_i$, and $(TV\colon F)=m$ then
57 $(D\colon C)\leq 2^{2k-2}m^2=4^{k-1}m^2$.\par % Need \par, for \Thm's sake.
58 \EndPage
```

| 29

Example 1g

Source:

*Transactions
of the
American
Mathematical
Society*

114, 216 (1965)

which proves that λ is in the center of D, i.e., $\lambda \in C$. It is evident from the definition that $s = \lambda r$, and (a) is proved.

To prove (b) it suffices to show that if $\sum_{i=1}^{k} r_i \alpha_i = 0$ then all $r_i = 0$. If this is not the case, let k (the number of elements of the last sum) be minimal; then for all $x \in R$:

$$0 = r_k x \Big(\sum r_i \alpha_i \Big) - \Big(\sum r_i \alpha_i \Big) x r_k = \sum_{i=1}^{k-1} (r_k x r_i - r_i x r_k) \alpha_i.$$

This element is of lower length. It follows therefore that $r_k x r_i - r_i x r_k = 0$ for $i = 1, \ldots k$. Hence, (a) yields that $r_i = \lambda_i r_k$, $\lambda_i \in C$. Thus, $\sum r_i \alpha_i = r_k \sum \lambda_i \alpha_i$. Now $r_k \neq 0$, by the minimality of k, and $\sum \lambda_i \alpha_i \in F$ which is a field from which we deduce that $\sum \lambda_i \alpha_i = 0$. But the α_i are a C-base, hence all $\lambda_i = 0$ which is impossible since in particular $\lambda_k = 1$.

THEOREM 7. *Let R be a dense ring of linear transformations of V_D and let F be a maximal commutative subfield D. If R_F contains a linear transformation of finite rank over F, then R contains also a linear transformation of finite rank over D, and $(D:C) < \infty$.*

Proof. It follows by Lemma 5 that R_F is a dense ring of linear transformations. Let $T \in R_F$ such that $(TV:F) < \infty$ and let $T = \sum_{i=1}^{k} r_i \alpha_i$ with $r_i \in R$ and $\{\alpha_i\}$ a C-base of F. Among all T with this property we choose T with k minimal. We note that for $x \in R$, $(r_k x T - T x r_k)V \subset (r_k x)TV + TV$. TV is of finite dimension and so is $r_k x TV$ since the latter is an F-homomorphic image of TV, where the homomorphism is obtained by the mapping $Tv \to r_k x Tv$. Consequently, $r_k x T - T x r_k = \sum (r_k x r_i - r_i x r_k) \alpha_i$ is of lower length, hence $r_k x r_i = r_i x r_k$ for all $x \in R$. It follows from (a) of the preceding lemma that $r_i = r_k \lambda_i$ with $\lambda_i \in C$ and, hence, $T = r_k \sum \lambda_i \alpha_i = r_k \alpha$ with $\alpha \in F$. Since $T \neq 0$, we have also $r_k \neq 0$ and $\alpha \neq 0$.

Consequently, $TV = r_k \alpha V = r_k V$ as α^{-1} exists in F. Now $r_k V$ is as well a D-space since $r_k R$ commutes with the elements of D. Hence $\infty > (r_k V:F) = (r_k V:D)(D:F)$ which yields that both $(r_k V:D) < \infty$ and $(D:F) < \infty$. The finiteness of the first proves the first part of the theorem and the finiteness of $(D:F)$ yields (e.g. [3, Chapter VII, Theorem 9.1, p. 175]) that $(D:C) < \infty$ and since F is maximal we also have $(D:C) = (D:F)^2$.

A bound for $(D:C)$ can be obtained as follows:

Let $T = \sum_{i=1}^{k} r_i \alpha_i$ be such that $(TV:F) = m$, then the preceding proof shows that either $TV = r_k V$ or there exists $T' = \sum r_i' \alpha_i$ of lower length and $(T'V:F) \leq 2m$. Continuing this way we get an $r \in R$ such that $(rV:F) \leq 2^{k-1}m$. Hence from the relation $(rV:F) = (rV:D)(D:F)$, and $(D:C) = (D:F)^2$ it follows that:

COROLLARY 8. *If $T = \sum_1^k r_i \alpha_i$, and $(TV:F) = m$ then $(D:C) \leq 2^{2k-2}m^2 = 4^{k-1}m^2$.*

```
1   \Example={1h}    \Volume 125, 38 (1966) % To place stuff in the right margin.
2   \Header{[October]}{Herbert Halpern} % Defined on page 14.
3   \pageno=38
4   \def\Wed{\mathord{\lower3pt\hbox{$\widehat{\hphantom{ab}}$}}}
5
6   \noindent The set $S$ is open and dense in $Z_1$; we set $N_1$ equal to the
7   complement of $S$ in~$Z_1$. The set~$N_1$ is nowhere dense in~$Z_1$ and thus
8   $N=\rho(N_1)$ is nowhere dense in~$Y$.
9
10  For each $\zeta\in Y-N$ we must prove that $f_\zeta$ satisfies properties (4)
11  and~(5). Let $P_0$ be the unique projection in $\{\,P_d\mid d\in D\,\}$ such
12  that $\hat P_0(\zeta_0)=1$ where $\zeta_0=\eta(\zeta)$. Since the algebra
13  $(E{\cal A}E)\cdot P_0$ is finite and homogeneous, there are equivalent
14  orthogonal abelian projections $E_1,E_2,\ldots,E_n$ such that $E_1+E_2+\cdots
15  +E_n=E$. Let $U_{jk}$ $(1\leq j, k\leq n)$ be partial isometric operators
16  in $(E{\cal A}E)\cdot P_0$ such that
17
18  (1) $U_{jk}U_{lm}=\delta_{mj}U_{lk}$, where $\delta$ is the Kronecker delta;
19
20  (2) $U^\ast_{jk}=U_{kj}$; and
21
22  (3) $U_{jj}=E_j$,
23
24  \noindent for all $1\leq j,k,l,m\leq n$. For each $A$ in $(E{\cal A}E)\cdot
25  P_0$, there are unique $B_{jk}$ $(1\leq j, k\leq n)$ in ${\cal L}_1P_0$
26  such that
27  $$ A=\sum_{j,k} B_{jk}U_{jk}. $$
28
29  We prove that the map $\Psi_{\zeta_0}=\Psi$ takes $(EAE)\cdot P_0$ onto the
30  set of all linear operators on an $n$-dimensional Hilbert space $H_n$. Let
31  $\Psi(U_{jk})=V_{jk}$ $(1\leq j, k\leq n)$. If $e_1,e_2,\ldots,e_n$ is an
32  orthonormal basis of $H_n$, define $V_{jk}e_l=\delta_{jl}e_k$. Then for each
33  $A$ in $(EAE)\cdot P_0$ we have
34  $$ \Psi(A) = \sum\alpha_{jk}V_{jk} $$
35  where $\alpha_{jk}=\hat B_{jk}(\zeta_0)$ if $A=\sum B_{jk}U_{jk}$, $B_{jk}\in
36  {\cal L}_1P_0$; thus $\Psi(A)$ is defined on~$H_n$. It is easy to see that
37  $\Psi((E{\cal A}E)\cdot P_0)$ is the set of all linear functionals on~$H_n$.
38
39  We have for each $A$ in $\cal A$ that
40  $$\eqalign{f_\zeta(A) & = (EAE)^{\#\Wed}(\zeta_0) \cr
41  & = (EAE\cdot P_0)^{\#\Wed}(\zeta_0) \cr
42  & = \sum_j(E_jAE_j)^{\#\Wed}(\zeta_0) \cr
43  & = \sum_j(\Psi(EAE)e_j,e_j). \cr} \leqno(1) $$
44  Now each functional $A\to (E_jAE_j)^{\#\Wed}(\zeta_0)$ is irreducible on
45  $E{\cal A}E$. Indeed, $E_j$ is abelian in $EAE$. Thus for each $A$ in $\cal A$
46  there is a $B$ in ${\cal L}_1P_0$ such that $E_jAE_j=BE_j$. So
47  $$\eqalign{(E_jAE_j)^\# & = (BE_j)^\# \cr
48  & = (BU_{kj}U_{jk})^\# \cr
49  & = (BU_{jk}U_{kj})^\# = (BE_k)^\#. \cr} $$
50  \EndPage
```

NOTES

4: The symbol constructed here will be used as an ordinary symbol (i.e., not as a binary operation, a relation, etc.); the use of \mathord will guarantee the correct spacing. Compare line 6 on page 20. The classification of mathematical symbols in TeX is discussed under classes in the Glossary.

Source:

*Transactions
of the
American
Mathematical
Society*
125, 38 (1966)

The set S is open and dense in Z_1; we set N_1 equal to the complement of S in Z_1. The set N_1 is nowhere dense in Z_1 and thus $N = \rho(N_1)$ is nowhere dense in Y.

For each $\zeta \in Y - N$ we must prove that f_ζ satisfies properties (4) and (5). Let P_0 be the unique projection in $\{ P_d \mid d \in D \}$ such that $\hat{P}_0(\zeta_0) = 1$ where $\zeta_0 = \eta(\zeta)$. Since the algebra $(EAE) \cdot P_0$ is finite and homogeneous, there are equivalent orthogonal abelian projections $E_1, E_2, \ldots, E_n$ such that $E_1 + E_2 + \cdots + E_n = E$. Let U_{jk} $(1 \le j, k \le n)$ be partial isometric operators in $(EAE) \cdot P_0$ such that
(1) $U_{jk}U_{lm} = \delta_{mj}U_{lk}$, where δ is the Kronecker delta;
(2) $U_{jk}^* = U_{kj}$; and
(3) $U_{jj} = E_j$,
for all $1 \le j, k, l, m \le n$. For each A in $(EAE) \cdot P_0$, there are unique B_{jk} $(1 \le j, k \le n)$ in $\mathcal{L}_1 P_0$ such that

$$A = \sum_{j,k} B_{jk} U_{jk}.$$

We prove that the map $\Psi_{\zeta_0} = \Psi$ takes $(EAE) \cdot P_0$ onto the set of all linear operators on an n-dimensional Hilbert space H_n. Let $\Psi(U_{jk}) = V_{jk}$ $(1 \le j, k \le n)$. If $e_1, e_2, \ldots, e_n$ is an orthonormal basis of H_n, define $V_{jk}e_l = \delta_{jl}e_k$. Then for each A in $(EAE) \cdot P_0$ we have

$$\Psi(A) = \sum \alpha_{jk} V_{jk}$$

where $\alpha_{jk} = \hat{B}_{jk}(\zeta_0)$ if $A = \sum B_{jk}U_{jk}$, $B_{jk} \in \mathcal{L}_1 P_0$; thus $\Psi(A)$ is defined on H_n. It is easy to see that $\Psi((EAE) \cdot P_0)$ is the set of all linear functionals on H_n.

We have for each A in $\mathcal{A}$ that

$$
\begin{aligned}
(1) \qquad f_\zeta(A) &= (EAE)^{\#\,\widehat{}}\,(\zeta_0) \\
&= (EAE \cdot P_0)^{\#\,\widehat{}}\,(\zeta_0) \\
&= \sum_j (E_j A E_j)^{\#\,\widehat{}}\,(\zeta_0) \\
&= \sum_j (\Psi(EAE)e_j, e_j).
\end{aligned}
$$

Now each functional $A \to (E_j A E_j)^{\#\,\widehat{}}\,(\zeta_0)$ is irreducible on EAE. Indeed, E_j is abelian in EAE. Thus for each A in $\mathcal{A}$ there is a B in $\mathcal{L}_1 P_0$ such that $E_j A E_j = B E_j$. So

$$
\begin{aligned}
(E_j A E_j)^{\#} &= (B E_j)^{\#} \\
&= (B U_{kj} U_{jk})^{\#} \\
&= (B U_{jk} U_{kj})^{\#} = (B E_k)^{\#}.
\end{aligned}
$$

```
 1 \Example={1i}    \Volume 169, 232 (1972) % To place stuff in the right margin.
 2 \Header{[July]}{D. J. Rodabaugh} % Defined on page 14.
 3 \pageno=232
 4 \def\Thm #1 #2 #3\par{\smallbreak{\bf #1} #2 {\it #3\par}} % See page 38.
 5 \def\Prf #1.{\smallskip{\bf #1.}} % See page 38.
 6
 7 \noindent $a$ in $N$ with $a^{n-k-1}\ne 0$. Let $c_1,\ldots, c_k$ be chosen
 8 so that they are not in $N_2$ and $a,\ldots,a^{n-k-1},c_1,\ldots c_k$ are a
 9 basis for $N$. (This is possible since $\dim N'_1=k+1$.) We know $a^2,\ldots,
10 a^{n-k-1}$ form a basis for $N_2$,
11 $$ ac_j=\sum_{i=2}^{n-k-1} \beta_{ij}a^i = a\left(\sum_{i=2}^{n-k-1}
12 \beta_{ij}a^{i-1}\right),\quad j=1,\ldots, k. $$
13 Define $b_j=c_j-\sum_{i=2}^{n-k-i}\beta_{ij}a^{i-1}$, $j=1,\ldots,k$. Clearly
14 $ab_j=0$, $j=1,\ldots, k$. Now, $b_j^2$ is in $N_2$ so
15 $b_j^2=\sum_{i=2}^{n-k-1} \gamma_{ij}a^i$ and
16 $$ 0 = (ab_j)b_j = ab_j^2 = \sum_{i=2}^{n-k-2} \gamma_{ij}a^{i+1}. $$
17 Hence, $\gamma_{ij}=0$ for $j=1,\ldots, k$ and $i=2,\ldots,n-k-2$. Defining
18 $\alpha_j=Y_{n-k-1,j}$ we have $b_j^2=a_ja^{n-k-1}$.
19
20 \Thm Lemma 5.6. If $\phi$ is an antiflexible map on an associative
21 commutative algebra $P$ of char.~$\neq2,3$ in which $ab=0$ then
22 $\phi(a^r,b^s)=0$ if $r>1$ or~$s>1$.
23
24 \Prf Proof. If $r>1$ then
25 $$ \phi(a^r,b^s) + \phi(b^sa,a^{r-1}) + \phi(b^sa^{r-1},a) = 0. $$
26 Since $b^sa=b^sa^{r-1}=0$, $\phi(a^r,b^s)=0$. The proof when $s>1$ is similar.
27
28 \Thm Theorem 5.6. Let $P=F\cdot1\oplus N$ where $N$ is an associative
29 commutative nil-algebra of type $(n-k,n,k+1)$ with $n-k>2$ over a field $F$
30 of char.~$\ne 2,3$. The algebra $P$ is nearly simple if and only if the
31 following hold:\EndLine
32 \indent {\rm(a)} N is spanned by $a,\ldots, a^{n-k-1},b_1,\ldots b_k$ where
33 $ab_i=b_i^2=b_ib_j=0$, $i,j=1,\ldots, k$.\EndLine
34 \indent {\rm(b)} Either $n-k={\rm char}\>F$ with $k$ even or
35 $n-k=m\,{\rm char}\>F$ for $m>1$.
36
37 \Prf Proof. By Theorem~5.5, there are elements $a,b_1,\ldots b_k$ with $N$
38 spanned by $a,\ldots a^{n-k-1},b_1,\ldots b_k$. Furthermore, $ab_i=0$,
39 $b_i^2=\alpha_i a^{n-k-1}$, $b_ib_j=\lambda_{ij}a^{n-k-1}$ for all $i,j$
40 where each $\alpha_i$, $\lambda_{ij}$ is in $F$. From this, it is clear that
41 $M$ is a subspace of the space spanned by $a^{n-k-1}$, $b_1,\ldots,b_k$.
42
43 Assume $P$ is nearly simple. Then there is a $\phi$ with $P(\phi)$ simple. We
44 first show that each $b_i$ is in $M$. To do this, it is necessary and
45 sufficient to prove that each $\alpha_i=o$ and each $\lambda_{ij}=0$. If
46 $x\ne 0$ is in $M$, Theorem~3.4 assures the existence of a $y$ in $N$ with
47 $\phi(x,y)\ne 0$. Thus, if $x$ in $M$ has the property that $\phi(x,y)=0$ for
48 all $y$ in $N$ then $x=0$. Since $a^{n-k-1}$ is in $M$, each $b_i^2$ and
49 each $b_ib_j$ are in $M$.
50
51 Lemma~5.6 implies $\phi(b_i^2,a^j)=0$ for all $i,j$. Since $n-k-1>1$,
52 $\phi(b_i^2,b_j)=\alpha_i\phi(a^{n-k-i},b_j)=0$ for all $i,j$ (also by
53 Lemma~5.6). Thus, for each $i$, $\phi(b_i^2,y)=0$ for each $y$ in $N$ so
54 $b_i^2=0$. If $p>1$, $a^p$ is in $N_2$ so $\phi(b_ib_j,a^p)=$ \dots
55 \EndPage
```

| 33

Example 1i

Source:
Transactions
of the
American
Mathematical
Society
169, 232 (1972)

a in N with $a^{n-k-1} \neq 0$. Let $c_1, \ldots, c_k$ be chosen so that they are not in N_2 and $a, \ldots, a^{n-k-1}, c_1, \ldots c_k$ are a basis for N. (This is possible since dim $N_1' = k+1$.) We know $a^2, \ldots, a^{n-k-1}$ form a basis for N_2,

$$ac_j = \sum_{i=2}^{n-k-1} \beta_{ij} a^i = a \left(\sum_{i=2}^{n-k-1} \beta_{ij} a^{i-1} \right), \quad j = 1, \ldots, k.$$

Define $b_j = c_j - \sum_{i=2}^{n-k-i} \beta_{ij} a^{i-1}$, $j = 1, \ldots, k$. Clearly $ab_j = 0$, $j = 1, \ldots, k$. Now, b_j^2 is in N_2 so $b_j^2 = \sum_{i=2}^{n-k-1} \gamma_{ij} a^i$ and

$$0 = (ab_j)b_j = ab_j^2 = \sum_{i=2}^{n-k-2} \gamma_{ij} a^{i+1}.$$

Hence, $\gamma_{ij} = 0$ for $j = 1, \ldots, k$ and $i = 2, \ldots, n-k-2$. Defining $\alpha_j = Y_{n-k-1, j}$ we have $b_j^2 = a_j a^{n-k-1}$.

Lemma 5.6. *If ϕ is an antiflexible map on an associative commutative algebra P of char. $\neq 2, 3$ in which $ab = 0$ then $\phi(a^r, b^s) = 0$ if $r > 1$ or $s > 1$.*

Proof. If $r > 1$ then

$$\phi(a^r, b^s) + \phi(b^s a, a^{r-1}) + \phi(b^s a^{r-1}, a) = 0.$$

Since $b^s a = b^s a^{r-1} = 0$, $\phi(a^r, b^s) = 0$. The proof when $s > 1$ is similar.

Theorem 5.6. *Let $P = F \cdot 1 \oplus N$ where N is an associative commutative nil-algebra of type $(n-k, n, k+1)$ with $n-k > 2$ over a field F of char. $\neq 2, 3$. The algebra P is nearly simple if and only if the following hold:*

(a) *N is spanned by $a, \ldots, a^{n-k-1}, b_1, \ldots, b_k$ where $ab_i = b_i^2 = b_i b_j = 0$, $i, j = 1, \ldots, k$.*

(b) *Either $n-k = $ char F with k even or $n-k = m$ char F for $m > 1$.*

Proof. By Theorem 5.5, there are elements $a, b_1, \ldots, b_k$ with N spanned by $a, \ldots, u^{n-k-1}, b_1, \ldots, b_k$. Furthermore, $ab_i = 0$, $b_i^2 = \alpha_i a^{n-k-1}$, $b_i b_j = \lambda_{ij} a^{n-k-1}$ for all i, j where each α_i, λ_{ij} is in F. From this, it is clear that M is a subspace of the space spanned by $a^{n-k-1}, b_1, \ldots, b_k$.

Assume P is nearly simple. Then there is a ϕ with $P(\phi)$ simple. We first show that each b_i is in M. To do this, it is necessary and sufficient to prove that each $\alpha_i = o$ and each $\lambda_{ij} = 0$. If $x \neq 0$ is in M, Theorem 3.4 assures the existence of a y in N with $\phi(x, y) \neq 0$. Thus, if x in M has the property that $\phi(x, y) = 0$ for all y in N then $x = 0$. Since a^{n-k-1} is in M, each b_i^2 and each $b_i b_j$ are in M.

Lemma 5.6 implies $\phi(b_i^2, a^j) = 0$ for all i, j. Since $n-k-1 > 1$, $\phi(b_i^2, b_j) = \alpha_i \phi(a^{n-k-i}, b_j) = 0$ for all i, j (also by Lemma 5.6). Thus, for each i, $\phi(b_i^2, y) = 0$ for each y in N so $b_i^2 = 0$. If $p > 1$, a^p is in N_2 so $\phi(b_i b_j, a^p) = \ldots$

```
1  \Example={1j}  \Volume 179, 314 (1973) % To place stuff in the right margin.
2  \Header{}{Michael A. Gauger} % Defined on page 14.
3  \pageno=314
4  \def\GL{\mathord{\rm GL}}
5  \def\Thm #1. #2\par{\bigbreak{\bf #1.\ }{\it #2\medskip}} % See page 38.
6
7  \line{\rlap{(7.1)}\hfil $\displaystyle \dim G_p =
8  \bigl({\textstyle g\choose 2}} - p\bigr)p; $ \hfil}
9  \bigskip \noindent hence
10 $$ \dim D_p = \bigl({\textstyle g\choose 2}} - p\bigr)p + 1. \leqno(7.2) $$
11 \indent Our aim now is to characterize those $g$ and $p$ such that
12 $G_p/\GL(V)$ is infinite, that is, such that there are infinitely many
13 $g$-generator, $p$-relation algebras. This is accomplished by comparing the
14 dimension of an orbit in $G_p$ to the dimension of $G_p$ itself. First,
15 however, we need to know the orbits are subvarieties. This requires a look at
16 the evaluation map $\GL(V)\times D_p\to D_p$.
17
18 We will show that the evaluation map $\alpha\colon\GL(V)\times D_p\to D_p$
19 given by $$ \alpha(\theta,w) = \wedge^p(\wedge^2\theta)(w) \leqno(7.3) $$
20 is a morphism of affine varieties. We proceed by showing that the
21 representation $\wedge^p\colon\GL(W)\to \GL(\wedge^pW)$ is a morphism for any
22 vector space~$W$.
23
24 Fix a basis $w_1,\ldots,w_n$ of $W$. Then all elements of the type
25 $w_{i_1}\wedge\ldots\wedge w_{i_p}$ where $1\leq i_1<i_2<\ldots<i_p\leq n$
26 form a basis of $\wedge^pW$. Let $I$ be the index set consisting of ordered
27 $p$-tuples of increasing integers between 1 and~$n$. We take $g_{ij}$,
28 $1\leq i, j\leq n$, as coordinate functions in $\GL(W)$ and $h_{\alpha\beta}$,
29 $\alpha,\beta\in I$ as coordinate functions in $\GL(\wedge^pW)$ determined
30 by the respective bases chosen above. If $\alpha$, $\beta\in I$, the $(\alpha,
31 \beta)$th coordinate function of $\wedge^p$ is the minor of $\vert g_{ij}
32 \vert$ determined by the rows $\alpha(1),\ldots,\alpha(p)$ and the columns
33 $\beta(1),\ldots,\beta(p)$. The coordinate ring of $\GL(\wedge^pW)$ is
34 generated by the $h_{\alpha\beta}$ together with $1/\det\vert h_{\alpha\beta}
35 \vert$, while that of $\GL(W)$ is generated by the various $g_{ij}$ together
36 with $1/\det\vert g_{ij}\vert$. The coordinate functions of $\wedge^p$ are
37 polynomials in the $g_{ij}$, so to show $\wedge^p$ is a morphism it suffices
38 to show that $1/\det\vert\wedge^p(g_{ij})\vert$ is a polynomial in $g_{ij}$
39 and $1/\det\vert g_{ij}\vert$. For this, the following lemma is useful.
40 \Thm Lemma 7.3. Every rational character of $\GL(W)$ is an integral power of
41 the determinant. \par
42 \Prf Proof. By rational is meant that the coordinate function is a sum of
43 quotients of polynomials in the coordinates of $W$. (See [6, p.~22].)
44
45 Now $\det\circ\wedge^p\colon\GL(W)\to K$ is a homogeneous polynomial
46 character, thus $\det\vert\wedge^p(g_{ij})\vert=(\det\vert g_{ij}\vert)^a$ for
47 some positive integer $a$, and we have proven
48 \Thm Lemma 7.4. Let $W$ be a $K$-vector space. The representation
49 $\wedge^p\colon\GL(W)\to\GL(\wedge^pW)$ is a morphism of affine varieties.\par
50 \medskip We mention without proof the following  \EndPage
```

NOTES

7–8: This was used to display Equation (7.1), because `$$...$$` leaves too much space above the equation when used at the very start of a page. Using `\rlap` ("right overlap") to place the equation number ensures that the equation is centered correctly on the page.

Space around an equation display can also be a problem if one is forcing a page break just after, or interfering in other ways with TeX's normal mechanisms. Left to itself, TeX usually does just fine: in fact, in keeping with high typographical tradition, it does not on its own start a page with a display.

MICHAEL A. GAUGER

| 35

Example 1j

Source:
*Transactions
of the
American
Mathematical
Society*
179, 314 (1973)

(7.1) $$\dim G_p = \left(\binom{g}{2} - p \right) p;$$

hence

(7.2) $$\dim D_p = \left(\binom{g}{2} - p \right) p + 1.$$

Our aim now is to characterize those g and p such that $G_p/\mathrm{GL}(V)$ is infinite, that is, such that there are infinitely many g-generator, p-relation algebras. This is accomplished by comparing the dimension of an orbit in G_p to the dimension of G_p itself. First, however, we need to know the orbits are subvarieties. This requires a look at the evaluation map $\mathrm{GL}(V) \times D_p \to D_p$.

We will show that the evaluation map $\alpha\colon \mathrm{GL}(V) \times D_p \to D_p$ given by

(7.3) $$\alpha(\theta, w) = \wedge^p(\wedge^2\theta)(w)$$

is a morphism of affine varieties. We proceed by showing that the representation $\wedge^p\colon \mathrm{GL}(W) \to \mathrm{GL}(\wedge^p W)$ is a morphism for any vector space W.

Fix a basis $w_1, \ldots, w_n$ of W. Then all elements of the type $w_{i_1} \wedge \ldots \wedge w_{i_p}$ where $1 \le i_1 < i_2 < \ldots < i_p \le n$ form a basis of $\wedge^p W$. Let I be the index set consisting of ordered p-tuples of increasing integers between 1 and n. We take g_{ij}, $1 \le i, j \le n$, as coordinate functions in $\mathrm{GL}(W)$ and $h_{\alpha\beta}$, $\alpha, \beta \in I$ as coordinate functions in $\mathrm{GL}(\wedge^p W)$ determined by the respective bases chosen above. If α, $\beta \in I$, the (α, β)th coordinate function of $\wedge^p$ is the minor of $|g_{ij}|$ determined by the rows $\alpha(1), \ldots, \alpha(p)$ and the columns $\beta(1), \ldots, \beta(p)$. The coordinate ring of $\mathrm{GL}(\wedge^p W)$ is generated by the $h_{\alpha\beta}$ together with $1/\det|h_{\alpha\beta}|$, while that of $\mathrm{GL}(W)$ is generated by the various g_{ij} together with $1/\det|g_{ij}|$. The coordinate functions of $\wedge^p$ are polynomials in the g_{ij}, so to show $\wedge^p$ is a morphism it suffices to show that $1/\det|\wedge^p(g_{ij})|$ is a polynomial in g_{ij} and $1/\det|g_{ij}|$. For this, the following lemma is useful.

Lemma 7.3. *Every rational character of* $\mathrm{GL}(W)$ *is an integral power of the determinant.*

Proof. By rational is meant that the coordinate function is a sum of quotients of polynomials in the coordinates of W. (See [6, p. 22].)

Now $\det \circ \wedge^p\colon \mathrm{GL}(W) \to K$ is a homogeneous polynomial character, thus $\det|\wedge^p(g_{ij})| = (\det|g_{ij}|)^a$ for some positive integer a, and we have proven

Lemma 7.4. *Let W be a K-vector space. The representation* $\wedge^p\colon \mathrm{GL}(W) \to \mathrm{GL}(\wedge^p W)$ *is a morphism of affine varieties.*

We mention without proof the following

```
1  \Example={1k}    \Volume 199, 370 (1974) % To place stuff in the right margin.
2  \Header{}{E. P. Kronstadt} % Defined on page 14.
3  \pageno=370
4  \def\Ma{{\scriptstyle\cal M}_A} % To imitate, badly, a symbol in the original.
5  \begingroup    % To confine some of the new settings made below.
6  \raggedright  \pretolerance2000 % These match the style of the original.
7
8  \noindent interpolating sequence w.r.t.\ every uniform algebra, we still say
9  $S$ is a general interpolating sequence.
10
11 If $S=\{a_i\}_{i=1}^\infty\subset D^n$ is an interpolating sequence, then it
12 is known that $S$ must be uniformly separated, i.e.\ there exists a constant
13 $M$ and functions $f_1,f_2,\cdots\in H^\infty(D^n)$ such that for all
14 $i$, $\Vert f_i\Vert\leq M$ and $f_i(a_i)=1$ while $f_i$ is zero on the
15 remaining points of $S$. (We will use $H^\infty(D^n)$ for $H^\infty(D^n,
16 {\bf C})$.) In 1958, L. Carleson [3] showed that for $S\subset D$ $(D=D^1)$,
17 uniform separation is a necessary and sufficient condition that $S$ be an
18 interpolating sequence (w.r.t.~{\bf C}). In this paper we attempt to extend
19 Carleson's results to sequences in~$D^n$ and to general interpolating
20 sequences. We succeed to the extent of showing that uniform separation
21 is sufficient in a large class of sequences in $D^n$ and~$D$.
22
23 After introducing notation in \S1, we review in \S2 some of the concepts
24 and results presented in~[10]. \S\S3 and~4 are devoted to a number of
25 techniques for constructing general interpolating sequences. \S5 contains the
26 main result of this paper demonstrating that uniformly separated sequences
27 in ``wedges'' of $D^n$ are general interpolating sequences. \S6 presents some
28 miscellaneous examples of general interpolating sequences, and the last
29 section indicates some directions in which all these results might be
30 generalized.
31 \medskip
32 1. Throughout this paper, $A$ will be a uniform algebra. We will let $\Ma$
33 denote the maximal ideal space or spectrum of $A$. $\Ma$ is of course a
34 subset of $A^\ast$, the dual space of $A$, and as such can be equipped with
35 the weak-$^\ast$ topology in which it is a compact Hausdorff space. We will
36 identify elements of $A$ with their Gelfand transforms, so we may consider
37 elements of $A$ as functions on $\Ma$.
38
39 It should be noted that $\Ma$ is naturally supplied with two topologies as a
40 subset of $A^\ast$---the weak-$^\ast$ topology and the norm topology. We will
41 sometimes have occasion to use both, and it may also happen that subsets
42 of $\Ma$ are naturally associated with other topologies. Throughout this
43 paper, if $Q\subset \Ma$, we will consistently use the symbol ${\rm Cl}(Q)$
44 to represent the weak-$^\ast$ closure of $Q$ and will explicitly mention all
45 other topologies. We will also use the symbol ${\rm Hk}(Q)$ for the hull of
46 the kernel of $Q$, i.e.
47 $$ {\rm Hk}(Q) = \{\,s\in\Ma : x(s)=0 \hbox{\ for every $x\in A$ for which
48 $x(Q)=0$}\,\}. $$
49
50 The norm topology for $\Ma$ is equivalent to the one induced by the
51 ``pseudo-hyperbolic'' metric:
52 $$ \rho(z,w)={\rm Sup}\{\,\vert x(z)\vert : x\in A, \Vert x\Vert \leq 1
53 \hbox{\ and\ }x(w)=0\,\}. $$
54
55 We will use $D$ to represent the open unit disk in the complex plane,~{\bf C},
56 and~$D^n$ for the open unit polydisk in $n$-dimensional complex
57 space~${\bf C}^n$. $T^n$ will refer to the essential boundary of $D^n$, i.e.
58 \par \endgroup \EndPage
```

Example 1k

Source:
*Transactions
of the
American
Mathematical
Society*
199, 370 (1974)

interpolating sequence w.r.t. every uniform algebra, we still say S is a general
interpolating sequence.

If $S = \{a_i\}_{i=1}^\infty \subset D^n$ is an interpolating sequence, then it is known that
S must be uniformly separated, i.e. there exists a constant M and functions
$f_1, f_2, \cdots \in H^\infty(D^n)$ such that for all i, $\|f_i\| \leq M$ and $f_i(a_i) = 1$ while f_i is
zero on the remaining points of S. (We will use $H^\infty(D^n)$ for $H^\infty(D^n, \mathbf{C})$.) In
1958, L. Carleson [3] showed that for $S \subset D$ ($D = D^1$), uniform separation
is a necessary and sufficient condition that S be an interpolating sequence
(w.r.t. $\mathbf{C}$). In this paper we attempt to extend Carleson's results to sequences
in D^n and to general interpolating sequences. We succeed to the extent of
showing that uniform separation is sufficient in a large class of sequences in
D^n and D.

After introducing notation in §1, we review in §2 some of the concepts and
results presented in [10]. §§3 and 4 are devoted to a number of techniques for
constructing general interpolating sequences. §5 contains the main result of
this paper demonstrating that uniformly separated sequences in "wedges"
of D^n are general interpolating sequences. §6 presents some miscellaneous
examples of general interpolating sequences, and the last section indicates
some directions in which all these results might be generalized.

1. Throughout this paper, A will be a uniform algebra. We will let $\mathcal{M}_A$
denote the maximal ideal space or spectrum of A. $\mathcal{M}_A$ is of course a subset
of A^*, the dual space of A, and as such can be equipped with the weak-*
topology in which it is a compact Hausdorff space. We will identify elements
of A with their Gelfand transforms, so we may consider elements of A as
functions on $\mathcal{M}_A$.

It should be noted that $\mathcal{M}_A$ is naturally supplied with two topologies as a
subset of A^*—the weak-* topology and the norm topology. We will sometimes
have occasion to use both, and it may also happen that subsets of $\mathcal{M}_A$ are
naturally associated with other topologies. Throughout this paper, if $Q \subset \mathcal{M}_A$,
we will consistently use the symbol $\mathrm{Cl}(Q)$ to represent the weak-* closure of Q
and will explicitly mention all other topologies. We will also use the symbol
$\mathrm{Hk}(Q)$ for the hull of the kernel of Q, i.e.

$$\mathrm{Hk}(Q) = \{\, s \in \mathcal{M}_A : x(s) = 0 \text{ for every } x \in A \text{ for which } x(Q) = 0 \,\}.$$

The norm topology for $\mathcal{M}_A$ is equivalent to the one induced by the
"pseudo-hyperbolic" metric:

$$\rho(z, w) = \mathrm{Sup}\{\, |x(z)| : x \in A, \|x\| \leq 1 \text{ and } x(w) = 0 \,\}.$$

We will use D to represent the open unit disk in the complex plane, $\mathbf{C}$,
and D^n for the open unit polydisk in n-dimensional complex space $\mathbf{C}^n$. T^n
will refer to the essential boundary of D^n, i.e.

```
1  \Example={11}    \Volume 226, 372 (1977) % To place stuff in the right margin.
2  \Header{}{P. W. Millar} % Defined on page 14.
3  \pageno=372
4  \def\Thm #1. #2\par{\smallskip{\Smallcaps #1\ }{\it #2\par}}
5  \def\Prf #1.{\smallskip{\Smallcaps #1.}}
6
7  \Thm Proposition 3.1. Assume that $-X$ is not a subordinator. The process
8  $\{X^\lambda(M+t)-I^\lambda, t>0\}$ is conditionally independent of
9  ${\cal F}^\lambda(M\,-)$, given $S^\lambda>M$: if
10 $A\in{\cal F}^\lambda(M\,-)$, $B\in\sigma\{X^\lambda(M+t)-I^\lambda,\,t>0\}$,
11 then
12 $$ P\{\,A\cap B\mid S^\lambda>M\,\} = P\{\,A\mid S^\lambda>M\,\}P
13 \{\,B\mid S^\lambda>M\,\}. $$
14
15 \Prf Proof. Since $-X$ is not a subordinator, $P\{S^\lambda>M\}>0$. Let
16 $\epsilon>0$ and define $T_0=\inf\{\,t>0:\vert X_t\vert >\epsilon\,\}$,
17 $T_n=\inf\{\,t>0:\vert X(t+T_n)-X(T_n)\vert >\epsilon\,\}$, $n\geq 1$. Let
18 $X_0=0$, $X_n=X(T_n)$, $n\geq1$. By the strong Markov property,
19 $\{X_n,n\geq 0\}$ is a random walk; $\{T_n,n\geq 0\}$ is a sequence of
20 independent, identically distributed random variables; and $(X_0,X_1,\ldots,
21 X_n,T_0,\ldots,T_{n-1})$ is independent of $\sigma\{X_{n+j}-X_n, j\geq1;
22 T_i, i\geq n\}$. The right continuous process $\{Y_t, t\geq 0\}$ defined by
23 $$\eqalign{ Y_t & = 0, \qquad t<T_0, \cr
24 & = X_n, \quad T_0+\cdots+T_{n-1} < t \leq T_0+\cdots+T_n, \cr}\leqno(3.3) $$
25 then converges as $\epsilon\to 0$ pathwise, uniformly to $\{X_t, t\geq 0\}$.
26 Define the process $\{Y_t^\lambda, t\geq o\}$ by
27 $$ Y_t^\lambda = Y_t \quad {\rm if}\;\; t<T_0+\cdots+T_N, \leqno(3.4) $$
28 where $N=\inf\{\,n:T_0+\cdots+T_n>S^\lambda\,\}$;
29 $$ Y_t^\lambda=\infty \quad {\rm if}\; t\geq T_0+\cdots+T_N. $$
30 Then as $\epsilon\to 0$, $Y^\lambda$ converges pathwise to $X^\lambda$, and
31 uniformly for $t\in [0,S)$. Let $Q$ be the index $i$ for which $X_i$ is the
32 (last) minimum of $Y^\lambda$, let $Y_i^\lambda$, $i\geq 0$, be the
33 successive values of $Y^\lambda$, and $T_i^\lambda$ the interjump times for
34 $Y^\lambda$. So if $\lambda>0$, there will exist an $i$ such that
35 $Y_i^\lambda=T_i^\lambda=\infty$. Notice that $Y_Q^\lambda$ is finite
36 with probability~1 and that as $\epsilon\to 0$, $Y_Q^\lambda$ converges to
37 $I^\lambda=\inf\nolimits_s X_s^\lambda$. Let $A$, $B$, $C$, $D$ be Borel
38 subsets of $(-\infty,\infty)$. Then, for example, if $i\geq 1$
39 $$\eqalign{P&\{Y_{Q-i}^\lambda \in A,\; T_{Q-i}^\lambda \in B,\;
40 Y_{Q+k}^\lambda - Y_Q^\lambda \in C,\; T_{Q+k}^\lambda \in D,\; N>Q>i\} \cr
41 &=\{\textstyle\sum\limits_{l>i}} P\{Y_{l-i}^\lambda \in A,\;
42 T_{l-i}^\lambda \in B,\; Y_{l+k}^\lambda - Y_l^\lambda \in C,\;
43 T_{l+k}^\lambda \in D,\; N>Q=l\}. \cr} \leqno(3.5) $$
44 Since $C$, $D$ do not contain $\infty$, a typical term in the summation
45 of~(3.5) may be written
46 $$\eqalign{{\textstyle\sum\limits_{n\geq l+k}} P \{ Q=l, Y_{l-i} \in A,\; &
47 T_{l-i} \in B,\; Y_{l+k} - Y_l \in C,\; T_{l+k} \in D, \cr
48 & T_0+\cdots+T_{n-1} \leq S < T_0+\cdots+T_n \}. \cr} \leqno(3.6) $$ \EndPage
```

NOTES ──

4, 7: \Thm, used in several previous examples—and defined slightly differently for each one—is a prototype for commands that automate the typesetting of Theorem statements (or of other special declarations). It is modelled after the command \proclaim in Plain TeX. When used, it reads everything that follows, until it sees a period immediately followed by a space, then sets that material in small capitals. It reads subsequent material, until the first paragraph-end (i.e, until, say, a blank line, or an occurrence of the command \par) and sets it in italics. Finally, it ends the paragraph itself. Other refinements are possible, and the definitions on earlier pages show a few (like arranging vertical spacing around the statement).

5: \Prf is similar to \Thm, but simpler.

| 39

Example 11

Source:

*Transactions
of the
American
Mathematical
Society*
226, 372 (1977)

PROPOSITION 3.1 *Assume that* $-X$ *is not a subordinator.   The process* $\{X^\lambda(M+t)-I^\lambda, t > 0\}$ *is conditionally independent of* $\mathcal{F}^\lambda(M-)$, *given* $S^\lambda > M$: *if* $A \in \mathcal{F}^\lambda(M-)$, $B \in \sigma\{X^\lambda(M+t) - I^\lambda, t > 0\}$, *then*

$$P\{A \cap B \mid S^\lambda > M\} = P\{A \mid S^\lambda > M\}P\{B \mid S^\lambda > M\}.$$

PROOF. Since $-X$ is not a subordinator, $P\{S^\lambda > M\} > 0$. Let $\epsilon > 0$ and define $T_0 = \inf\{t > 0 : |X_t| > \epsilon\}$, $T_n = \inf\{t > 0 : |X(t+T_n)-X(T_n)| > \epsilon\}$, $n \geq 1$. Let $X_0 = 0$, $X_n = X(T_n)$, $n \geq 1$. By the strong Markov property, $\{X_n, n \geq 0\}$ is a random walk; $\{T_n, n \geq 0\}$ is a sequence of independent, identically distributed random variables; and $(X_0, X_1, \ldots, X_n, T_0, \ldots, T_{n-1})$ is independent of $\sigma\{X_{n+j} - X_n, j \geq 1; T_i, i \geq n\}$. The right continuous process $\{Y_t, t \geq 0\}$ defined by

$$(3.3) \quad \begin{aligned} Y_t &= 0, \qquad t < T_0, \\ &= X_n, \quad T_0 + \cdots + T_{n-1} < t \leq T_0 + \cdots + T_n, \end{aligned}$$

then converges as $\epsilon \to 0$ pathwise, uniformly to $\{X_t, t \geq 0\}$. Define the process $\{Y_t^\lambda, t \geq o\}$ by

$$(3.4) \qquad Y_t^\lambda = Y_t \quad \text{if } t < T_0 + \cdots + T_N,$$

where $N = \inf\{n : T_0 + \cdots + T_n > S^\lambda\}$;

$$Y_t^\lambda = \infty \quad \text{if } t \geq T_0 + \cdots + T_N.$$

Then as $\epsilon \to 0$, Y^λ converges pathwise to X^λ, and uniformly for $t \in [0, S)$. Let Q be the index i for which X_i is the (last) minimum of Y^λ, let Y_i^λ, $i \geq 0$, be the successive values of Y^λ, and T_i^λ the interjump times for Y^λ. So if $\lambda > 0$, there will exist an i such that $Y_i^\lambda = T_i^\lambda = \infty$. Notice that Y_Q^λ is finite with probability 1 and that as $\epsilon \to 0$, Y_Q^λ converges to $I^\lambda = \inf_s X_s^\lambda$. Let A, B, C, D be Borel subsets of $(-\infty, \infty)$. Then, for example, if $i \geq 1$

$$(3.5) \quad \begin{aligned} &P\{Y_{Q-i}^\lambda \in A, T_{Q-i}^\lambda \in B, Y_{Q+k}^\lambda - Y_Q^\lambda \in C, T_{Q+k}^\lambda \in D, N > Q > i\} \\ &= \sum_{l>i} P\{Y_{l-i}^\lambda \in A, T_{l-i}^\lambda \in B, Y_{l+k}^\lambda - Y_l^\lambda \in C, T_{l+k}^\lambda \in D, N > Q = l\}. \end{aligned}$$

Since C, D do not contain ∞, a typical term in the summation of (3.5) may be written

$$(3.6) \quad \begin{aligned} \sum_{n \geq l+k} P\{Q = l, Y_{l-i} \in A, T_{l-i} \in B, Y_{l+k} - Y_l \in C, T_{l+k} \in D, \\ T_0 + \cdots + T_{n-1} \leq S < T_0 + \cdots + T_n\}. \end{aligned}$$

```
 1 \BookTitle={Semi-Riemannian Geometry}
 2 \Author={Barrett O'Neill}
 3 \CopyrightDate={1983}
 4 \PublishInfo={Academic Press}
 5
 6 \NewExample
 7 Matrices crop up everywhere, and they often pose special typesetting
 8 problems. The input for the partitioned matrices below illustrates some of the
 9 workarounds necessary in Plain~\TeX\ for unusual matrix effects. Example~17
10 uses \LamSTeX\ to show how that package handles such arrangemerts. Only
11 fragments from the source are displayed here; no attempt is made to match
12 the page layout.
13
14 \def\Vrule{\smash{\vrule height7pt depth\baselineskip}}
15 \def\LVrule{\smash{\vrule height1.5\baselineskip depth1pt}}
16 \def\Squeeze{\noalign{\vskip-.5\baselineskip}}
17 \def\Hrule #1{\Squeeze\multispan#1\hrulefill}
18 \def\CompressMatrices{\ifmmode \def\quad{\hskip.5em\relax}\fi}
19 \def\M{\hphantom{-}}
20 \font\Bigrm=cmr12
21
22 \ShortExample <318>
23 \textindent{(1)} {\it The automorphism $\sigma$  of $SO(n+1)$}. By the column
24 vector conventions, $\zeta$ can be regarded as the diagonal matrix with
25 entries 1, $-1$, \dots, $-1$. Thus by Lemma~28, if $a\in SO(n+1)$,
26 $${\CompressMatrices
27 \sigma(a) = \zeta a \zeta =
28 \left[\matrix{\M a_{00}&\Vrule&-a_{01}&\cdots&-a_{0n} \cr
29       \Hrule5\cr
30       -a_{10}&\Vrule \cr
31       \vdots&\Vrule&&(a_{ij}) \cr
32       -a_{n0}&\LVrule \cr }\right] }
33 \quad (1\le i,\; j\le n). $$
34 So ${\rm Fix}(\sigma)$ is $S(O(1)\times O(n))$, and $F_0$ is the isotropy
35 group $1\times SO(n)\approx SO(n)$.
36 \EndShortExample
37
38 \ShortExample <261>
39 \smallskip
40 $${\CompressMatrices
41 \pmatrix{&\Vrule\cr \Squeeze
42         \matrix{\M a&b\cr -b&a} & \Vrule & \hbox{\Bigrm0} \cr
43         \Squeeze &\Vrule\cr \Hrule3\cr
44         \M\hbox{\Bigrm0} & \LVrule & D_{n-2}\cr}} \quad (b\ne0); $$
45 \EndShortExample \EndPage
```

NOTES

1–4, 6, 22, 36, 38, 45: Formatting commands from the file OutForm.tex. See Chapter 6.

14–20: New commands aimed at formatting partitioned matrices.

14–15: These give two types of vertical rules; the first extends into the line below, the second into the line above. \smash hides their true vertical size from TeX's processing.

16: Gives negative vertical space.

17: Gives a horizontal rule spanning a specified number of columns (see lines 29 and 43).

18: This is a trick to work around the fact that a \quad of intercolumn spacing is hardwired into TeX's \matrix command. A check is made to see if math mode is being used, and if it is, \quad is redefined to half its normal size. This is living dangerously, but it works, and it has no effect on anything outside the formula where it is used.

19: Leaves a space the width of a minus sign, since such a space is often needed in aligning entries.

20: Gives the large typeface needed in some places.

26–33, 40–44: These lines show how the new commands are used.

Plain TeX

———————— **Example 2** ————————

From *Semi-Riemannian Geometry*
BARRETT O'NEILL
Academic Press, 1983

Matrices crop up everywhere, and they often pose special typesetting problems. The input for the partitioned matrices below illustrates some of the workarounds necessary in Plain TeX for unusual matrix effects. Example 17 uses LAMS-TeX to show how that package handles such arrangements. Only fragments from the source are displayed here; no attempt is made to match the page layout.

(Page 318)

(1) *The automorphism σ of $SO(n+1)$.* By the column vector conventions, ζ can be regarded as the diagonal matrix with entries $1, -1, \ldots, -1$. Thus by Lemma 28, if $a \in SO(n+1)$,

$$\sigma(a) = \zeta a \zeta = \left[\begin{array}{c|ccc} a_{00} & -a_{01} & \cdots & -a_{0n} \\ \hline -a_{10} & & & \\ \vdots & & (a_{ij}) & \\ -a_{n0} & & & \end{array}\right] \quad (1 \leq i,\, j \leq n).$$

So $\mathrm{Fix}(\sigma)$ is $S(O(1) \times O(n))$, and F_0 is the isotropy group $1 \times SO(n) \approx SO(n)$.

(Page 261)

$$\left(\begin{array}{cc|c} a & b & \\ -b & a & 0 \\ \hline & 0 & D_{n-2} \end{array}\right) \quad (b \neq 0);$$

41

```
 1  \BookTitle={Geometry of Manifolds}
 2  \Author={Richard L.~Bishop \& Richard J.~Crittenden}
 3  \CopyrightDate={1964}
 4  \PublishInfo={Academic Press}
 5
 6  \ExampleType{\TeX\ and Bookmaking}%
 7  \NewExample
 8  This example illustrates how another fairly straightforward piece of
 9  mathematical text can be typeset with \TeX. It also illustrates how some of
10  the book-formatting commands developed here may be used to give a number of
11  ..........................................................................
12  that comes with Plain \TeX, and it is obviously inadequate. Later in the
13  present book, better examples will be shown; these examples make use of
14  special arrow fonts (see Examples~17 and~18).
15  \EndPage
16
17  % Formatting commands for the 'output' part of the right-hand pages:
18  \LMarginOdd=.75 in  % Sets the left margin for odd-numbered pages.
19  \LMarginEven=1 in   % Sets the left margin for even-numbered pages.
20  \TopLRPno           % Puts page numbers on the top, at the left or right.
21  \hsize 4.25 in  \vsize 6.65 in % Sets a page size.
22
23  % Now, select fonts (see Chapter 5 for an explanation of the names):
24  \ChapterFonts{cmssbx10 scaled \magstephalf}{cmssi12 scaled\magstep2}
25  \SectionFont{cmssbx10}
26  \SubSectionFont{cmss10}
27  \PropositionFonts{cmssbx10}{cmr10}
28  \ProofFont{cmss10}
29
30  % Utility definitions:
31  \def\IP #1{\langle#1\rangle}
32  \def\<#1>{\ifmmode \IP{#1}\else $\IP{#1}$\fi}
33  \def\PHI {{\mit \Phi}}
34  \def\Gd {\ifmmode G_{d,2}\else $G_{d,2}$\fi}
```

NOTES

1–4, 6, 7, 15: Formatting commands from the file `OutForm.tex`; see Chapter 6.

18–28: Some book-formatting commands from the file `BookForm.tex`.

18–19: The left margin on the odd- and even-numbered pages is specified in this way; right margins are not specified, because the normal reference line for horizontal positioning in TeX is the left edge of the page. The actual left margins on the facing page will be slightly, but systematically, different from the specified values depending on how exactly the present book is "trimmed" by the printer and bound.

24–28: These select the typefaces to be used for the titles of chapters, for section headings, for stating theorems, etc. Chapter 5 gives information on typeface names.

31–34: Utility definitions set up to ease the burden of some of the typing for this example.

T_EX *and Bookmaking*

────────────── **Example 3** ──────────────

From *Geometry of Manifolds*
RICHARD L. BISHOP & RICHARD J. CRITTENDEN
Academic Press, 1964

This example illustrates how another fairly straightforward piece of mathematical text can be typeset with T_EX. It also illustrates how some of the book-formatting commands developed here may be used to give a number of effects: running headlines that vary from page to page, automatic numbering and positioning of chapter and section headings (using specially selected typefaces) and so forth. The left and top margins will have approximately the same sizes as in the original.

The definitions of the new commands are all given in the "book format" file, `BookForm.tex`, listed in Chapter 6. There are, as well, brief discussions there on why each command is set up the way it is. Commands of this kind allow much of the dull routine of document preparation to be automated and thus, as far as users are concerned, abolished.

The source for this Example has been followed fairly closely, with two exceptions. Some formulas in the original contain the character "o" taken from a special font called fraktur: later examples (starting with Example 5) will use this font, but for now boldface is used as a substitute. The other liberty was taken with the diagram on page 57 (page 167 of the source). The diagonal arrow used there is the longest that comes with Plain T_EX, and it is obviously inadequate. Later in the present book, better examples will be shown; these examples make use of special arrow fonts (see Examples 17 and 18).

43

```
1  % First set initial values for page numbers, the chapter number, etc.:
2  \StartBook    % This initializes everything.
3  \pageno=161   \ChapNo=8 % Current page number and previous chapter number.
4
5  \Chapter{Riemannian Curvature}
6
7  The main properties of the Riemannian curvature are established, including
8  direct, but for the most part impractical, method of computing curvature.
9  Following a selection of examples, the Jacobi equation is established for
10 vector fields associated with rectangles with geodesic longitudinals
11 and a number of local and global consequences are derived. In particular, it
12 is shown that for a complete Riemannian manifold with nonpositive curvature,
13 the exponential is a covering map [{\sl 24, 33, 50, 83\/}].
14
15 \Section{Riemannian Curvature}
16
17 Let $M$ be a $d$-dimensional Riemannian manifold with metric \<\;,\;>
18 and curvature transformation $R_{st}$, $s$, $t$ tangents to $M$. A
19 {\it plane section $P$ at $m\in M$} is a 2-dimensional subspace of $M_m$.
20
21 Let $P$ be a plane section at $m$, and let $s$, $t\in M_m$ be two vectors
22 spanning $P$.
23
24 The {\it Riemannian\/} (or {\it sectional\/}) {\it curvature of}
25 $P$\negthinspace, $K(P)$,  is defined~by
26 $$ K(P)={\<R_{st}s,t> \over A(s,t)^2} $$
27 where $A(s,t)=(\Vert s\Vert^2\;\Vert t\Vert^2-\<s,t>^2)^{1/2}$ is the area of
28 the parallelogram spanned by $s$ and $t$.
29
30 The first aim is to prove that $K(P)$ depends on $P$ alone and not on the
31 particular choice of $s$ and $t$ spanning $P$. Simultaneously, it will be
32 proved that $K(P)$ determines $R_{st}$, and so nothing is lost by considering
33 the Riemannian curvature instead of the curvature form on $F(M)$.
34
```

NOTES ───

2, 3, 5, 15: Apart from **\pageno**, these are the special book-formatting commands defined in the file **BookForm.tex**. See Chapter 6.

26, 27: The \<...,...> combination was defined on the previous page.

The rest of the page is straightforward.

Source:

*Geometry of
Manifolds*
RICHARD
L. BISHOP
& RICHARD
J. CRITTENDEN
© 1964
Academic Press

CHAPTER 9

Riemannian Curvature

The main properties of the Riemannian curvature are established, including direct, but for the most part impractical, method of computing curvature. Following a selection of examples, the Jacobi equation is established for vector fields associated with rectangles with geodesic longitudinals and a number of local and global consequences are derived. In particular, it is shown that for a complete Riemannian manifold with nonpositive curvature, the exponential is a covering map [*24, 33, 50, 83*].

9.1 Riemannian Curvature

Let M be a d-dimensional Riemannian manifold with metric $\langle \, , \, \rangle$ and curvature transformation R_{st}, s, t tangents to M. A *plane section* P *at* $m \in M$ is a 2-dimensional subspace of M_m.

Let P be a plane section at m, and let s, $t \in M_m$ be two vectors spanning P.

The *Riemannian* (or *sectional*) *curvature of* P, $K(P)$, is defined by

$$K(P) = \frac{\langle R_{st}s, t \rangle}{A(s,t)^2}$$

where $A(s,t) = (\|s\|^2 \|t\|^2 - \langle s,t \rangle^2)^{1/2}$ is the area of the parallelogram spanned by s and t.

The first aim is to prove that $K(P)$ depends on P alone and not on the particular choice of s and t spanning P. Simultaneously, it will be proved that $K(P)$ determines R_{st}, and so nothing is lost by considering the Riemannian curvature instead of the curvature form on $F(M)$.

```
1  Note that, if $\dim M=2$, there is only one plane section at each $m\in M$,
2  and so $K$ is a real-valued function on $M$, called the {\it Gaussian
3  curvature}.
4
5  \ProofsEtc Problem 1. Let $f\colon M\to N$ be a local isometry. Show that
6  $f$ preserves curvature. Prove that the $d$-dimensional Riemannian sphere
7  has constant curvature, and hence also the $d$-dimensional real projective
8  space.
9
10 \Proposition Lemma 1. For $x$, $y$, $z$, $w\in M_m$, the curvature tensor $R$
11 satisfies the following properties:
12
13 \smallskip {\baselineskip18 pt \settabs 2\columns
14 \+\quad (a) $R_{xy}=-R{yx}$,
15 &\quad (b) $\<R_{xy}z,w>=-\<z,R_{xy}w>$,\cr
16 \+\quad (c) $R_{xy}z+R_{zx}y+R_{yz}x=0$,
17 &\quad (d) $\<R_{xy}z,w>=\<R_{zw}x,y>$.\cr}
18 \medskip
19
20 One way of interpreting these properties is to view $R$ as a linear
21 transformation of the second Grassmann space $G^2_m$, that is, the space of
22 bivectors. To do this we define $R_{xy}$ to be the bivector which satisfies
23 $\<R_{xy},zw> = \<R_{xy}z,w>$ for all decomposable
24 bivectors $zw$. Then (b) says that $R_{xy}$ is a well-defined bivector,
25 (a) says that it depends only on the bivector $xy$ and not on $x$ and $y$
26 individually, so that $xy\to R_{xy}$ can be extended linearly to an
27 endomorphism of all bivectors; (d) says that $R$ is a symmetric
28 transformation of bivectors, so the corresponding quadratic form determines
29 the transformation. Furthermore, (c) says that the quadratic form is
30 determined by its values on the decomposable elements alone. (See corollary~2
31 below.)
32
33 If $x_1,\ldots,x_d$ is a coordinate system at $m$, then the classical object
34 $R_{ijkl}$ is given by
35 $$R_{ijkl}=\<R_{X_iX_j}X_k,X_l>$$
36 where $X_i=D_{x_i}$. The above formulas then correspond to the classical ones,
37 namely:
38
39 \smallskip {\baselineskip18 pt \settabs 2\columns
40 \+\quad (a${}'$) $R_{ijkl}=-R_{jikl}$,
41 &\quad (b${}'$) $R_{ijkl}=-R_{ijlk}$,\cr
42 \+\quad (c${}'$) $R_{ijkl}+R_{kijl}+R_{jkil}=0$,
43 &\quad (d${}'$) $R_{ijkl}=R_{klij}$.\cr}
44 \medskip
45
46 \ProofsEtc Proof of lemma 1. Let $x,y\in M_m$, $b\in F(M)$, $\bar x, \bar y$
47 \in F(M)_b$ such that $d\pi\;\bar x=x$, $d\pi\;\bar y=y$. Then by 6.1.5, if
48 $z\in M_m$,
49 $$R_{xy}z=-b\PHI(\bar x,\bar y)b^{-1}z,$$
50 where $b$ is regarded as a map $R^d\to M_m$.
51
```

NOTES _____

5, 10, 46: These lines show how various formatting commands from `BookForm.tex` are used. Observe how the typefaces selected on page 42, lines 27–28 show up in the output here.

15, 17, 23, 35, 49: The commands \<...,...> and \PHI were defined on page 42.

Example 3

Source:

*Geometry of
Manifolds*
RICHARD
L. BISHOP
& RICHARD
J. CRITTENDEN
© 1964
Academic Press

Note that, if $\dim M = 2$, there is only one plane section at each $m \in M$, and so K is a real-valued function on M, called the *Gaussian curvature*.

Problem 1. Let $f: M \to N$ be a local isometry. Show that f preserves curvature. Prove that the d-dimensional Riemannian sphere has constant curvature, and hence also the d-dimensional real projective space.

Lemma 1. For $x, y, z, w \in M_m$, the curvature tensor R satisfies the following properties:

(a) $R_{xy} = -Ryx$,

(b) $\langle R_{xy}z, w \rangle = -\langle z, R_{xy}w \rangle$,

(c) $R_{xy}z + R_{zx}y + R_{yz}x = 0$,

(d) $\langle R_{xy}z, w \rangle = \langle R_{zw}x, y \rangle$.

One way of interpreting these properties is to view R as a linear transformation of the second Grassmann space G_m^2, that is, the space of bivectors. To do this we define R_{xy} to be the bivector which satisfies $\langle R_{xy}, zw \rangle = \langle R_{xy}z, w \rangle$ for all decomposable bivectors zw. Then (b) says that R_{xy} is a well-defined bivector, (a) says that it depends only on the bivector xy and not on x and y individually, so that $xy \to R_{xy}$ can be extended linearly to an endomorphism of all bivectors; (d) says that R is a symmetric transformation of bivectors, so the corresponding quadratic form determines the transformation. Furthermore, (c) says that the quadratic form is determined by its values on the decomposable elements alone. (See corollary 2 below.)

If $x_1, \ldots, x_d$ is a coordinate system at m, then the classical object R_{ijkl} is given by

$$R_{ijkl} = \langle R_{X_i X_j} X_k, X_l \rangle$$

where $X_i = D_{x_i}$. The above formulas then correspond to the classical ones, namely:

(a') $R_{ijkl} = -R_{jikl}$,

(b') $R_{ijkl} = -R_{ijlk}$,

(c') $R_{ijkl} + R_{kijl} + R_{jkil} = 0$,

(d') $R_{ijkl} = R_{klij}$.

Proof of lemma 1. Let $x, y \in M_m$, $b \in F(M)$, $\bar{x}, \bar{y} \in F(M)_b$ such that $d\pi\,\bar{x} = x$, $d\pi\,\bar{y} = y$. Then by 6.1.5, if $z \in M_m$,

$$R_{xy}z = -b\Phi(\bar{x}, \bar{y})b^{-1}z,$$

where b is regarded as a map $R^d \to M_m$.

48

```
1   (a) then follows from the fact that $\PHI$ is a 2-form, and hence is
2   alternating, while (b) follows since $\PHI$ is ${\bf o}(d)$-valued,
3   ${\bf o}(d)$ consisting of skew-symmetric transformations of $R^d$.
4
5   In order to prove (c), we first notice that for $a$, $b$, $c\in R^d$,
6   $$\eqalignno{
7   H[[E(a),E(b)],E(c)]&=-H[\lambda\PHI(E(a),E(b)),E(c)]\qquad\qquad
8   &\rm (6.1.4)\qquad\cr
9   &\!=-E(\PHI(E(a),E(b))c)
10  &\rm (6.2.1).\quad\qquad\cr}$$
11
12  Choosing $\bar x$, $\bar y$ above in a particular way, we have
13  $$R_{xy}z=-b\PHI(E(b^{-1}(x))(b),E(b^{-1}(y))(b))b^{-1}(z),$$
14  so that
15  $$\eqalignno{
16  E(b^{-1}R_{xy}z)(b)&=-E(\PHI(E(b^{-1}(x))(b),E(b^{-1}(y))(b))b^{-1}
17  z)(b)&\cr
18  &\!=H[[E(b^{-1}(x)),E(b^{-1}(y))],E(b^{-1}(z))](b).&\cr}$$
19  But then the Jacobi identity gives
20  $$E(b^{-1}(R_{xy}z+R_{zx}y+R_{yz}x))(b)=0,$$
21  which proves (c), since $E$ and $b$ are one-to-one.
22
23  (d) follows from (a), (b), (c) by taking inner products of equation~(c)
24  with~$w$, then cyclicly permuting $x$, $y$, $z$, $w$. The four equations thus
25  obtained are then added, and proper use of (a) and~(b) will yield~(d). Details
26  are left to the reader.
27
28  \Proposition Corollary 1. K(P) is well defined.
29
30  \ProofsEtc Proof. For $x$, $y\in M_m$, let $K(x,y) = \<R_{xy}x,y>
31  /A(x,y)^2$. We point out that
32  $$\leqalignno{\qquad\qquad&K(x,y)=K(y,x),&\qquad\llap{\rm (i)}\cr
33  &K(ax,by)=K(x,y),\qquad \hbox{if $ab\neq0$},&\qquad\llap{\rm (ii)}\cr
34  &K(x+cy,y)=K(x,y).&\qquad\llap{\rm (iii)}\cr}$$
35  It then follows that if $x'=ax+by$, $y'=cx+dy$, $ad-bc\ne 0$, then
36  $K(x',y)=K(x,y)$, since it is well known that the transformation from $(x,y)$
37  to $(x',y')$ can be obtained by a sequence of the types indicated in (i),
38  (ii), and (iii).
39
```

NOTES

The command `\PHI` occurs in several places on this page; it is defined on page 42.

6–10: The `\quad` and `\qquad` commands next to the equation numbers are there only to match the style of the original. Without them, `\eqalignno`—one of TeX's basic formula aligning commands—would put equation numbers flush right. Incidentally, it is more common to align equations exactly along the '=' signs than the way they appear on the

right-hand page. The slight shift of the second line to the left is achieved by using the negative space given by `\!`.

28, 30: Commands from `BookForm.tex`, or defined on page 42.

32–34: The `\llap` commands are used to make the equation numbers line up correctly—they cause the numbers to overlap the `\qquad` of space from the left.

| 49

Example 3

Source:

Geometry of Manifolds

RICHARD
L. BISHOP
& RICHARD
J. CRITTENDEN

© 1964
Academic Press

(a) then follows from the fact that Φ is a 2-form, and hence is alternating, while (b) follows since Φ is $\mathbf{o}(d)$-valued, $\mathbf{o}(d)$ consisting of skew-symmetric transformations of R^d.

In order to prove (c), we first notice that for $a, b, c \in R^d$,

$$H[[E(a), E(b)], E(c)] = -H[\lambda\Phi(E(a), E(b)), E(c)] \qquad (6.1.4)$$
$$= -E(\Phi(E(a), E(b))c) \qquad (6.2.1).$$

Choosing $\bar{x}, \bar{y}$ above in a particular way, we have

$$R_{xy}z = -b\Phi(E(b^{-1}(x))(b), E(b^{-1}(y))(b))b^{-1}(z),$$

so that

$$E(b^{-1}R_{xy}z)(b) = -E(\Phi(E(b^{-1}(x))(b), E(b^{-1}(y))(b))b^{-1}z)(b)$$
$$= H[[E(b^{-1}(x)), E(b^{-1}(y))], E(b^{-1}(z))](b).$$

But then the Jacobi identity gives

$$E(b^{-1}(R_{xy}z + R_{zx}y + R_{yz}x))(b) = 0,$$

which proves (c), since E and b are one-to-one.

(d) follows from (a), (b), (c) by taking inner products of equation (c) with w, then cyclicly permuting x, y, z, w. The four equations thus obtained are then added, and proper use of (a) and (b) will yield (d). Details are left to the reader.

Corollary 1.   K(P) is well defined.

Proof.   For $x, y \in M_m$, let $K(x, y) = \langle R_{xy}x, y \rangle / A(x, y)^2$. We point out that

(i) $K(x, y) = K(y, x),$

(ii) $K(ax, by) = K(x, y), \qquad$ if $ab \neq 0,$

(iii) $K(x + cy, y) = K(x, y).$

It then follows that if $x' = ax + by$, $y' = cx + dy$, $ad - bc \neq 0$, then $K(x', y) = K(x, y)$, since it is well known that the transformation from (x, y) to (x', y') can be obtained by a sequence of the types indicated in (i), (ii), and (iii).

\Proposition Corollary 2. The \<R_{xy}x,y> determines the
curvature transformations.

\ProofsEtc Proof. More precisely, \<R_{xy}z,w> is the only 4-linear function
satisfying the properties of lemma~1 which restricts to \<R_{xy}z,y>. So we
assume that we have two 4-linear functions, f and f', on M_m which
satisfy the conditions corresponding to (a)--(d) and such that $f(x,y,x,y)$
=f'(x,y,x,y)$, all x, $y\in M_m$. Letting $g=f-f'$, we see that g
satisfies these same conditions corresponding to (a)--(d). Replacing x by
$x+z$ in $g(x,y,x,y)=0$ we get
$$ g(x,y,x,y)+g(z,y,x,y)+g(x,y,z,y)=0 $$
and hence,
$$ g(x,y,z,y)+g(z,y,x,y)=0. $$
By (d)
$$ g(x,y,z,y)=0. $$
Replacing y by $y+w$ and following the same procedure gives
$$ g(x,y,z,w)+g(x,y,z,w)=0. $$

By (d), then (a), we get
$$g(x,y,z,w)=g(y,z,x,w), $$
so $g(.,\, .,\, ., w)$ is invariant under cyclic permutation of the three
entries. But the sum over such permutations is 0 by (c), so $g=0$. QED

\ProofsEtc Remarks. (1) Sometimes it is more convenient to deal with
curvature instead of curvature transformations and this corollary assures
us that this will not lose information.

(2) If M has two Riemannian structures such that at a single point the inner
product and curvature are the same, then the curvature transformations are the
same.

(3) It is not correct to say that the curvature determines the curvature
transformation, for two different Riemannian structures, with different
curvature transformations, may give rise to the same curvature. For example,
let $f\colon S^2\to S^2$ be any diffeomorphism of the Riemannian 2-sphere.
Viewing f as an isometry gives two Riemannian structures on S^2 with
different curvature transformations but the same (constant) curvature.

Example 3

Source:

*Geometry of
Manifolds*

RICHARD
L. BISHOP
& RICHARD
J. CRITTENDEN

© 1964
Academic Press

Corollary 2.   The $\langle R_{xy}x, y\rangle$ determines the curvature transformations.

Proof.   More precisely, $\langle R_{xy}z, w\rangle$ is the only 4-linear function satisfying the properties of lemma 1 which restricts to $\langle R_{xy}z, y\rangle$. So we assume that we have two 4-linear functions, f and f', on M_m which satisfy the conditions corresponding to (a)–(d) and such that $f(x, y, x, y) = f'(x, y, x, y)$, all $x, y \in M_m$. Letting $g = f - f'$, we see that g satisfies these same conditions corresponding to (a)–(d). Replacing x by $x + z$ in $g(x, y, x, y) = 0$ we get

$$g(x, y, x, y) + g(z, y, x, y) + g(x, y, z, y) = 0$$

and hence,

$$g(x, y, z, y) + g(z, y, x, y) = 0.$$

By (d)

$$g(x, y, z, y) = 0.$$

Replacing y by $y + w$ and following the same procedure gives

$$g(x, y, z, w) + g(x, y, z, w) = 0.$$

By (d), then (a), we get

$$g(x, y, z, w) = g(y, z, x, w),$$

so $g(.,\,.,\,.,w)$ is invariant under cyclic permutation of the three entries. But the sum over such permutations is 0 by (c), so $g = 0$. QED

Remarks.   (1) Sometimes it is more convenient to deal with curvature instead of curvature transformations and this corollary assures us that this will not lose information.

(2) If M has two Riemannian structures such that at a single point the inner product and curvature are the same, then the curvature transformations are the same.

(3) It is not correct to say that the curvature determines the curvature transformation, for two different Riemannian structures, with different curvature transformations, may give rise to the same curvature. For example, let $f: S^2 \to S^2$ be any diffeomorphism of the Riemannian 2-sphere. Viewing f as an isometry gives two Riemannian structures on S^2 with different curvature transformations but the same (constant) curvature.

```
 1  \ProofsEtc Problem 2. Let $q(x,y)=f(x,y,x,y)$, $x$, $y\in M_m$. Establish the
 2  following explicit formula for $f$ in terms of $q$:
 3  $$\eqalign{& 6f(x,y,z,w) = q(x+z,y+w) - q(x+w,y+z) \cr
 4  &\qquad + q(x,y+z) - q(x,y+w) - q(y,x+z) + q(y,x+w) \cr
 5  &\qquad - q(z,y+w) + q(z,x+w) - q(w,x+z) + q(w,y+z) \cr
 6  &\qquad + q(x,w) - q(x,z) + q(y,w). \cr}$$
 7
 8  \ProofsEtc Problem 3. Use the following outline to prove
 9  {\it Schur's theorem\/} [{\sl 17\/}]: If $K$ is constant on every fibre of
10  $\Gd(M)$, then $K$ is constant on $\Gd(M)$, for $d>2$.
11
12  (a) This hypothesis is equivalent to: for every $x$, $y\in R^d$,\break
13  \<\PHI(E(x),E(y))x,y> is constant on fibres of $F(M)$.
14
15  (b) Since the functions depending on $x$, $y$ in (a) determine the
16  functions \<\PHI(E(x),E(y))z,w>, the hypothesis is equivalent to
17  $$\PHI(E(x),E(y))\quad \hbox{is constant on fibres of $F(M)$.}$$
18  \indent (c) If $F\PHI(E(x),E(y))=0$ for every vertical $F$, $x$, $y\in R^d$,
19  then $E(z)\PHI(E(x),E(y))=0$ for every $x$, $y$, $z\in R^d$, and hence
20  $\PHI(E(x),E(y))$ is constant on $F(M)$, $K$ constant on $\Gd(M)$.
21  \smallskip
22  \noindent [{\it Hint\/}: Use the Bianchi identity
23  $$\displaylines{
24  D\PHI(E(x),E(y),E(z)) = E(x)\PHI(E(y),E(z)) + E(y)\PHI(E(z),E(x))\hfill\cr
25  \hfill {} + E(z)\PHI(E(x),E(y)) = 0,\cr}$$
26  and the fact that
27  $$ [\bar A, E(x)] = \bar A E(x) - E(x)\bar A = E(Ax) \quad {\rm for} \quad
28  A\in {\bf o}(d),$$
29  so $E(x)\bar A + E(Ax) = \bar A E(x)$.]

31  \Section{Computation of the Riemannian Curvature}

33  We indicate briefly how the Riemannian curvature can be computed in terms
34  of the metric coefficients $g_{ij}$. In particular, we show the connection
35  between the curvature transformation and the metric.

37  By 6.4.3, if $X$, $Y$ are vector fields, then
38  $$R_{XY} = \nabla_{[X,Y]} - [\nabla_X,\nabla_Y],$$
```

NOTES _____

The commands \Gd, \PHI and \<... ,... > were all defined on page 42.

Other non-standard commands (their names start with capital letters) are from BookForm.tex; see Chapter 6.

31: Observe that the \Section command (from BookForm.tex) does many things. It places the section title flush left, automatically numbers it, and automatically changes the headline as well.

Example 3

Source:

*Geometry of
Manifolds*

RICHARD
L. BISHOP
& RICHARD
J. CRITTENDEN

© 1964
Academic Press

Problem 2.   Let $q(x, y) = f(x, y, x, y)$, x, $y \in M_m$. Establish the following explicit formula for f in terms of q:

$$6f(x, y, z, w) = q(x + z, y + w) - q(x + w, y + z)$$
$$+ q(x, y + z) - q(x, y + w) - q(y, x + z) + q(y, x + w)$$
$$- q(z, y + w) + q(z, x + w) - q(w, x + z) + q(w, y + z)$$
$$+ q(x, w) - q(x, z) + q(y, w).$$

Problem 3.   Use the following outline to prove *Schur's theorem* [17]: If K is constant on every fibre of $G_{d,2}(M)$, then K is constant on $G_{d,2}(M)$, for $d > 2$.

(a) This hypothesis is equivalent to: for every x, $y \in R^d$, $\langle \Phi(E(x), E(y))x, y \rangle$ is constant on fibres of $F(M)$.

(b) Since the functions depending on x, y in (a) determine the functions $\langle \Phi(E(x), E(y))z, w \rangle$, the hypothesis is equivalent to

$$\Phi(E(x), E(y)) \quad \text{is constant on fibres of } F(M).$$

(c) If $F\Phi(E(x), E(y)) = 0$ for every vertical F, x, $y \in R^d$, then $E(z)\Phi(E(x), E(y)) = 0$ for every $x, y, z \in R^d$, and hence $\Phi(E(x), E(y))$ is constant on $F(M)$, K constant on $G_{d,2}(M)$.

[*Hint*: Use the Bianchi identity

$$D\Phi(E(x), E(y), E(z)) = E(x)\Phi(E(y), E(z)) + E(y)\Phi(E(z), E(x))$$
$$+ E(z)\Phi(E(x), E(y)) = 0,$$

and the fact that

$$[\bar{A}, E(x)] = \bar{A}E(x) - E(x)\bar{A} = E(Ax) \quad \text{for} \quad A \in \mathbf{o}(d),$$

so $E(x)\bar{A} + E(Ax) = \bar{A}E(x)$.]

9.2 Computation of the Riemannian Curvature

We indicate briefly how the Riemannian curvature can be computed in terms of the metric coefficients g_{ij}. In particular, we show the connection between the curvature transformation and the metric.

By 6.4.3, if X, Y are vector fields, then

$$R_{XY} = \nabla_{[X,Y]} - [\nabla_X, \nabla_Y],$$

```
 1  where $\nabla_X$ is the covariant derivative in the direction of $X$.
 2  However, by [{\sl 66}, p.~77] if $X$, $Y$, $Z$ are vector fields,
 3  $$\displaylines{
 4  2\<\nabla_XY,Z> = X\<Y,Z> + Y\<X,Z> - Z\<X,Y> \qquad \cr
 5  \qquad \qquad + \<[X,Y],Z> + \<[Z,X],Y> + \<X,[Z,Y]>. \cr}$$
 6  These two formulas give the desired connection.
 7
 8  \ProofsEtc Problem 4. The above formula depends on the following facts:
 9
10  (1) Torsion zero if and only if $[X,Y]=\nabla_XY-\nabla_YX$, $X$, $Y$ any
11  $C^\infty$ vector fields.
12
13  (2) Parallel translation preserves the inner product if and only if $X\<Y,Z>
14  = \<\nabla_XY,Z> + \<Y,\nabla_XZ>$. Prove these statements and the formula.
15  Derive an explicit formula for $K(D_{x_i},D_{x_j})$ in terms of the~$g_{ij}$.
16
17  \ProofsEtc Problem 5. Use this formula to obtain an alternate proof for
18  problem~7.18.
19
20  \Section{Continuity of the Riemannian curvature}
21  \def\tmprel {\lower3pt\hbox{$\buildrel\approx\over\to$}} % For temporary use.
22
23  $K$ is not a function on $M$, the Riemannian manifold, but it is a function
24  on the Grassman bundle of 2-planes of $M$ [3.3(5)], and in fact a continuous
25  function. From this it will follow that the curvature on a compact subset
26  of $M$ is bounded.
27
28  Let \Gd\ be the Grassman manifold of plane sections (two-\break
29  dimensional subspaces) of $R^d$. (See problem 7.30.) So
30  $$\Gd=O(d)/O(2) \times O'(d-2).$$
31  We denote by $\Gd(M)$ the bundle with fibre \Gd\ associated to the frame
32  bundle $F(M)$, where $M$ is a Riemannian manifold. Thus \break
33  $\Gd(M)=F(M)\times
34  {}_{O(d)}\Gd$. If $m\in M$, we write $\Gd(m)$ for the fibre of $\Gd(M)$
35  over $m$. If $b\in F(M)$ such that $\pi(b)=m$, then $b\colon\Gd\tmprel\Gd(m)$
36  by: $P\to \{(b,P)\}=(b,P)O(d)$, the equivalence class of $(b,P)$ in $\Gd(M)$.
37  But we know that $b\colon R^d\tmprel M_m$, so $b\colon\Gd\tmprel\allowbreak
38  \{\hbox{plane sections at }m\}$, and the resulting identification of $\Gd(m)$
39  with $\{\{$plane sections at~$m\}\}$ is independent of $b$. Hence, the Riemannian
40  curvature $K$ can be viewed as a real-valued function defined on $\Gd(M)$.
41  (We are here using the notation $F(M) \times {}_{O(d)}\Gd$ for the space
42  $(F(M) \times \Gd)/O(d)$ of 3.3.)
43
```

NOTES

21: This shows how symbols can be built "on the fly" for temporary use.

35: This is how the new relation, `\tmprel`, is used.

Source:

*Geometry of
Manifolds*

RICHARD
L. BISHOP
& RICHARD
J. CRITTENDEN

© 1964
Academic Press

where ∇_X is the covariant derivative in the direction of X. However, by [66, p. 77] if X, Y, Z are vector fields,

$$2\langle \nabla_X Y, Z \rangle = X\langle Y, Z \rangle + Y\langle X, Z \rangle - Z\langle X, Y \rangle$$
$$+ \langle [X, Y], Z \rangle + \langle [Z, X], Y \rangle + \langle X, [Z, Y] \rangle.$$

These two formulas give the desired connection.

Problem 4.   The above formula depends on the following facts:

(1) Torsion zero if and only if $[X, Y] = \nabla_X Y - \nabla_Y X$, X, Y any C^∞ vector fields.

(2) Parallel translation preserves the inner product if and only if $X\langle Y, Z \rangle = \langle \nabla_X Y, Z \rangle + \langle Y, \nabla_X Z \rangle$. Prove these statements and the formula. Derive an explicit formula for $K(D_{x_i}, D_{x_j})$ in terms of the g_{ij}.

Problem 5.   Use this formula to obtain an alternate proof for problem 7.18.

9.3 Continuity of the Riemannian curvature

K is not a function on M, the Riemannian manifold, but it is a function on the Grassman bundle of 2-planes of M [3.3(5)], and in fact a continuous function. From this it will follow that the curvature on a compact subset of M is bounded.

Let $G_{d,2}$ be the Grassman manifold of plane sections (two-dimensional subspaces) of R^d. (See problem 7.30.) So

$$G_{d,2} = O(d)/O(2) \times O'(d-2).$$

We denote by $G_{d,2}(M)$ the bundle with fibre $G_{d,2}$ associated to the frame bundle $F(M)$, where M is a Riemannian manifold. Thus $G_{d,2}(M) = F(M) \times_{O(d)} G_{d,2}$. If $m \in M$, we write $G_{d,2}(m)$ for the fibre of $G_{d,2}(M)$ over m. If $b \in F(M)$ such that $\pi(b) = m$, then $b: G_{d,2} \xrightarrow{\approx} G_{d,2}(m)$ by: $P \to \{(b, P)\} = (b, P)O(d)$, the equivalence class of (b, P) in $G_{d,2}(M)$. But we know that $b: R^d \xrightarrow{\approx} M_m$, so $b: G_{d,2} \xrightarrow{\approx}$ {plane sections at m}, and the resulting identification of $G_{d,2}(m)$ with {plane sections at m} is independent of b. Hence, the Riemannian curvature K can be viewed as a real-valued function defined on $G_{d,2}(M)$. (We are here using the notation $F(M) \times_{O(d)} G_{d,2}$ for the space $(F(M) \times G_{d,2})/O(d)$ of 3.3.)

```
1  \Proposition Proposition 1. The function $K\colon\Gd(M)\to R$ is $C^\infty$,
2  and hence, in particular, continuous.
3
4  \ProofsEtc Proof. Consider the diagram
5  $$\matrix{\matrix{
6  F(M)\times O(d)\cr \downarrow\,p\cr F(M)\times\Gd\cr \downarrow\,q}
7  &\searrow\; K\circ q\circ p\cr
8  F(M)\times {}_{O(d)}\Gd
9  &{\buildrel K\over \longrightarrow}\quad R\hfill}$$
10 $$\scriptstyle p,\; q\; \hbox{\sevenrm identification maps}$$
11 \medskip
12 It is only necessary to show that $K\circ q\circ p$ is $C^\infty$. To define
13 the map $p$ we must first choose an element, say $P_0$, of \Gd. Then $p(b,g)
14 = (b,gP_0)$. Hence, $K\circ q\circ p(b,g) = K(b(gP_0))$, remembering that
15 $b\colon\Gd\to\{\hbox{plane sections at }m\}$. Let $P_0$ be spanned by
16 orthonormal vectors $x$, $y\in R^d$. Then
17 $$\eqalign{
18 K(bgP_0)&= \<R_{b(gx)b(gy)}b(gx),b(gy)> \cr
19 &=-\<b\PHI(E(gx)(b),E(gy)(b))gx,gy>,\cr}$$
20 which is clearly $C^\infty$ in $b$ and~$g$. QED
21
22 Since \Gd\ is compact, we have the:
23
24 \Proposition Corollary. If $C\subset M$ is compact, then there exist
25 $H$, $L\in M$ such that for any plane section $P$ at any point $m\in C$,
26 $H\leq K(P)\leq L$.
27
28 % Need the blank line above, or an explicit '\par' command.
29 \EnoughAlready
30
```

NOTES ⎯⎯⎯⎯⎯⎯⎯⎯⎯⎯⎯⎯⎯⎯⎯⎯⎯⎯⎯⎯⎯⎯⎯⎯⎯⎯⎯

27: The blank line is needed because the command \Proposition expects a paragraph-end to mark the end of the statement.

29: Defined in the file OutForm.tex; see Chapter 6. The example ends here.

| 57

Example 3

Source:

Geometry of Manifolds

RICHARD L. BISHOP & RICHARD J. CRITTENDEN

© 1964 Academic Press

Proposition 1.   The function $K: G_{d,2}(M) \to R$ is C^∞, and hence, in particular, continuous.

Proof.   Consider the diagram

$$
\begin{array}{c}
F(M) \times O(d) \\
\downarrow p \\
F(M) \times G_{d,2} \qquad \searrow \; K \circ q \circ p \\
\downarrow q \\
F(M) \times_{O(d)} G_{d,2} \xrightarrow{\;K\;} R
\end{array}
$$

p, q identification maps

It is only necessary to show that $K \circ q \circ p$ is C^∞. To define the map p we must first choose an element, say P_0, of $G_{d,2}$. Then $p(b, g) = (b, gP_0)$. Hence, $K \circ q \circ p(b, g) = K(b(gP_0))$, remembering that $b: G_{d,2} \to \{$plane sections at $m\}$. Let P_0 be spanned by orthonormal vectors $x, y \in R^d$. Then

$$
\begin{aligned}
K(bgP_0) &= \langle R_{b(gx)b(gy)} b(gx), b(gy) \rangle \\
&= -\langle b\Phi(E(gx)(b), E(gy)(b))gx, gy \rangle,
\end{aligned}
$$

which is clearly C^∞ in b and g. QED

Since $G_{d,2}$ is compact, we have the:

Corollary.   If $C \subset M$ is compact, then there exist $H, L \in M$ such that for any plane section P at any point $m \in C$, $H \le K(P) \le L$.

$\vdots$

```
 1 │ \BookTitle={Partial Differential Equations in Physics}
 2 │ \Author={Arnold Sommerfeld}
 3 │ \CopyrightDate={1964}
 4 │ \PublishInfo={Academic Press}
 5 │
 6 │ \NewExample
 7 │ For the most part, all that the typesetting of differential equations involves
 8 │ .........................................................................
 9 │ the bottom right-hand corner of each page of the original was located. As
10 │ with the previous example, the top and left margins here will only
11 │ approximately match those of the source.
12 │
13 │ \EndPage
14 │ % Formatting commands for the full right-hand pages:
15 │ \MarkPageCorner     % This gives a bottom right-hand corner 'cropmark'.
16 │ \BotMargin=.375in   % Sets the bottom margin.
17 │
18 │ % Formatting commands for the book displayed on the right-hand pages:
19 │ \LMarginOdd=.625 in  % Sets the left margin for odd-numbered pages.
20 │ \LMarginEven=.625 in % Sets the left margin for even-numbered pages.
21 │ \hsize 4.25 in  \vsize 6.875 in  % Sets the page size.
22 │ \voffset-.125in       % Sets the vertical position.
23 │ \parindent15pt        % Sets the paragraph indentation.
24 │ \FootnoteSize{.2\vsize} % At most, this fraction of the page for footnotes.
25 │ \SmallFootnotes        % Chooses small type for footnotes.
26 │ \TopLRpno              % Puts page numbers at the top left and right.
27 │ \CenteredTitles        % Centers all titles.
28 │ \NumberFootnotes       % Starts automatic footnote numbering.
29 │
30 │ \EveryChapter={Chapter \RomanNumeral{\the\ChapNo}} % Ditto for chapter heads.
31 │ \EverySection={\S\ \the\SectNo.} % Sets repeated material for section heads.
32 │ \EverySubSection={} % Nothing is chosen here for subsections.
33 │ \DontResetSectNos     % The section numbers will accumulate through chapters.
34 │
35 │ % Select fonts (see Chapter 5 for an explanation of the names):
36 │ \ChapterFonts{cmcsc10}{cmbx12}
37 │ \SectionFont{cmbx10}
38 │ \SubSectionFont{cmcsc10}
39 │
40 │ % Choose spacings:
41 │ \StartofChapSkip=30pt plus4pt minus2pt % Gap to be left above the title.
42 │ \InChapTitleSkip=\bigskipamount        % Gap between lines in the title.
43 │ \AfterChapTitleSkip=\InChapTitleSkip   % Gap to be left below the title.
44 │ \BetweenSectionSkip=\InChapTitleSkip   % Gap to be left between sections.
45 │ \AfterSectionTitleSkip=\medskipamount  % Gap to be left after section titles.
46 │
47 │ % Set the headline format:
48 │ \LeftRunningHead={\hfil {\Smallcaps partial differential equations}\hfil
49 │          \llap{\bf \S\thinspace\the\SectNo\FirstEquationOnPage}}
50 │ \RightRunningHead={\rlap{\bf\S\thinspace\the\SectNo\LastEquationOnPage}\hfil
51 │                  {\Smallcaps \Lcase{\the\ChapName}}\hfil}
```

NOTES

1–4, 6, 13, 15–16: Commands from `OutForm.tex` (see Chapter 6) that set the output format.

19–51: These set the book format for this example. See the file `BookForm.tex` in Chapter 6 for the definitions. The commands are fairly flexible; slightly different settings from the previous example (see page 42) give a very different look to this one. Some of the commands used here were not needed there since that example matched many of the default settings of `BookForm.tex`.

TEX and Bookmaking

————————— **Example 4** —————————

From *Partial Differential Equations in Physics*
ARNOLD SOMMERFELD
Academic Press, 1964

For the most part, all that the typesetting of differential equations involves is a knowledge of how to use the \over command (or \frac in $\mathcal{AMS}$-TEX) and of where to look up such symbols as ∂ or ∇. Occasionally, there are also slightly subtle, but again not terribly complicated, issues of spacing involved. This example illustrates some of these issues.

The chief novelties here are, however, the side shows (or, more correctly, the top and bottom shows). The footnotes in Sommerfeld's book are set in eight-point type, they involve a fair amount of mathematics (including displays) and they often stretch over several pages. For the most part, the footnotes are distinguished from the main page solely by the small type and the narrow intervening strip of white space. In a few cases, mainly where the material just above the footnote appears similar to that in it, the compositor of the original added a horizontal rule. Since this seems best done "by eye," the reproduction shown here provides a mechanism to switch on and off such a rule as needed (after a trial run).

The headlines in the original contain equation numbers. This example introduces automatic equation numbering and a mechanism whereby these numbers can find their way to the headline.

Another typographically interesting feature here is the diagram on page 71 (page 75 of the source). Later examples will discuss more systematically the question of including figures in TEX documents; right now the point being made is that certain classes of simple figures can be drawn—with some effort—purely from within Plain TEX.

The edition of Sommerfeld's book that has been used as the source is the inexpensive paperback edition (which explains the small overall page size) from 1964. A "crop mark" at the bottom right indicates where the bottom right-hand corner of each page of the original was located. As with the previous example, the top and left margins here will only approximately match those of the source.

```
 1  \StartBook              % Sets the initial values of page numbers, etc.
 2  \Chapter{Fourier Series and Integrals}
 3
 4  Fourier's Th\'eorie analytique de la chaleur\footnote{Jean Baptiste Fourier,
 5  1768--1830. His book on the conduction of heat appeared in 1822 in Paris.
 6  Fourier also distinguished himself as an algebraist, engineer, and writer
 7  on the history of Egypt, where he had accompanied Napoleon.\hfil\break
 8  \indent The influence of his book even outside France is illustrated by the
 9  following quotation: ''Fourier's incentive kindled the spark in (the then
10  16-year-old) William Thomson as well as in Franz Neumann.'' (F. Klein,
11  Vorlesungen \"uber die Geschichte der Mathematik im 19.\ Jahrhundert, v.~I,
12  p.~223.)}
13  is the bible of the mathematical physicist. It contains not only an
14  exposition of the trigonometric series and integrals named after Fourier, but
15  the general boundary value problem is treated in an exemplary fashion for the
16  typical case of heat conduction.
17
18  In mathematical lectures on Fourier series emphasis is usually put on the
19  concept of arbitrary function, on its continuity properties and its
20  singularities (accumulation points of an infinity of maxima and minima).
21  This point of view becomes immaterial in the physical applications. For, the
22  initial or boundary values of functions considered here, partially because
23  of the atomist nature of matter and of interaction, must always be taken as
24  smoothed mean values, just as the partial differential equations in which
25  they enter arise from a statistical averaging of much more complicated
26  elementary laws. Hence we are concerned with relatively simple idealized
27  functions and with their approximation with ''least possible error.'' What
28  is meant by the latter is explained by Gauss in his ''Method of Least
29  Squares.'' We shall see that it opens a simple and rigorous approach not
30  only to Fourier series but to all other series expansions of mathematical
31  physics in spherical and cylindrical harmonics, or generally in
32  eigenfunctions.
33
34  \Section{Fourier Series}
35  Let an arbitrary function $f(x)$ be given in the interval $-\pi\leq x\leq
36  +\pi$; this function may, e.g., be an empirical curve determined by
37  sufficiently many and sufficiently accurate measurements. We want to
38  approximate it by the sum of $2n+1$ trigonometric terms
39  $$\eqalign{
40  S_n(x)=A_0&{} + A_1\cos x + A_2\cos 2x +\cdots+ A_n\cos nx\cr
41          &{} + B_1\sin x + B_2\sin 2x +\cdots+ B_n\sin nx}\leqno\EqLbl{}$$
```

NOTES

Most of the page is straightforward. Observe that the **\footnote** command now automatically numbers footnotes (see page 58, line 28).
41: **\EqLbl{}** is also from **BookForm.tex**: it numbers equations. If an argument is specified, it will place the argument next to the equation number. For example, if **\EqLbl{a}** had been used above, the equation on the right hand page would have been labelled "(1a)".

Source:

*Partial
Differential
Equations
in Physics*
ARNOLD
SOMMERFELD
© 1964
Academic Press

CHAPTER I

Fourier Series and Integrals

Fourier's Théorie analytique de la chaleur[1] is the bible of the mathematical physicist. It contains not only an exposition of the trigonometric series and integrals named after Fourier, but the general boundary value problem is treated in an exemplary fashion for the typical case of heat conduction.

In mathematical lectures on Fourier series emphasis is usually put on the concept of arbitrary function, on its continuity properties and its singularities (accumulation points of an infinity of maxima and minima). This point of view becomes immaterial in the physical applications. For, the initial or boundary values of functions considered here, partially because of the atomist nature of matter and of interaction, must always be taken as smoothed mean values, just as the partial differential equations in which they enter arise from a statistical averaging of much more complicated elementary laws. Hence we are concerned with relatively simple idealized functions and with their approximation with "least possible error." What is meant by the latter is explained by Gauss in his "Method of Least Squares." We shall see that it opens a simple and rigorous approach not only to Fourier series but to all other series expansions of mathematical physics in spherical and cylindrical harmonics, or generally in eigenfunctions.

§ 1. Fourier Series

Let an arbitrary function $f(x)$ be given in the interval $-\pi \leq x \leq +\pi$; this function may, e.g., be an empirical curve determined by sufficiently many and sufficiently accurate measurements. We want to approximate it by the sum of $2n + 1$ trigonometric terms

$$(1) \quad \begin{aligned} S_n(x) = {} & A_0 + A_1 \cos x + A_2 \cos 2x + \cdots + A_n \cos nx \\ & + B_1 \sin x + B_2 \sin 2x + \cdots + B_n \sin nx \end{aligned}$$

---

[1] Jean Baptiste Fourier, 1768–1830. His book on the conduction of heat appeared in 1822 in Paris. Fourier also distinguished himself as an algebraist, engineer, and writer on the history of Egypt, where he had accompanied Napoleon.

The influence of his book even outside France is illustrated by the following quotation: "Fourier's incentive kindled the spark in (the then 16-year-old) William Thomson as well as in Franz Neumann." (F. Klein, Vorlesungen über die Geschichte der Mathematik im 19. Jahrhundert, v. I, p. 223.)

```
 1 | By what criteria shall we choose the coefficients $A_k,B_k$ at our disposal?
 2 | We shall denote the error term $f(x)-S_n(x)$ by $\epsilon_n(x)$; thus
 3 | $$f(x)=S_n(x)+\epsilon_n(x).\leqno\EqLbl{}$$
 4 | Following Gauss we consider the {\it mean square error
 5 | $$M={1\over 2\pi}\int\limits_{-\pi}^{+\pi}\epsilon_n^2 dx\leqno\EqLbl{}$$
 6 | and reduce $M$ to a minimum through the choice of the $A_k,B_k$}.
 7 |
 8 | To this we further remark that the corresponding measure of the total error
 9 | formed with the first power of $\epsilon_n$ would not be suitable, since
10 | arbitrarily large positive and negative errors could then cancel each other
11 | and would not count in the total error. On the other hand the use of the
12 | absolute value $\vert\epsilon_n\vert$ under the integral sign in place of
13 | $\epsilon_n^2$ would be inconvenient because of its non-analytic
14 | character.\footnote{A completely different approach is taken by the
15 | great Russian mathematician Tchebycheff in the approximation named after him.
16 | He considers not the {\it mean\/} but the {\it maximal\/}
17 | $\vert\epsilon_n\vert$ appearing in the interval of integration, and makes
18 | this a minimum through the choice of the coefficients at his
19 | disposal.}
20 |
21 | The requirement that (\PrevEq) be a minimum leads to the equations
22 | $$\eqalign{
23 | -{\partial M\over \partial A_k} =& {1\over \pi} \int\limits_{-\pi}^{+\pi}
24 | \{f(x)-S_n(x)\}\cos kx dx = 0,\quad k=0,1,2,\ldots, n \cr
25 | -{\partial M\over \partial B_k} =& {1\over \pi} \int\limits_{-\pi}^{+\pi}
26 | \{f(x)-S_n(x)\}\sin kx dx = 0,\quad k=0,1,2,\ldots, n.\cr} \leqno\EqLbl{}$$
27 | \indent
28 | These are exactly the $2n+1$ equations for the determination of the $2n+1$
29 | unknowns $A,B$. A favorable feature here is that each individual coefficient
30 | $A$ or $B$ is determined directly and is not connected recursively with the
31 | other $A,B$. We owe this to the {\it orthogonality relations\/} that exist
32 | among trigonometric functions:%
33 | \footnote{Here and below all integrals
34 | are to be taken from $-\pi\;\hbox{to}\;+\pi$. In order to justify the word
35 | ``orthogonality'' we recall that two vectors {\bf u,v} which are orthogonal
36 | in Euclidean three-dimensional, or for that matter $n$-dimensional space,
37 | satisfy the condition that their scalar product
38 | $$(\hbox{\bf u v})=\sum_1^N\hbox{\bf u}_i\hbox{\bf v}_i = 0$$
39 | vanishes. The integrals appearing in (\PrevEqs{0}) can be considered as
40 | sums of this same type with infinitely many terms. See the remarks in \S 26
41 | about so-called ``Hilbert space.''}
42 | $$\displaylines{\textstyle
43 | \EqLbl{}\hphantom{a}\qquad\qquad \int \cos kx \sin lx dx = 0, \hfill\cr
44 | \noalign{\smallskip}
45 | \ShiftEquationNo{-1}  % To keep the same equation number as before.
46 | \!\!\left.\matrix{\textstyle
47 | \EqLbl{a}\qquad\qquad \int \cos kx \cos lx dx \cr
48 | \noalign{\medskip}
49 | \EqLbl{b}\qquad\qquad \int \sin kx \sin lx dx}\right\rbrace =0,
50 | \> k\ne l.\hfill \cr}$$
51 | \smallbreak
```

NOTES ────────────────

\EqLbl occurs at several places here; see the previous page for a discussion.

21: \PrevEq allows symbolic references to the previous equation number.

Example 4

Source:

*Partial
Differential
Equations
in Physics*
ARNOLD
SOMMERFELD
© 1964
Academic Press

By what criteria shall we choose the coefficients A_k, B_k at our disposal? We shall denote the error term $f(x) - S_n(x)$ by $\epsilon_n(x)$; thus

$$(2) \qquad\qquad f(x) = S_n(x) + \epsilon_n(x).$$

Following Gauss we consider the *mean square error*

$$(3) \qquad\qquad M = \frac{1}{2\pi} \int\limits_{-\pi}^{+\pi} \epsilon_n^2\, dx$$

and reduce M to a minimum through the choice of the A_k, B_k.

To this we further remark that the corresponding measure of the total error formed with the first power of ϵ_n would not be suitable, since arbitrarily large positive and negative errors could then cancel each other and would not count in the total error. On the other hand the use of the absolute value $|\epsilon_n|$ under the integral sign in place of ϵ_n^2 would be inconvenient because of its non-analytic character.[2]

The requirement that (3) be a minimum leads to the equations

$$
\begin{aligned}
(4) \qquad &-\frac{\partial M}{\partial A_k} = \frac{1}{\pi} \int\limits_{-\pi}^{+\pi} \{f(x) - S_n(x)\} \cos kx\, dx = 0, \quad k = 0, 1, 2, \ldots, n \\[2mm]
&-\frac{\partial M}{\partial B_k} = \frac{1}{\pi} \int\limits_{-\pi}^{+\pi} \{f(x) - S_n(x)\} \sin kx\, dx = 0, \quad k = 0, 1, 2, \ldots, n.
\end{aligned}
$$

These are exactly the $2n + 1$ equations for the determination of the $2n + 1$ unknowns A, B. A favorable feature here is that each individual coefficient A or B is determined directly and is not connected recursively with the other A, B. We owe this to the *orthogonality relations* that exist among trigonometric functions:[3]

$$(5) \qquad\qquad \int \cos kx \sin lx\, dx = 0,$$

$$(5a) \qquad\qquad \left. \int \cos kx \cos lx\, dx \right\}$$
$$(5b) \qquad\qquad \left. \int \sin kx \sin lx\, dx \right\} = 0,\ k \neq l.$$

[2] A completely different approach is taken by the great Russian mathematician Tchebycheff in the approximation named after him. He considers not the *mean* but the *maximal* $|\epsilon_n|$ appearing in the interval of integration, and makes this a minimum through the choice of the coefficients at his disposal.

[3] Here and below all integrals are to be taken from $-\pi$ to $+\pi$. In order to justify the word "orthogonality" we recall that two vectors $\mathbf{u}, \mathbf{v}$ which are orthogonal in Euclidean three-dimensional, or for that matter n-dimensional space, satisfy the

```
 1  In order to prove them it is not necessary to write down the cumbersome
 2  addition formulae of trigonometric functions, but to think rather of their
 3  connection with the exponential functions $e^{\pm ikx}$ and $e^{\pm ilx}$.
 4  The integrands of (\PrevEq a,b) consist then of only four terms of the form
 5  $\exp \{\pm i(k+l)x\}$ or $\exp \{\pm i(k-l)x\}$, all of which vanish upon
 6  integration unless $l=k$. This proves (\PrevEq a,b). The fact that (\PrevEq)
 7  is valid even without this restriction follows from the fact that for $l=k$
 8  it reduces~to
 9  $${1\over 4i} \int (e^{2ikx} - e^{-2ikx}) dx = 0. $$
10  In a similar manner one obtains the values of (\PrevEq a,b) for $l=k>0$ (only
11  the product of $\exp(ikx)$ and $\exp(-ikx)$ contributes to it): this value
12  simply becomes equal to $\pi$; for $l=k=0$ the value of the integral in
13  (\PrevEq a) obviously equals $2\pi$. We therefore can replace (\PrevEq a,b)
14  by the single formula which is valid also for $l=k>0$
15  $${1\over \pi}\int \cos kx \cos lx dx={1\over \pi}\int \sin kx \sin lx =
16  \delta_{kl}\leqno\EqLbl{}$$
17  with the usual abbreviation
18  $$\delta_{kl}=\cases{0\ldots l\ne k \cr 1\ldots l=k.\cr}$$
19  Equation (\PrevEq) for $k=l$ is called the {\it normalizing condition}. It is
20  to be augmented for the exceptional case $l=k=0$ by the trivial statement
21  $$\ShiftEquationNo{-1}
22  {1\over 2\pi}\int dx=1.\leqno\EqLbl{a}$$
23  \indent
24  If we now substitute (\PrevEqs2), (\PrevEq) and (\PrevEq a) in (\PrevEqs3)
25  then in the integrals with $S_n$ all terms except the $k$-th vanish, and we
26  obtain directly {\it Fourier's representation of coefficients\/}:
27  \OnFootnoterule{\hsize}
28  $$\matrix{
29  \!\!\left.\matrix{\displaystyle A_k={1\over \pi}\int f(x) \cos kx dx \cr
30       \noalign{\medskip}
31  \displaystyle B_k={1\over \pi}\int f(x) \sin kx dx \cr}\right\rbrace k>0,
32  \hfill\cr
33  \noalign{\medskip}
34  \displaystyle A_0={1\over 2\pi}\int f(x) dx.\hfill \cr}\leqno\EqLbl{}$$
```

NOTES

There are again several occurrences of \PrevEq and \EqLbl on this page. See pages 60 and 62.

21: Backs up the equation number, to match the numbering in the source.

24: \PrevEqs is used like \PrevEq, except that one has to specify how far back one is referring.

27: Switches on a "footnote rule." It wasn't needed here, but the original has one. (The continuation footnote there began with the displayed equation—an act from which TEX abstains—and something was needed to distinguish it from the display at the end of the main page.)

Example 4

Source:

*Partial
Differential
Equations
in Physics*
ARNOLD
SOMMERFELD

© 1964
Academic Press

In order to prove them it is not necessary to write down the cumbersome addition formulae of trigonometric functions, but to think rather of their connection with the exponential functions $e^{\pm ikx}$ and $e^{\pm ilx}$. The integrands of (5a,b) consist then of only four terms of the form $\exp\{\pm i(k+l)x\}$ or $\exp\{\pm i(k-l)x\}$, all of which vanish upon integration unless $l = k$. This proves (5a,b). The fact that (5) is valid even without this restriction follows from the fact that for $l = k$ it reduces to

$$\frac{1}{4i} \int (e^{2ikx} - e^{-2ikx})dx = 0.$$

In a similar manner one obtains the values of (5a,b) for $l = k > 0$ (only the product of $\exp(ikx)$ and $\exp(-ikx)$ contributes to it): this value simply becomes equal to π; for $l = k = 0$ the value of the integral in (5a) obviously equals 2π. We therefore can replace (5a,b) by the single formula which is valid also for $l = k > 0$

$$(6) \qquad \frac{1}{\pi} \int \cos kx \cos lx dx = \frac{1}{\pi} \int \sin kx \sin lx dx = \delta_{kl}$$

with the usual abbreviation

$$\delta_{kl} = \begin{cases} 0 \dots l \neq k \\ 1 \dots l = k. \end{cases}$$

Equation (6) for $k = l$ is called the *normalizing condition*. It is to be augmented for the exceptional case $l = k = 0$ by the trivial statement

$$(6a) \qquad \frac{1}{2\pi} \int dx = 1.$$

If we now substitute (5), (6) and (6a) in (4) then in the integrals with S_n all terms except the k-th vanish, and we obtain directly *Fourier's representation of coefficients*:

$$(7) \qquad \left. \begin{aligned} A_k &= \frac{1}{\pi} \int f(x) \cos kx dx \\ B_k &= \frac{1}{\pi} \int f(x) \sin kx dx \end{aligned} \right\} k > 0,$$

$$A_0 = \frac{1}{2\pi} \int f(x)dx.$$

---

condition that their scalar product

$$(\mathbf{u} \ \mathbf{v}) = \sum_1^N \mathbf{u}_i \mathbf{v}_i = 0$$

vanishes. The integrals appearing in (5) can be considered as sums of this same type with infinitely many terms. See the remarks in §26 about so-called "Hilbert space."

```
1  \indent
2  Our approximation $S_n$ is hereby determined completely. If, e.g., $f(x)$ were
3  given empirically then the integrations (\PrevEq) would have to be carried
4  out numerically or by machine.\footnote{Integrating machines that serve in
5  Fourier analysis are called ``harmonic analyzers.'' The most perfect of these
6  is the machine of Bush and Caldwell; it can be used also for the integration
7  of arbitrary simultaneous differential equations; see {\it Phys.\ Rev.}
8  {\bf 38}, 1898 (1931).}
9
10 From (\PrevEq) one sees directly that for an even function $f(-x)=f(+x)$,
11 all $B_k$ vanish, whereas for an odd function, $f(-x)=-f(+x)$, all $A_k$,
12 including $A_0$, vanish. Hence the former is approximated by a {\it pure
13 cosine series}, the latter by a {\it pure sine series}.
14 \OffFootnoterule
15 \EnoughAlready
16
17 %%% These will be used later:
18 \newdimen\DOne    \newdimen\DTwo    \newbox\BZ
19 \DOne=.1in        \DTwo=.4in        \setbox\BZ=\hbox{$\smash{+\atop\bullet}$}
20 \def\EqBox #1#2{\noindent$\vcenter{\hsize #1 \leftline{$#2$}}$}
21 \def\JagVrule{$\vcenter to32pt{\leaders\vbox to 8pt{\hsize.4pt \vfil\noindent
22              \vrule width.4pt height3pt depth3pt\par\vfil}\vfill}$}%
23 \def\DotHrule #1#2#3{\dimen255=#2\advance\dimen255 by\wd\BZ
24  \vrule height.4pt depth0pt width#1
25  \raise2pt\rlap{$\smash{#3\atop\bullet}$}\vrule height.4pt depth0pt
26  width\dimen255}%
27 \def\Ls #1{\DotHrule{\DOne}{\DTwo}{#1}}%
28 \def\Rs #1{\DotHrule{\DTwo}{\DOne}{#1}}%
29 \def\MainBox #1#2#3#4#5{$\vcenter to .5in{\hbox{\JagVrule\Ls{#1}\JagVrule
30  \Rs{#2}\JagVrule\Ls{#3}\JagVrule\Rs{#4}\JagVrule
31  \DotHrule{\DOne}{.15in}{#5}}\vfil$}%
32 \def\Lbls #1{\dimen255=\DOne\advance\dimen255 by\DTwo
33  \advance\dimen255 by.4pt \advance\dimen255 by\wd\BZ
34  \vbox{\hsize\dimen255 \leftline{$\scriptstyle#1$}}}%
35 \def\LabelBox #1#2#3#4#5{$\setbox1=
36    \hbox{\Lbls{#1}\Lbls{#2}\Lbls{#3}\Lbls{#4}\Lbls{#5}}\ht1=0pt
37    \wd1=0pt \dp1=4pt
38    \dimen255=\DOne\advance\dimen255 by\DTwo\advance\dimen255 by\wd\BZ
39    \vcenter to24pt{\vfil\moveleft.12in\box1
40    \hbox{\JagVrule\hskip\dimen255\JagVrule
41        \lower10pt\hbox to\dimen255{$\leftarrow$\hfil
42        $\scriptstyle\xi$\hfil$\rightarrow$\hskip\DOne\hskip\wd\BZ}\JagVrule
43        \hskip\dimen255\JagVrule\hskip\dimen255\JagVrule\hskip\DOne
44        \hskip\wd\BZ \hskip.15in}}$}%
45
```

NOTES

14: Switches the footnote rule off. The placement of this command is tricky: placed too high, it is read by TeX before the previous footnote is set and it takes effect too soon; placed too low, it isn't read soon enough to affect the footnote it is intended to alter. The placement here was made after trial runs. **15**: From `OutForm.tex`; see Chapter 6.

18–44: Commands that were composed to aid in typesetting the diagram on page 71. They put together a set of boxes, each containing some ingredient of the final picture. The main ones are `\EqBox`, which is used to place a formula in a given space, and `\MainBox`, which typesets the main diagram.

| 67

Example 4

Source:

*Partial
Differential
Equations
in Physics*
ARNOLD
SOMMERFELD

© 1964
Academic Press

Our approximation S_n is hereby determined completely. If, e.g., $f(x)$ were given empirically then the integrations (7) would have to be carried out numerically or by machine.[4]

From (7) one sees directly that for an even function $f(-x) = f(+x)$, all B_k vanish, whereas for an odd function, $f(-x) = -f(+x)$, all A_k, including A_0, vanish. Hence the former is approximated by a *pure cosine series*, the latter by a *pure sine series*.

$\vdots$

---

[4] Integrating machines that serve in Fourier analysis are called "harmonic analyzers." The most perfect of these is the machine of Bush and Caldwell; it can be used also for the integration of arbitrary simultaneous differential equations; see *Phys. Rev.* **38**, 1898 (1931).

```
1  \ChapName={Introduction to Partial Differential Equations}
2  \pageno=33  \SectNo=7   \EquationNo=5  \FootnoteNo=1
3  \def\Pd #1 #2/#3 {{\partial^{#1}#2\over \partial#3^{#1}}}
4
5  $$\sqcap\llap{$\sqcup$}u=0 \qquad \hbox{with} \qquad \rlap{$\sqcap$}\sqcup =
6  \sum_{k=1}^4 \Pd2 /x_k \leqno\EqLbl{}$$
7  by introducing the fourth coordinate $x_4$ (or $x_0$) $=ict$ in addition
8  to the three spatial coordinates $x_1,x_2,x_3$. For an oscillating membrane
9  we have (\PrevEqs2) with two spatial dimensions, for an oscillating string we
10 have one spatial dimension. In the latter case we write
11 $$\Pd2 u/x = {1\over c^2}\Pd2 u/t \qquad \hbox{or sometimes} \quad
12 \hbox{(\PrevEqs{0}a)}\quad \Pd2 u/x - \Pd2 u/y = 0,\leqno\EqLbl{}$$
13 setting, for the time being, $y=ct$ (not $y=ict$). Neither membrane nor
14 string has a proper elasticity; the constant $c$ is computed from the tension
15 imposed from outside and from the density per unit of area or of length.
16
17 In the general theory of elasticity one has, as a special case, the
18 differential equation for the transverse vibrations of a thin disc
19 $${\mit\Delta\Delta}=-{1\over c^2}\Pd2 u/t ,\qquad\qquad
20 {\mit\Delta\Delta}=\Pd4 /x + 2{\partial^4\over \partial x^2 \partial y^2}
21 + \Pd4 /y ;\leqno\EqLbl{}$$
22 for reasons of dimensionality $c$ here does not stand for the velocity of
23 sound in the elastic material as it does in acoustics, but is computed from
24 the elasticity, density, and thickness of the disc. Analogously, the
25 differential equation of an oscillating elastic rod is
26 $$\Pd4 u/x = -{1\over c^2}\Pd2 u/t .\leqno\EqLbl{}$$
27 This will be derived in exercise II.1, where the resulting characteristic
28 frequencies will be compared with the acoustic frequencies of open and of
29 covered pipes.
30 \EnoughAlready
```

NOTES

A jump forward is made, to page 33 in the source. The page begins with a displayed equation, something that TeX would not have done on its own, were it typesetting the full book. In any case, starting here with $$...$$ leads to a lot of white space at the top. Page 34 suggests a way of dealing with such occurrences.

1–2: Several variables are reset, since the new page falls in a new chapter.

3: An abbreviation is introduced in order to reduce the typing needed for the many partial derivatives on the page. The spaces in the definition are significant: they tell the command, for instance, exactly what material goes in the denominator.

5: Since standard TeX fonts don't supply a box operator, a crude one is constructed here. Also see \square and box operator in the Glossary.

6, 11, 12, 19, etc.: The lines show how \Pd is to be used; observe carefully the use of spaces. On the whole, because it is tricky to use such commands correctly, the use of spaces in this way isn't to be engaged in lightly.

Example 4

Source:

*Partial
Differential
Equations
in Physics*
ARNOLD
SOMMERFELD

© 1964
Academic Press

$$(5) \qquad \Box u = 0 \qquad \text{with} \qquad \Box = \sum_{k=1}^{4} \frac{\partial^2}{\partial x_k^2}$$

by introducing the fourth coordinate x_4 (or x_0) $= ict$ in addition to the three spatial coordinates x_1, x_2, x_3. For an oscillating membrane we have (4) with two spatial dimensions, for an oscillating string we have one spatial dimension. In the latter case we write

$$(6) \quad \frac{\partial^2 u}{\partial x^2} = \frac{1}{c^2} \frac{\partial^2 u}{\partial t^2} \qquad \text{or sometimes} \quad (6a) \quad \frac{\partial^2 u}{\partial x^2} - \frac{\partial^2 u}{\partial y^2} = 0,$$

setting, for the time being, $y = ct$ (not $y = ict$). Neither membrane nor string has a proper elasticity; the constant c is computed from the tension imposed from outside and from the density per unit of area or of length.

In the general theory of elasticity one has, as a special case, the differential equation for the transverse vibrations of a thin disc

$$(7) \quad \Delta\Delta = -\frac{1}{c^2} \frac{\partial^2 u}{\partial t^2}, \qquad \Delta\Delta = \frac{\partial^4}{\partial x^4} + 2\frac{\partial^4}{\partial x^2 \partial y^2} + \frac{\partial^4}{\partial y^4};$$

for reasons of dimensionality c here does not stand for the velocity of sound in the elastic material as it does in acoustics, but is computed from the elasticity, density, and thickness of the disc. Analogously, the differential equation of an oscillating elastic rod is

$$(8) \qquad \frac{\partial^4 u}{\partial x^4} = -\frac{1}{c^2} \frac{\partial^2 u}{\partial t^2}.$$

This will be derived in exercise II.1, where the resulting characteristic frequencies will be compared with the acoustic frequencies of open and of covered pipes.

$$\vdots$$

```
 1  \ChapName={Boundary Value Problems in Heat Conduction}
 2  \pageno=75  \SectNo=16
 3  \def\Int {\int\limits_0^1\!\!}
 4  \def\SmFrac #1/#2 {{\textstyle {#1\over #2}}}
 5  \def\Half {\SmFrac 1/2 }
 6
 7  {\bf a)\ \ a)}\par
 8  {\offinterlineskip
 9  \line{\EqBox{1.25in}{u=0\hbox{\ \ for }}\cases{x=0\cr x=1}}
10      \hfil\LabelBox{x=-1}{x=0}{x=1}{x=21}{x=31}\ }%
11  \line{\EqBox{1.25in}{} \hfil \MainBox -+-+- }%
12  \setbox3=\hbox to\hsize{\EqBox{1.25in}{}\hfil \dimen255=\DTwo
13          \advance\dimen255 by\DOne
14          \advance\dimen255 by\wd\BZ\hskip\dimen255 \hskip1pt
15          \hbox to \dimen255{$\longleftarrow\hfil{\scriptstyle l}\hfil
16          \longrightarrow$}\hskip 2\dimen255 \hskip1.4pt \hskip\wd\BZ
17          \hskip\DOne\hskip.15in\ }\ht3=-5pt % The neg. height causes a shift up.
18  \box3  % This places the 'l' label crudely, 'by hand'.
19  \medskip
20  \EqBox{\hsize}{\eqalign{f(x)&=\textstyle\sum B_n \sin\pi n \SmFrac x/l ,\;\;
21      B_n=\SmFrac 2/1 \Int f(x)\sin\pi n \SmFrac x/l dx\cr
22      G&=\vartheta\left(\SmFrac x-\xi/2l \mid\tau\right)-
23      \vartheta\left(\SmFrac x+\xi/2l \mid\tau\right).\cr}}
24  \par}
25
26  \bigskip {\bf b)\ \ b)}\smallskip
27  \line{\EqBox{1.25in}{\Pd{} u/x = 0
28      \hbox{\ \ for }\cases{x=0\cr x=1}}\hfil\MainBox +++++ }
29  \EqBox{\hsize}{\eqalign{f(x)&=\textstyle\sum A_n \cos\pi n \SmFrac x/l ,\;\;
30      A_n=\SmFrac 2/1 \Int f(x)\cos\pi n \SmFrac x/l dx,\;\;
31      A_0=\SmFrac 1/1 \Int f(x)dx,\cr
32      G&=\vartheta\left(\SmFrac x-\xi/2l \mid\tau\right)+
33      \vartheta\left(\SmFrac x+\xi/2l \mid\tau\right).\cr}}
34
35  \bigskip{\bf a)\ \ b)}\smallskip
36  \line{\EqBox{1.25in}{\eqalign{u=0&\;\hbox{\ for }\;x=0\cr
37      \textstyle \Pd{} u/x = 0&\;\hbox{\ \ ''\ \ }\;x=1\cr}}\hfil\MainBox -++-- }
38  \EqBox{\hsize}{\eqalign{f(x)&=\textstyle\sum B_n \sin\pi(n+\Half)\SmFrac x/l ,
39      \;\;B_n=\SmFrac 2/1 \Int f(x)\sin\pi(n+\Half)\SmFrac x/l dx,\cr
40      G&=\vartheta\left(\SmFrac x-\xi/4l \mid\tau\right) -
41      \vartheta\left(\SmFrac x+\xi/4l \mid\tau\right) +
42      \vartheta\left(\SmFrac x+\xi-2l/4l \mid\tau\right) -
43      \vartheta\left(\SmFrac x-\xi-2l/4l \mid\tau\right).\cr}}
44  \EnoughAlready
```

NOTES

This page shows applications of the commands set up on page 66.

3–5: Useful abbreviations.

7–24: The first picture is the worst of the lot, since it involves extra ingredients. Ignoring those for the moment, the pattern in all three pictures is this: a line, given by \line, is filled with two boxes. The first is called \EqBox; it typesets a formula in a specified space. The second is \MainBox, which typesets a sequence of five specified positive and negative charges along a line with dashed vertical lines in between. Below \line comes another \EqBox, this one the width of the page. For the first picture, the extra ingredients are the \line with the labels on the top (input lines 9–10) and the convoluted placement of the label l under the diagram.

27–33, 36–43: These follow the pattern described above; most of the messiness comes from the equations under the diagrams being messy.

Example 4

Source:

*Partial
Differential
Equations
in Physics*
ARNOLD
SOMMERFELD
© 1964
Academic Press

a) a)

$$u = 0 \ \text{for} \ \begin{cases} x = 0 \\ x = l \end{cases}$$

$x=-l$ $x=0$ $x=l$ $x=2l$ $x=3l$

$\leftarrow \xi \rightarrow$

$-$ $+$ $-$ $+$ $-$

$\longleftarrow l \longrightarrow$

$$f(x) = \sum B_n \sin \pi n \tfrac{x}{l}, \quad B_n = \tfrac{2}{l} \int_0^l f(x) \sin \pi n \tfrac{x}{l} dx$$

$$G = \vartheta\left(\tfrac{x-\xi}{2l} \mid \tau\right) - \vartheta\left(\tfrac{x+\xi}{2l} \mid \tau\right).$$

b) b)

$$\tfrac{\partial u}{\partial x} = 0 \ \text{for} \ \begin{cases} x = 0 \\ x = l \end{cases}$$

$+$ $+$ $+$ $+$ $+$

$$f(x) = \sum A_n \cos \pi n \tfrac{x}{l}, \quad A_n = \tfrac{2}{l} \int_0^l f(x) \cos \pi n \tfrac{x}{l} dx, \quad A_0 = \tfrac{1}{l} \int_0^l f(x) dx,$$

$$G = \vartheta\left(\tfrac{x-\xi}{2l} \mid \tau\right) + \vartheta\left(\tfrac{x+\xi}{2l} \mid \tau\right).$$

a) b)

$$u = 0 \ \text{for} \ x = 0$$
$$\tfrac{\partial u}{\partial x} = 0 \ \text{"} \ \ x = l$$

$-$ $+$ $+$ $-$ $-$

$$f(x) = \sum B_n \sin \pi (n + \tfrac{1}{2}) \tfrac{x}{l}, \quad B_n = \tfrac{2}{l} \int_0^l f(x) \sin \pi (n + \tfrac{1}{2}) \tfrac{x}{l} dx,$$

$$G = \vartheta\left(\tfrac{x-\xi}{4l} \mid \tau\right) - \vartheta\left(\tfrac{x+\xi}{4l} \mid \tau\right) + \vartheta\left(\tfrac{x+\xi-2l}{4l} \mid \tau\right) - \vartheta\left(\tfrac{x-\xi-2l}{4l} \mid \tau\right).$$

$\vdots$

```
 1 | \ChapName={Cylinder and Sphere Problems}
 2 | \pageno=116  \SectNo=21  \EquationNo=22
 3 |
 4 | \SubSection{D. Generalization of the Saddle-Point Method According to Debye}
 5 | Although in later applications we shall in general apply only the
 6 | asymptotic limiting value of the Bessel functions as determined at the end of
 7 | \S 19, we wish to discuss here certain more general expansions due to Hankel,
 8 | which progress according to negative powers of $\rho$ and in which the first
 9 | term is the above mentioned asymptotic limiting value. Actually these series
10 | are {\it divergent}, being developments at an essential singularity, but they
11 | are frequently called {\it semi-convergent}. The first terms decrease rapidly,
12 | but from a certain term on they increase to infinity. We obviously must break
13 | off at that term in order to obtain approximation formulas.
14 |
15 | The shortest way of obtaining these series is from the differential equation
16 | for the Hankel function, by substituting formal power series, and then
17 | computing the coefficients by setting the factor of each power equal to zero
18 | (this is obviously not completely rigorous). Considering (19.55) and (19.56)
19 | we write
20 | $$H_n^{1,2}(\rho)=\sqrt{2\over \pi\rho}e^{\pm i(\rho-(n+{1\over 2})\pi/2)}
21 | \Bigl(a_0+{a_1\over\rho} + {a_2\over\rho^2} + \cdots + {a_m\over\rho^m} +
22 | \cdots \Bigr)\leqno\EqLbl{}$$
23 | and after dividing out the factor $\sqrt{2/\pi}\exp\{\pm i(\rho -
24 | (n+{1\over 2})\pi/2)\}$ from the differential equation (19.11) we find the
25 | terms with $\rho^{-m-{3\over 2}}$ to be
26 | $$\matrix{- & a_{m+1} & \mp & 2i(m+{1\over 2})a_m & + &
27 | (m+{1\over 2})(m-{1\over 2})a_{m-i},\vphantom{\Big(}\cr
28 | &&&\hfill \pm\; i a_m &&\hfill -(m-{1\over 2})a_{m-1},\vphantom{\Big(}\cr
29 | + & a_{m+1} &&&&\hfill -n^2a_{m-1}.\vphantom{\Big(}\cr}$$
30 | where the consecutive rows correspond to the consecutive terms
31 | $$\textstyle {d^2Z\over d\rho^2},\qquad {1\over \rho}{dZ\over d\rho}, \qquad
32 | \bigl(1-{n^2\over \rho^2}\bigr)Z$$
33 | in (19.11). Summing the three rows we get the following {\it first order
34 | recursion formula}
35 | \EnoughAlready
```

NOTES

1–2: The commands that set things up correctly for the new page (in a new chapter).

4: The \SubSection command can gracefully handle long titles, continuing them onto a second line when necessary.

26–29: \matrix is used here to align bits of the formula. The spacing is a little wide, but it can be reduced, if desired, by using the \CompressMatrices command introduced on page 40.

| 73

Example 4

Source:

*Partial
Differential
Equations
in Physics*
ARNOLD
SOMMERFELD
© 1964
Academic Press

D. Generalization of the Saddle-Point Method According to Debye

Although in later applications we shall in general apply only the asymptotic limiting value of the Bessel functions as determined at the end of §19, we wish to discuss here certain more general expansions due to Hankel, which progress according to negative powers of ρ and in which the first term is the above mentioned asymptotic limiting value. Actually these series are *divergent*, being developments at an essential singularity, but they are frequently called *semi-convergent*. The first terms decrease rapidly, but from a certain term on they increase to infinity. We obviously must break off at that term in order to obtain approximation formulas.

The shortest way of obtaining these series is from the differential equation for the Hankel function, by substituting formal power series, and then computing the coefficients by setting the factor of each power equal to zero (this is obviously not completely rigorous). Considering (19.55) and (19.56) we write

$$(22) \quad H_n^{1,2}(\rho) = \sqrt{\frac{2}{\pi\rho}} e^{\pm i(\rho-(n+\frac{1}{2})\pi/2)} \left(a_0 + \frac{a_1}{\rho} + \frac{a_2}{\rho^2} + \cdots + \frac{a_m}{\rho^m} + \cdots \right)$$

and after dividing out the factor $\sqrt{2/\pi} \exp\{\pm i(\rho - (n + \frac{1}{2})\pi/2)\}$ from the differential equation (19.11) we find the terms with $\rho^{-m-\frac{3}{2}}$ to be

$$- \quad a_{m+1} \quad \mp \quad 2i(m + \tfrac{1}{2})a_m \quad + \quad (m + \tfrac{1}{2})(m - \tfrac{1}{2})a_{m-i},$$
$$\pm\, ia_m \qquad\qquad -(m - \tfrac{1}{2})a_{m-1},$$
$$+ \quad a_{m+1} \qquad\qquad\qquad -n^2 a_{m-1}.$$

where the consecutive rows correspond to the consecutive terms

$$\frac{d^2 Z}{d\rho^2}, \qquad \frac{1}{\rho}\frac{dZ}{d\rho}, \qquad \left(1 - \frac{n^2}{\rho^2}\right)Z$$

in (19.11). Summing the three rows we get the following *first order recursion formula*

$$\vdots$$

```
 1  \input amssym.def      % This file calls up the AMS symbol fonts.
 2  \input amssym.tex      % This file provides names for the symbols.
 3
 4  \BookTitle={Foundations of Modern Analysis}
 5  \Author={J.~Dieudonn\'e}
 6  \CopyrightDate={1969}
 7  \PublishInfo={Academic Press}
 8
 9  \NewExample
10  At the start of Dieudonn\'e's book there is a list of the notations that are
11  used in it. The list covers a variety of symbols, and it seemed like a good
12  ..........................................................
13  to understand. Two different ways of achieving this are shown, both
14  briefly discussed in the notes.
15  \EndPage
16
17  % Formatting commands for the book shown on the right-hand pages:
18  \LMarginOdd=.625 in   % Left margin for odd-numbered pages.
19  \LMarginEven=.875 in  % Left margin for even-numbered pages.
20  \hsize 4.5 in  \vsize 7.125 in  % Sets the page size.
21  \voffset-.1in          % Sets the vertical offset.
22  \LeftTitles            % Puts titles on the left.
23  \EveryChapter={}       % Want nothing automatically appearing at the start.
24  \TopLRPno              % Sets page numbers on the top.
25  % Select fonts (see Chapter 5 for an explanation of the names):
26  \ChapterFonts{cmssbx10}{cmss12}
27  \LoadEight             % To be able to use 8-point type.
28  % Choose spacings:
29  \StartofChapSkip=60pt plus4pt minus2pt % Space above the chapter title.
30  \InChapTitleSkip=\medskipamount        % Space between lines of the title.
31  \AfterChapTitleSkip=100pt              % Space after the title.
32  % Set the headline format:
33  \LeftRunningHead={{\Sansserif\qquad\quad\the\ChapName}\hfil}
34  \RightRunningHead={\hfil {\Sansserif\the\ChapName\quad\qquad}}
35
36  % Special formatting commands for this example:
37  \newdimen\EntryWidth  \EntryWidth=1.375in
38  \def\EntryBox #1{\setbox0=\hbox{$\displaystyle#1$\qquad}
39    \ifdim\wd0>\EntryWidth \EntryWidth=\wd0 \box0
40      \else \hbox to\EntryWidth{\unhbox0\hfil}\fi}
41  \def\DescriptionBox #1{\vtop{\advance\hsize by -\EntryWidth
42    \parindent-1em #1\strut}}
43  \def\X #1[#2]{\leftline{\strut\EntryBox{#1}\DescriptionBox{#2}}
44              \vskip\parskip}
45  \def\StartList {\begingroup \def\par {\endgraf \expandafter\X }}
46  \def\EndList {[]\endgroup \vskip-\baselineskip}
```

NOTES

4–34: The standard setup for a new example.
37: Introduces a new dimensional variable; it will control the standard width of the "entry column."
38–40: The command that typesets the entry, in a box of width at least \EntryWidth: it first checks the natural width of the entry (plus a \qquad of space), then adjusts \EntryWidth if needed.
41–42: This typesets the description of the entry, in a box of width equal to the page width (\hsize) minus \EntryWidth. \vtop is used because the top line is required to line up with the entry box. A negative paragraph indentation is picked in order to make the first line stick out to the left.
43: One way to make new entries is to explicitly use the command \X; see the next page for examples.
45–46: These allow entries to be listed with no visible formatting commands. (\X is smuggled in, inside the redefinition of \par; \expandafter postpones the substitution of \X by the replacement text in its definition.)

TEX and Bookmaking

——————— Example 5 ———————

From *Foundations of Modern Analysis*
J. DIEUDONNÉ
Academic Press, 1969

At the start of Dieudonné's book there is a list of the notations that are used in it. The list covers a variety of symbols, and it seemed like a good way to give some mild exercise to TEX's symbolic muscles.

Some of the symbols shown here are not part of the Plain TEX distribution; they come, instead, from sets of fonts distributed with $\mathcal{AMS}$-TEX. The fonts may be used even if one is not using the full $\mathcal{AMS}$-TEX package; a smaller file, `amssym.def`, does all the necessary work (see the first input line on the facing page). Chapter 3 gives more information on $\mathcal{AMS}$-TEX, Chapter 5 on fonts in general, and the inside back cover on where to get such fonts.

The original source hasn't been followed slavishly: for example, capital letters in mathematical formulas appear in math italics here, even though they are in regular, upright roman in the original. (It is always possible to get roman letters in formulas by asking for them explicitly with `\rm`—or `\roman` if one is using $\mathcal{AMS}$-TEX—but that would unnecessarily clutter the input for this example without making any significant point.) Further, none of the standard sets of typefaces that are available contain a really curly script face, as is used in the source, so TEX's calligraphic letters have been used as an approximation.

As a side issue, the table-like format that is reproduced here has its own interest. Symbols are tabulated in the first column and a description (possibly multiline) in the second. The description is set with hanging indentation, in that the first line is out- rather than indented. And, as a minor twist, the first column is allowed to occasionally intrude into the space of the second. The notes below the input box on the facing page comment briefly on the commands set up here that achieve all of this.

Despite the slightly unusual format of these pages, the input has been kept largely uncluttered—after a few new commands are defined initially. It is often a problem with markup languages—systems that mark the input text with explicit typesetting instructions—like TEX, that the input file gets so filled with typesetting commands as to make its content almost undecipherable at first glance. This example shows how it is possible to hide most of the explicit formatting, leaving the content of the input file easy to understand. Two different ways of achieving this are shown, both briefly discussed in the notes.

```
 1  \def\Abs #1{\left\vert #1\right\vert}  % For absolute values.
 2  \def\Norm #1{\left\Vert #1\right\Vert} % For norms.
 3  \def\Smpr{\mathop{\rm pr}\nolimits} % Imitates the definitions of `\cos', etc.
 4  \def\CircleAccent{\mathaccent"7017 } % This is how new accents are defined.
 5
 6  \StartBook
 7  \Chapter{NOTATIONS}     \pageno=-15
 8  \footline{\hfil\folio} % Puts the page number on the bottom right.
 9
10  {\Eightpoint In the following definitions the first digit refers to the number
11  of the chapter in which the notation occurs and the second to the section
12  within the chapter. \vskip2\bigskipamount}
13
14  % Start with the blank line, or `invisible formatting', approach:
15  \StartList
16
17  = [equals: 1.1]
18
19  \ne [is different from: 1.1]
20
21  \in [is an element of, belongs to: 1.1]
22
23  \notin [is not an element of: 1.1]
24
25  \subset [is a subset of, is contained in: 1.1]
26
27  \supset [contains: 1.1]
28
29  \EndList
30
31  % The rest of the input will use a more `explicit formatting' approach:
32  \X \mathrel{\subset\mkern-9mu\mid} [is not contained in: 1.1]
33  \X \{x\in X\mid P(x)\} [the set of elements of $X$ having property $P$: 1.1]
34  \X \varnothing [the empty set: 1.1]  % AMS symbol
35  \X \{a\} [the set having $a$ as unique element: 1.1]
36  \X {\frak B}(X) [the set of subsets of $X$: 1.1] % AMS symbol
37  \X X-Y,\, \complement_X Y,\, \complement Y % AMS symbol
38     [the complement of $Y$ in $X$: 1.2]
39  \X \cup [union: 1.2]
40  \X \cap [intersection: 1.2]
41  \X (a,b) [ordered pair: 1.3]
42  \X \Smpr_1c,\, \Smpr_2c [first and second projection: 1.3] % Defined above.
43  \X G(x),\, G^{-1}(y) [cross sections of $G \subset X\times Y$: 1.3]
44  \X X\times Y [product of two sets: 1.3]
45  \X X_1\times X_2\times \ldots \times X_n [product of $n$ sets: 1.3]
46  \X \Smpr_i z [$i$th projection: 1.3] % Defined above.
47  \X \Smpr_{i_1 i_2 \ldots i_k}(z) [partial projection: 1.3] % Defined above.
48  \X X^n [product of $n$ sets equal to $X$: 1.3]
49  \X F(x) [value of the mapping $F$ at $x$: 1.4]
50  \X Y^X,\, {\cal F}(X,Y) [set of mappings of $X$ into $Y$: 1.4]
```

NOTES

1–4: Convenience commands. \mathop is used (line 3) in order to tell TeX that the symbol being constructed is an operator. See the use of \mathrel on page 20 (and line 32 of this page) and that of \mathord on page 30. Line 4 shows how new accents are defined; Chapter 5 gives a full explanation

of the code numbers involved.

15–29: If using this approach, blank lines must be left between entries.

32...: \X can be used to start each entry, with (as above) the description encased in brackets.

Example 5

Source:
*Foundations
of Modern
Analysis*
J. DIEUDONNÉ
© 1969
Academic Press

NOTATIONS

In the following definitions the first digit refers to the number of the chapter in which the notation occurs and the second to the section within the chapter.

$=$	equals: 1.1
$\neq$	is different from: 1.1
$\in$	is an element of, belongs to: 1.1
$\notin$	is not an element of: 1.1
$\subset$	is a subset of, is contained in: 1.1
$\supset$	contains: 1.1
$\not\subset$	is not contained in: 1.1
$\{x \in X \mid P(x)\}$	the set of elements of X having property P: 1.1
$\varnothing$	the empty set: 1.1
$\{a\}$	the set having a as unique element: 1.1
$\mathfrak{B}(X)$	the set of subsets of X: 1.1
$X - Y$, $\complement_X Y$, $\complement Y$	the complement of Y in X: 1.2
$\cup$	union: 1.2
$\cap$	intersection: 1.2
(a, b)	ordered pair: 1.3
$\mathrm{pr}_1 c$, $\mathrm{pr}_2 c$	first and second projection: 1.3
$G(x)$, $G^{-1}(y)$	cross sections of $G \subset X \times Y$: 1.3
$X \times Y$	product of two sets: 1.3
$X_1 \times X_2 \times \ldots \times X_n$	product of n sets: 1.3
$\mathrm{pr}_i z$	ith projection: 1.3
$\mathrm{pr}_{i_1 i_2 \ldots i_k}(z)$	partial projection: 1.3
X^n	product of n sets equal to X: 1.3
$F(x)$	value of the mapping F at x: 1.4
Y^X, $\mathcal{F}(X, Y)$	set of mappings of X into Y: 1.4

```
 1 \X 1_X [identity mapping of $X$: 1.4]
 2 \X x\to T(x) [mapping: 1.4]
 3 \X F(A) [direct image: 1.5]
 4 \X F^{-1}(A) [inverse image: 1.5]
 5 \X F^{-1}(y) [inverse image of a one element set $\{y\}$: 1.5]
 6 \X F(.,y),\, F(x,.)
 7   [partial mappings of a mapping $F$ of $A\subset X\times Y$ into $Z$: 1.5]
 8 \X j_A [natural injection: 1.6]
 9 \X F^{-1} [inverse mapping of a bijective mapping: 1.6]
10 \X G\circ F [composed mapping: 1.7]
11 \X (x_\lambda)_{\lambda \in L} [family: 1.8]
12 \X {\bf N} [set of natural integers: 1.8]
13 \X \{x_1,\ldots,x_n\} [set of elements of a finite sequence: 1.8]
14 \X \textstyle\bigcup\limits_{\lambda\in L} A_\lambda,\,
15   \bigcup\limits_\lambda A_\lambda [union of a family of sets: 1.8]
16 \X \textstyle\bigcap\limits_{\lambda\in L} A_\lambda,\,
17   \bigcap\limits_\lambda A_\lambda [intersection of a family of sets: 1.8]
18 \X X/R [quotient set of a set $X$ by an equivalence relation $R$: 1.8]
19 \X \textstyle\prod\limits_{\lambda\in L} X_\lambda
20   [product of a family of sets: 1.8]
21 \X \Smpr_{\bf J} [projection on a partial product: 1.8] % Defined above.
22 \X (u_\lambda) [mapping into a product of sets: 1.8]
23 \X {\bf R} [set of real numbers: 2.1]
24 \X x+y [sum of real numbers: 2.1]
25 \X xy [product of real numbers: 2.1]
26 \X 0 [element of {\bf R}: 2.1]
27 \X -x [opposite of a real number: 2.1]
28 \X 1 [element of {\bf R}: 2.1]
29 \X x^{-1},\, 1/x [inverse in {\bf R}: 2.1]
30 \X x\leqslant y,\, y\geqslant x [order relation in {\bf R}: 2.1] % AMS symbol
31 \X x<y,\, y>x [relation in {\bf R}: 2.1]
32 \X \left\rbrack a,b\right\lbrack\!,\, \lbrack a,b\rbrack,\,
33   \left\lbrack a,b\right\lbrack\!,\, \left\rbrack a,b\right\rbrack
34   % NOTE: `\left' & `\right' are needed for proper spacing.
35   [intervals in {\bf R}: 2.1]
36 \X {\bf R}_+,\, {\bf R}_+^\ast
37   [set of real numbers $\geqslant0$ (resp.~$>0$): 2.2]
38 \X \Abs x,\, x^+,\, x^-  % Defined above.
39   [absolute value, positive and negative part of a real number: 2.2]
40 \X {\bf Q} [set of rational numbers: 2.2]
41 \X {\bf Z} [set of positive or negative integers: 2.2]
42 \X {\rm l.u.b.}\;X,\, \sup X [least upper bound of a set: 2.3]
43 \X {\rm g.l.b.}\;X,\, \inf X [greatest lower bound of a set: 2.3]
44 \X \sup_{x\in A} f(x),\, \inf_{x\in A} f(x)
45   [supremum and infimum of $f$ in $A$: 2.3]
46 \X \bar{\bf R} [extended real line: 3.3]
```

Example 5

Source:

*Foundations
of Modern
Analysis*

J. DIEUDONNÉ

© 1969
Academic Press

1_X	identity mapping of X: 1.4
$x \to T(x)$	mapping: 1.4
$F(A)$	direct image: 1.5
$F^{-1}(A)$	inverse image: 1.5
$F^{-1}(y)$	inverse image of a one element set $\{y\}$: 1.5
$F(.,y), F(x,.)$	partial mappings of a mapping F of $A \subset X \times Y$ into Z: 1.5
j_A	natural injection: 1.6
F^{-1}	inverse mapping of a bijective mapping: 1.6
$G \circ F$	composed mapping: 1.7
$(x_\lambda)_{\lambda \in L}$	family: 1.8
$\mathbf{N}$	set of natural integers: 1.8
$\{x_1, \ldots, x_n\}$	set of elements of a finite sequence: 1.8
$\bigcup_{\lambda \in L} A_\lambda, \bigcup_\lambda A_\lambda$	union of a family of sets: 1.8
$\bigcap_{\lambda \in L} A_\lambda, \bigcap_\lambda A_\lambda$	intersection of a family of sets: 1.8
X/R	quotient set of a set X by an equivalence relation R: 1.8
$\prod_{\lambda \in L} X_\lambda$	product of a family of sets: 1.8
$\mathrm{pr}_\mathbf{J}$	projection on a partial product: 1.8
(u_λ)	mapping into a product of sets: 1.8
$\mathbf{R}$	set of real numbers: 2.1
$x + y$	sum of real numbers: 2.1
xy	product of real numbers: 2.1
0	element of $\mathbf{R}$: 2.1
$-x$	opposite of a real number: 2.1
1	element of $\mathbf{R}$: 2.1
$x^{-1}, 1/x$	inverse in $\mathbf{R}$: 2.1
$x \leqslant y, y \geqslant x$	order relation in $\mathbf{R}$: 2.1
$x < y, y > x$	relation in $\mathbf{R}$: 2.1
$]a,b[, [a,b], [a,b[,]a,b]$	intervals in $\mathbf{R}$: 2.1
$\mathbf{R}_+, \mathbf{R}_+^*$	set of real numbers $\geqslant 0$ (resp. > 0): 2.2
$|x|, x^+, x^-$	absolute value, positive and negative part of a real number: 2.2
$\mathbf{Q}$	set of rational numbers: 2.2
$\mathbf{Z}$	set of positive or negative integers: 2.2
l.u.b. X, sup X	least upper bound of a set: 2.3
g.l.b. X, inf X	greatest lower bound of a set: 2.3
$\sup_{x \in A} f(x), \inf_{x \in A} f(x)$	supremum and infimum of f in A: 2.3
$\bar{\mathbf{R}}$	extended real line: 3.3

```
80   1 │ \X +\infty, -\infty [points at infinity in $\bar{\bf R}$: 3.3]
     2 │ \X x\leqslant y,\, y\geqslant x  % AMS symbol
     3 │    [order relation in $\bar{\bf R}$: 3.3]
     4 │ \X d(A,B) [distance of two sets: 3.4]
     5 │ \X B(a;r),\, B'(a;r),\, S(a;r)
     6 │    [open ball, closed ball, sphere of center $a$ and radius $r$: 3.4]
     7 │ \X \delta(A) [diameter: 3.4]
     8 │ \X \CircleAccent A [interior: 3.7] % Defined above.
     9 │ \X \bar A [closure: 3.8]
    10 │ \X {\rm fr}(A) [frontier: 3.8]
    11 │ \X \lim_{x\to A,\, x\in A} f(x) [limit of  a function: 3.13]
    12 │ \X \lim_{n\to \infty} x_n [limit of a sequence: 3.13]
    13 │ \X \Omega(a;f) [oscillation of a function: 3.14]
    14 │ \X \log_a x [logarithm of a real number: 4.3]
    15 │ \X a^x [exponential of base $a$ ($x$ real): 4.3]
    16 │ \X {\bf C} [set of complex numbers: 4.4]
    17 │ \X z+z',\, zz' [sum, product of complex numbers: 4.4]
    18 │ \X 0,\, 1,\, i [elements of {\bf C}: 4.4]
    19 │ \X {\cal R}z,\, {\cal I}z [real and imaginary parts: 4.4]
    20 │ \X \bar z [conjugate of a complex number: 4.4]
    21 │ \X \Abs z [absolute value of a real number: 4.4] % Defined above.
    22 │ \X x+y,\, \lambda x,\, x [sum and product by a scalar in a vector space: 5.1]
    23 │ \X 0 [element of a vector space: 5.1]
    24 │ \X \Norm x [norm: 5.1] % Defined above.
    25 │ \X \textstyle\sum\limits_{n=0}^\infty x_n [sum of a series, series: 5.2]
    26 │ \X \textstyle\sum\limits_{\alpha\in A} x_\alpha
    27 │    [sum of an absolutely summable family: 5.3]
    28 │ \X (c_0) [space of sequences tending to 0: 5.3, prob.~5]
    29 │ \X {\cal L}(E;F) [space of linear continuous mappings: 5.7]
    30 │ \X \Norm u [norm of a linear continuous mapping: 5.7] % Defined above.
    31 │ \X {\cal L}(E_1,\ldots, E_n; F)
    32 │    [space of multilinear continuous mappings: 5.7]
    33 │ \X l^1 [space of absolutely convergent series: 5.7, prob.~1]
    34 │ \X l^\infty [space of bounded sequences: 5.7, prob~1]
    35 │ \X (x\mid y) [scalar product: 6.2]
    36 │ \X P_{F} [orthogonal projection: 6.3]
    37 │ \X l^2,\, l_{\bf R}^2,\, l_{\bf C}^2 [Hilbert spaces of sequences: 6.5]
    38 │ \X {\cal B}_F(A),\, {\cal B}_{\bf R}(A),\, {\cal B}_{\bf C}(A)
    39 │    [spaces of bounded mappings: 7.1]
    40 │ \X {\cal C}_F(E) [space of continuous mappings: 7.2]
    41 │ \X {\cal C}_F^\infty(E) [space of bounded continuous mappings: 7.2]
    42 │ \X f(x+),\, f(x-) [limits to the right, to the left: 7.6]
    43 │ \X f'(x_0),\, Df(x_0) [(total) derivative at $x_0$: 8.1]
```

Example 5

Source:

*Foundations
of Modern
Analysis*

J. DIEUDONNÉ

© 1969
Academic Press

$+\infty, -\infty$	points at infinity in $\bar{\mathbf{R}}$: 3.3
$x \leqslant y,\ y \geqslant x$	order relation in $\bar{\mathbf{R}}$: 3.3
$d(A, B)$	distance of two sets: 3.4
$B(a; r),\ B'(a; r),\ S(a; r)$	open ball, closed ball, sphere of center a and radius r: 3.4
$\delta(A)$	diameter: 3.4
$\mathring{A}$	interior: 3.7
$\bar{A}$	closure: 3.8
$\mathrm{fr}(A)$	frontier: 3.8
$\lim_{x \to A,\, x \in A} f(x)$	limit of a function: 3.13
$\lim_{n \to \infty} x_n$	limit of a sequence: 3.13
$\Omega(a; f)$	oscillation of a function: 3.14
$\log_a x$	logarithm of a real number: 4.3
a^x	exponential of base a (x real): 4.3
$\mathbf{C}$	set of complex numbers: 4.4
$z + z',\ zz'$	sum, product of complex numbers: 4.4
$0, 1, i$	elements of $\mathbf{C}$: 4.4
$\mathcal{R}z,\ \mathcal{I}z$	real and imaginary parts: 4.4
$\bar{z}$	conjugate of a complex number: 4.4
$|z|$	absolute value of a real number: 4.4
$x + y,\ \lambda x,\ x$	sum and product by a scalar in a vector space: 5.1
0	element of a vector space: 5.1
$\|x\|$	norm: 5.1
$\sum_{n=0}^{\infty} x_n$	sum of a series, series: 5.2
$\sum_{\alpha \in A} x_\alpha$	sum of an absolutely summable family: 5.3
(c_0)	space of sequences tending to 0: 5.3, prob. 5
$\mathcal{L}(E; F)$	space of linear continuous mappings: 5.7
$\|u\|$	norm of a linear continuous mapping: 5.7
$\mathcal{L}(E_1, \ldots, E_n; F)$	space of multilinear continuous mappings: 5.7
l^1	space of absolutely convergent series: 5.7, prob. 1
l^∞	space of bounded sequences: 5.7, prob 1
$(x \mid y)$	scalar product: 6.2
P_F	orthogonal projection: 6.3
$l^2,\ l^2_{\mathbf{R}},\ l^2_{\mathbf{C}}$	Hilbert spaces of sequences: 6.5
$\mathcal{B}_F(A),\ \mathcal{B}_{\mathbf{R}}(A),\ \mathcal{B}_{\mathbf{C}}(A)$	spaces of bounded mappings: 7.1
$\mathcal{C}_F(E)$	space of continuous mappings: 7.2
$\mathcal{C}_F^\infty(E)$	space of bounded continuous mappings: 7.2
$f(x+),\ f(x-)$	limits to the right, to the left: 7.6
$f'(x_0),\ Df(x_0)$	(total) derivative at x_0: 8.1

```
 1 | \X f',\, Df [derivative (as a function): 8.1]
 2 | \X f'(\alpha),\, D_+f(\alpha) [derivative on the right: 8.4]
 3 | \X f_g'(\beta),\, D_-f(\beta) [derivative on the left: 8.4]
 4 | \X \textstyle\int_\alpha^\beta f(\xi)\, d\xi [integral: 8.7]
 5 | \X e,\, \exp(x),\, \log x [($x$ real): 8.8]
 6 | \X D_1f(a_1,a_2),\, D_2f(a_1,a_2) [partial derivatives: 8.9]
 7 | \X f_{\xi_i}'(\xi_1,\ldots, \xi_n),\,
 8 |    {\partial \;\over \partial\xi_i} f(\xi_1,\ldots,\xi_n)
 9 |    [partial derivatives: 8.10]
10 | \X {D(f_1,\ldots, f_n) \over D(\xi_1,\ldots, \xi_n)},\,
11 |    {\partial(f_1,\ldots, f_n) \over \partial(\xi_1,\ldots, \xi_n)}
12 |    [jacobian: 8.10]
13 | \X f''(x_0),\, D^2f(x_0),\, f^{(p)}(x_0),\, D^pf(x_0)
14 |    [higher derivatives: 8.12]
15 | \X f\ast \rho [regularization: 8.12, prob.~2]
16 | \X {\cal E}_F^{(p)}(A)
17 |    [space of $p$ times continuously differentiable mappings: 8.13]
18 | \X \Abs\alpha,\, M_\alpha,\, D^\alpha,\, D_{M_\alpha} % Defined above.
19 |    [($\alpha$ composite index): 8.13]
20 | \X e^z,\, \exp(z) [($z$ complex): 9.5]
21 | \X \sin z,\, \cos z [sine and cosine: 9.5]
22 | \X \pi [9.5]
23 | \X \log z,\, {\rm am}(z),\, {t\choose n},\, (1+z)^t
24 |    [($z$, $t$ complex numbers): 9.5, prob.~8]
25 | \X \gamma^0 [opposite path: 9.6]
26 | \X \gamma_1\vee\gamma_2 [juxtaposition of paths: 9.6]
27 | \X \textstyle\int_\gamma f(z)\,dz [integral along a road: 9.6]
28 | \X j(a;\gamma) [index with respect to a circuit: 9.8]
29 | \X E(z,p) [primary factor: 9.12, prob.~1]
30 | \X {\mit\Gamma}(z) [gamma function: 9.12, prob.~2]
31 | \X \gamma [Euler's constant: 9.12, prob.~2]
32 | \X \textstyle\int_\gamma f(z)\,dz
33 |    [integral along an endless road: 9.12, prob.~3]
34 | \X \omega(a;f),\, \omega(a) [order of a function at a point: 9.15]
35 | \X {\cal L}(E) [algebra of operators: 11.1]
36 | \X uv [composed operator: 11.1]
37 | \X {\rm sp}(u) [spectrum: 11.1]
38 | \X E(\zeta),\, E(\zeta;u) [eigenspace: 11.1]
39 | \X \tilde u [continuous extension: 11.2]
40 | \X N(\lambda),\, N(\lambda;u),\, F(\lambda),\, F(\lambda;u)
41 |    [subspaces attached to an eigenvalue of a \break compact operator: 11.4]
42 | \X k(\lambda),\, k(\lambda;u) [order of an eigenvalue: 11.4]
43 | \X u^\ast [adjoint operator: 11.5]
44 | \EndPage  \bye
```

Example 5

Source:

*Foundations
of Modern
Analysis*

J. DIEUDONNÉ
© 1969
Academic Press

f', Df	derivative (as a function): 8.1
$f'(\alpha)$, $D_+f(\alpha)$	derivative on the right: 8.4
$f'_g(\beta)$, $D_-f(\beta)$	derivative on the left: 8.4
$\int_\alpha^\beta f(\xi)\,d\xi$	integral: 8.7
e, $\exp(x)$, $\log x$	(x real): 8.8
$D_1f(a_1,a_2)$, $D_2f(a_1,a_2)$	partial derivatives: 8.9
$f'_{\xi_i}(\xi_1,\ldots,\xi_n)$, $\dfrac{\partial}{\partial\xi_i}f(\xi_1,\ldots,\xi_n)$	partial derivatives: 8.10
$\dfrac{D(f_1,\ldots,f_n)}{D(\xi_1,\ldots,\xi_n)}$, $\dfrac{\partial(f_1,\ldots,f_n)}{\partial(\xi_1,\ldots,\xi_n)}$	jacobian: 8.10
$f''(x_0)$, $D^2f(x_0)$, $f^{(p)}(x_0)$, $D^pf(x_0)$	higher derivatives: 8.12
$f * \rho$	regularization: 8.12, prob. 2
$\mathcal{E}_F^{(p)}(A)$	space of p times continuously differentiable mappings: 8.13
$|\alpha|$, M_α, D^α, D_{M_α}	(α composite index): 8.13
e^z, $\exp(z)$	(z complex): 9.5
$\sin z$, $\cos z$	sine and cosine: 9.5
π	9.5
$\log z$, $\mathrm{am}(z)$, $\dbinom{t}{n}$, $(1+z)^t$	(z, t complex numbers): 9.5, prob. 8
γ^0	opposite path: 9.6
$\gamma_1 \vee \gamma_2$	juxtaposition of paths: 9.6
$\int_\gamma f(z)\,dz$	integral along a road: 9.6
$j(a;\gamma)$	index with respect to a circuit: 9.8
$E(z,p)$	primary factor: 9.12, prob. 1
$\Gamma(z)$	gamma function: 9.12, prob. 2
γ	Euler's constant: 9.12, prob. 2
$\int_\gamma f(z)\,dz$	integral along an endless road: 9.12, prob. 3
$\omega(a;f)$, $\omega(a)$	order of a function at a point: 9.15
$\mathcal{L}(E)$	algebra of operators: 11.1
uv	composed operator: 11.1
$\mathrm{sp}(u)$	spectrum: 11.1
$E(\zeta)$, $E(\zeta;u)$	eigenspace: 11.1
$\tilde{u}$	continuous extension: 11.2
$N(\lambda)$, $N(\lambda;u)$, $F(\lambda)$, $F(\lambda;u)$	subspaces attached to an eigenvalue of a compact operator: 11.4
$k(\lambda)$, $k(\lambda;u)$	order of an eigenvalue: 11.4
u^*	adjoint operator: 11.5

```
 1 | % A new file begins here; to be processed using German hyphenation patterns.
 2 | \input OutForm.tex     % Input the 'output format' file.
 3 | \input BookForm.tex    % Input the 'book format' file.
 4 | \input amssym.def      % The definitions of AMS symbol invoking commands.
 5 | \input amssym.tex      % The names for AMS symbols.
 6 | \input german.sty      % Commands to simplify input, use German styles, etc.
 7 |
 8 | \RealPno=85
 9 | \ExampleNumber=5    % The number of the example that this one follows.
10 |
11 | \BookTitle={Mengenlehre}
12 | \Author={F.~Hausdorff}
13 | \CopyrightDate={1927}
14 | \PublishInfo={Walter de Gruyter \& Co.}
15 |
16 | \ExampleType={\TeX\ and Languages}%
17 | \NewExample
18 | This example reproduces some pages from a classic German mathematics book. In
19 | addition to the interest of seeing how \TeX\ does German, the original also
20 | ...........................................................................
21 | Readers interested in languages in general may also want to look at the
22 | short discussions of Cyrillic fonts and 256-character fonts in Chapter~5.
23 | \EndPage
24 |
25 | % Formatting commands for the 'output' part of the right-hand pages:
26 | \LMarginEven=.9125 in  % Left margin for even-numbered pages.
27 | \LMarginOdd=.625in     % Left margin for odd-numbered pages.
28 | \hsize 4.4375 in  \vsize 6.8125 in  % Size. NOTE: pages will appear offset.
29 | \parindent20pt         % Set paragraph indentation.
30 | \font\Eightrm=cmr8     % Used in headlines.
31 | \CenteredTitles
32 | \EverySection={\S\ \the\SectNo.}
33 | \EveryChapter={}
34 | \EverySubSection={}
35 |
36 | % Select fonts (see Chapter 5 for an explanation of the names):
37 | \SectionFont{cmbx10}
38 |
39 | % Choose spacings:
40 | \AfterSectionTitleSkip=\medskipamount
41 | \abovedisplayshortskip=1pt plus 1pt
42 | \abovedisplayskip=2pt plus 2pt
43 | \belowdisplayshortskip=1pt plus 1pt
44 | \belowdisplayskip=2pt plus 2pt
45 |
46 | % Set headline format:
47 | \LeftRunningHead={\hfil {\Eightrm Zweites Kapital.\ \ Kardinalzahlen.}\hfil}
48 | \RightRunningHead={\hfil {\Eightrm\S\ \the\SectNo.\ \ \the\SectionName}\hfil}
49 | \TopLRpno
```

NOTES

2–3: Both the files needed for formatting book examples are brought in here (see page 14).

4–5: These allow the use of extra symbols, fraktur fonts, etc. See lines 1–2, page 74.

6: A German style file that allows common input to be entered simply; e.g., it allows the use of " to get an umlaut, in place of the standard \", and "s to get ß, in place of \ss. Further, it permits TeX to hyphenate after certain accents, and it provides utilities for typesetting dates, quotation marks, etc., in accepted styles.

8–49: The (by now, standard) set-up for a new example.

T_EX and Languages

─────────── **Example 6** ───────────

From *Mengenlehre*
F. HAUSDORFF
Walter de Gruyter & Co., 1927

This example reproduces some pages from a classic German mathematics book. In addition to the interest of seeing how T_EX does German, the original also involves some moderately interesting alignments in its displayed equations.

Correctly doing a language involves using the correct hyphenation patterns. A *pattern* is a set of instructions that tells the program how to go about finding in a systematic way the potential hyphenation points of words. When a format file is created (see § 1.2), hyphenation patterns may be built in through a command called \patterns. Exceptions to the rules may be explicitly listed by using a command called \hyphenation. T_EX allows the use of \patterns only when a format file is being made, but exceptions can be added at any time.

Versions of T_EX from 3.0 on can handle multiple hyphenation patterns in a powerful way. A command called \language has been added that allows users to switch between patterns, even in mid-paragraph. In order to enjoy the full benefits of this feature, one needs to be using a format that has had such multiple hyphenation patterns built into it. Hyphenation exceptions can, however, continue to be specified at any time for any language. The file below illustrates the procedure (it should be copied as is and run through T_EX):

```
\hsize 75pt
\language0 \hyphenation{xxx-xxx}   % Hyphenation exception for language 0.
\language1 \hyphenation{yyy-yyy}   % Hyphenation exception for language 1.
% Begin with language 1 (since we are already in it):
xxxxxx xxxxxx xxxxxx xxxxxx \par  yyyyyy yyyyyy yyyyyy yyyyyy \par
% Switch to language 0 and use the same 'words':
\language0
xxxxxx xxxxxx xxxxxx xxxxxx \par  yyyyyy yyyyyy yyyyyy yyyyyy \par
\bye
```

If this experiment is done, it will be seen that T_EX can hyphenate "yyyyyy" in language 1 but not "xxxxxx," with the reverse holding for language 0.

The present example is typeset with a German T_EX, i.e., a version of T_EX where the format file has German hyphenation patterns built into it, and with a German style file (input line 6); see PARTL [1988]. This file contains, among other things, commands that allow input to be entered more simply.

Readers interested in languages in general may also want to look at the short discussions of Cyrillic fonts and 256-character fonts in Chapter 5.

```
 1 | \SectionName={Mengenvergleichung.}
 2 | \SectNo=5  \pageno=27
 3 |
 4 | I. {\it Jede unendliche Menge enth"alt eine abz"ahlbare Teilmenge.}
 5 |
 6 | Es sei $a_1$ ein Element der unendlichen Menge A, $a_2$ ein Element der
 7 | (immer noch) unendlichen Menge $A-\{a_1\}$, $a_3$ ein Element der (immer noch)
 8 | unendlichen Menge $A-\{a_1,a_2\}$ usw. Die paarweise verschiedenen Elemente
 9 | $a_1$, $a_2$, $a_3$, \dots\ bilden eine abz"ahlbare Teilmenge von $A$.
10 |
11 | II. {\it Jede unendliche Menge ist mit einer echten Teilmenge "aquivalent.}
12 |
13 | Indem man aus der unendlichen Menge $A$ eine abz"ahlbare Teilmenge
14 | herauszieht und deren Komplement in $A$ mit $B$ bezeichnet, erh"alt man
15 | $$ \eqalignno{ A = &\{a_1, a_2, \ldots, \rlap{$a_n$}\hphantom{a_{n+1}},
16 | \ldots\,\} + B; \cr
17 | \noalign{\hbox{dies ist z.\ B. mit}}
18 | &\{a_2, a_3, \ldots, a_{n+1},\ldots\,\} + B \cr} $$
19 | "aquivalent (man ordne jedem $a_n$ als Bild $a_{n+1}$, jedem $b\in B$ als
20 | Bild $b$ selbst~zu).
21 |
22 | Umgekehrt kann eine mit einer echten Teilmenge "aquivalente Menge nicht
23 | endlich sein. Diese Eigenschaft charakterisiert also die unendlichen Megen
24 | (und ist von R. Dedekind zur Definition der unendlichen Mengen verwendet
25 | worden).
26 |
27 | III. ("Aquivalenzsatz von F. Bernstein.) {\it Zwei Mengen, deren jede mit
28 | einer Teilmenge der andern "aquivalent ist, sind selbst "aquivalent}.
29 |
30 | Es sei $A\sim B_1$, $B\sim A_1$, $A_1$ Teilmenge von $A$, und zwar echte, da
31 | sonst nichts zu beweisen ist, also $A_1\subset A$, $B_1\subset B$. Durch die
32 | eineindeutige Abbildung von $B$ auf $A_1$ wird auch eine (echte) Teilmenge
33 | $A_2$ von $A_1$ abgebildet, also
34 | $$ A\supset A_1\supset A_2, \quad A\sim B_1\sim A_2. $$
35 | \indent Demnach ist der Staz auf folgenden zur"uckgef"uhrt:
36 | $$ \hbox{wenn}\quad A\supset A_1\supset A_2, \quad A\sim A_2, \qquad
37 | \hbox{so ist } A\sim A_1, $$
38 | d.\ h.\ {\it wenn eine Menge zwischen zwei "aquivalenten Mengen liegt, so
39 | ist sie mit diesen "aquivalent}.
40 |
41 | Bei der eineindeutigen Abbildung von $A$ auf $A_2$ werde $A_1$ auf $A_3$,
42 | $A_2$ auf $A_4$, $A_3$ auf $A_5$ usw.\ abgebildet, wobei also
43 | $$ A\supset A_1\supset A_2\supset A_3\supset A_4\supset A_5\supset \cdots. $$
44 | Sei
45 | $$ D = AA_1A_2\ldots $$
46 | der Durchschnitt der Mengen $A_n$, dann ist
47 | $$\eqalign{
48 | A\hphantom{_1}&= D + (A-A_1) + (A_1-A_2) + (A_2-A_3) + (A_3-A_4) + \cdots \cr
49 | A_1&=D\hphantom{{}+(A-A_1)} {} + (A_1-A_2) + (A_2-A_3) + (A_3-A_4) +
50 | \cdots \cr} $$
51 | Denn, um z.\ B. die erste Formel zu beweisen: ein Element $a\in A$ geh"ort
52 | entweder allen $A_n$ an, also zu $D$, oder es gibt ein erstes $A_n$, dem es
53 | nicht angeh"ort, w"ahrend es noch zu $A_{n-1}$ geh"ort, also
```

Example 6

Source:
Mengenlehre
F. HAUSDORFF
© 1927
Walter de
Gruyter & Co.

§ 5. Mengenvergleichung. 27

I. *Jede unendliche Menge enthält eine abzählbare Teilmenge.*

Es sei a_1 ein Element der unendlichen Menge A, a_2 ein Element der (immer noch) unendlichen Menge $A - \{a_1\}$, a_3 ein Element der (immer noch) unendlichen Menge $A - \{a_1, a_2\}$ usw. Die paarweise verschiedenen Elemente $a_1, a_2, a_3, \ldots$ bilden eine abzählbare Teilmenge von A.

II. *Jede unendliche Menge ist mit einer echten Teilmenge äquivalent.*

Indem man aus der unendlichen Menge A eine abzählbare Teilmenge herauszieht und deren Komplement in A mit B bezeichnet, erhält man

$$A = \{a_1, a_2, \ldots, a_n \quad, \ldots\} + B;$$

dies ist z. B. mit

$$\{a_2, a_3, \ldots, a_{n+1}, \ldots\} + B$$

äquivalent (man ordne jedem a_n als Bild a_{n+1}, jedem $b \in B$ als Bild b selbst zu).

Umgekehrt kann eine mit einer echten Teilmenge äquivalente Menge nicht endlich sein. Diese Eigenschaft charakterisiert also die unendlichen Megen (und ist von R. Dedekind zur Definition der unendlichen Mengen verwendet worden).

III. (Äquivalenzsatz von F. Bernstein.) *Zwei Mengen, deren jede mit einer Teilmenge der andern äquivalent ist, sind selbst äquivalent.*

Es sei $A \sim B_1$, $B \sim A_1$, A_1 Teilmenge von A, und zwar echte, da sonst nichts zu beweisen ist, also $A_1 \subset A$, $B_1 \subset B$. Durch die eineindeutige Abbildung von B auf A_1 wird auch eine (echte) Teilmenge A_2 von A_1 abgebildet, also

$$A \supset A_1 \supset A_2, \quad A \sim B_1 \sim A_2.$$

Demnach ist der Staz auf folgenden zurückgeführt:

wenn $\quad A \supset A_1 \supset A_2, \quad A \sim A_2, \quad$ so ist $A \sim A_1$,

d. h. *wenn eine Menge zwischen zwei äquivalenten Mengen liegt, so ist sie mit diesen äquivalent.*

Bei der eineindeutigen Abbildung von A auf A_2 werde A_1 auf A_3, A_2 auf A_4, A_3 auf A_5 usw. abgebildet, wobei also

$$A \supset A_1 \supset A_2 \supset A_3 \supset A_4 \supset A_5 \supset \cdots.$$

Sei

$$D = A A_1 A_2 \ldots$$

der Durchschnitt der Mengen A_n, dann ist

$$A \ = D + (A - A_1) + (A_1 - A_2) + (A_2 - A_3) + (A_3 - A_4) + \cdots$$
$$A_1 = D \qquad\qquad + (A_1 - A_2) + (A_2 - A_3) + (A_3 - A_4) + \cdots$$

Denn, um z. B. die erste Formel zu beweisen: ein Element $a \in A$ gehört entweder allen A_n an, also zu D, oder es gibt ein erstes A_n, dem es nicht

```
 1  $a\in A_{n-1}-A_n$ ($A_0=A$ gesetzt). Da nun
 2  $$ A - A_1 \sim A_2 - A_3 \sim A_4 - A_5 \sim \cdots, $$
 3  so ist auch
 4  $$ (A-A_1) + (A_2-A_3) + \cdots \sim (A_2-A_3) + (A_4-A_5) + \cdots, $$
 5  die links stehende Menge wird auf rechts stehende abgebildet. Indem man die
 6  "ubrigen Elemente von $A$ sich selbst als Bilder zuordnet, erh"alt man eine
 7  eineindeutige Abbildung von $A$ auf $A_1$, also $A\sim A_1$.
 8
 9  Bis jetzt haben wir nur von "Aquivalenz, also von Gleichheit zweier
10  Kardinalzahlen gesprochen. Die n"achste Frage w"are nun: wenn zwei Mengen
11  nicht "aquivalent, also ihre Kardinalzahlen ungleich sind, l"a"st sich
12  dann in nat"urlicher Weise die eine Kardinalzahl als die gr"o"sere, die
13  andere als die kleinere definieren? Kurz gesagt: haben die Kardinalzahlen
14  Gr"o"sencharakter? Sind sie vergleichbar?
15
16  Die Antwort scheint vorl"aufig verneinend auszufallen. Bedeuten $A$ und $B$
17  zwei Mengen, $A_1$ und $B_1$ irgendwelche Teilmengen von ihnen, so bestehen
18  vier M"oglichkeiten:
19  \halign{\indent(#)\quad\hfil&Es gibt #ein\ \hfil&$A_1\sim B$#\ \hfil&#ein\
20  \hfil&$B_1\sim A$.#\hfil\cr
21  1&&&und &\cr
22  2&k&,&aber &\cr
23  3&&,&aber k&\cr
24  4&k&&und k&\cr}
25
26  Im Fall~(1) ist nach dem "Aquivalenzsatz $\frak a=\frak b$; im Fall~(2)
27  wird man naturgem"a"s $\frak a < \frak b$, im Fall~(3) $\frak a >
28  \frak b$ zu definieren haben. Im Fall~(4) k"onnen wir offenbar, wegen seiner
29  Symmetrie in bezug auf beide Mengen, weder $\frak a < \frak b$ noch
30  $\frak a > \frak b$  definieren, da sonst beides zugleich gelten m"u"ste;
31  ebensowenig ist $\frak a=\frak b$ zul"assig, was mit der bisherigen
32  nat"urlichen Definition der Gleichheit in Widerspruch treten w"urde. Wir
33  haben hier also eine vierte Relation, die wir $\frak a \parallel \frak b$
34  schreiben und als {\it Unvergleichbarkeit\/} von $\frak a$ mit $\frak b$
35  bezeichnen, w"ahrend in den drei ersten F"allen $\frak a$ und $\frak b$
36  vergleichbar hei"sen sollen. Also:
37  $$\displaylines{\hskip1.3125in
38  \left.\matrix{(1)\; \frak a = \frak b\cr
39                (2)\; \frak a < \frak b\cr
40                (3)\; \frak a > \frak b\cr}\right\} \frak a,\; \frak b
41  \hbox{ vergleichbar} \hfill \cr
42  \noalign{\vskip-2.5pt}
43  \hskip1.3125in \left.\matrix{(4)\; \frak a \parallel \frak b\cr}\right.
44  \;\frak a,\> \frak b \hbox{\ \ unvergleichbar}.\hfill \cr} $$
45
46  Die Vergleichbarkeit w"are also nur zu retten, wenn man zeigen k"onnte,
47  da"s der vierte Fall tats"achlich nicht eintreten kann. Bei zwei endlichen
48  Mengen ist dies in der Tat der Fall; denn ist, mit Numerierung der Elemente,
49  $A = \{a_1,a_2,\ldots, a_m\}$, $B=\{b_1,b_2,\ldots,b_n\}$ und bildet man
50  behufs eineindeutiger Zuordnung die Paare $(a_1,b_1)$, $(a_2,b_2)$ usw., so
51  kommt man zu Ende, sobald eine der Mengen verbraucht ist. (Auch eine endliche
52  und eine unendliche M"achtigkeit sind stets vergleichbar, jene ist die
```

Example 6

Source:
Mengenlehre
F. HAUSDORFF
© 1927
Walter de
Gruyter & Co.

angehört, während es noch zu A_{n-1} gehört, also $a \in A_{n-1} - A_n$ ($A_0 = A$ gesetzt). Da nun

$$A - A_1 \sim A_2 - A_3 \sim A_4 - A_5 \sim \cdots,$$

so ist auch

$$(A - A_1) + (A_2 - A_3) + \cdots \sim (A_2 - A_3) + (A_4 - A_5) + \cdots,$$

die links stehende Menge wird auf rechts stehende abgebildet. Indem man die übrigen Elemente von A sich selbst als Bilder zuordnet, erhält man eine eineindeutige Abbildung von A auf A_1, also $A \sim A_1$.

Bis jetzt haben wir nur von Äquivalenz, also von Gleichheit zweier Kardinalzahlen gesprochen. Die nächste Frage wäre nun: wenn zwei Mengen nicht äquivalent, also ihre Kardinalzahlen ungleich sind, läßt sich dann in natürlicher Weise die eine Kardinalzahl als die größere, die andere als die kleinere definieren? Kurz gesagt: haben die Kardinalzahlen Größencharakter? Sind sie vergleichbar?

Die Antwort scheint vorläufig verneinend auszufallen. Bedeuten A und B zwei Mengen, A_1 und B_1 irgendwelche Teilmengen von ihnen, so bestehen vier Möglichkeiten:

(1)   Es gibt ein $A_1 \sim B$ und ein $B_1 \sim A$.
(2)   Es gibt kein $A_1 \sim B$, aber ein $B_1 \sim A$.
(3)   Es gibt ein $A_1 \sim B$, aber kein $B_1 \sim A$.
(4)   Es gibt kein $A_1 \sim B$ und kein $B_1 \sim A$.

Im Fall (1) ist nach dem Äquivalenzsatz $\mathfrak{a} = \mathfrak{b}$; im Fall (2) wird man naturgemäß $\mathfrak{a} < \mathfrak{b}$, im Fall (3) $\mathfrak{a} > \mathfrak{b}$ zu definieren haben. Im Fall (4) können wir offenbar, wegen seiner Symmetrie in bezug auf beide Mengen, weder $\mathfrak{a} < \mathfrak{b}$ noch $\mathfrak{a} > \mathfrak{b}$ definieren, da sonst beides zugleich gelten müßte; ebensowenig ist $\mathfrak{a} = \mathfrak{b}$ zulässig, was mit der bisherigen natürlichen Definition der Gleichheit in Widerspruch treten würde. Wir haben hier also eine vierte Relation, die wir $\mathfrak{a} \parallel \mathfrak{b}$ schreiben und als *Unvergleichbarkeit* von $\mathfrak{a}$ mit $\mathfrak{b}$ bezeichnen, während in den drei ersten Fällen $\mathfrak{a}$ und $\mathfrak{b}$ vergleichbar heißen sollen. Also:

$$\left.\begin{array}{l}(1)\ \mathfrak{a} = \mathfrak{b} \\ (2)\ \mathfrak{a} < \mathfrak{b} \\ (3)\ \mathfrak{a} > \mathfrak{b}\end{array}\right\}\ \mathfrak{a},\ \mathfrak{b}\ \text{vergleichbar}$$

$$(4)\ \mathfrak{a} \parallel \mathfrak{b}\quad \mathfrak{a},\ \mathfrak{b}\ \text{unvergleichbar}.$$

Die Vergleichbarkeit wäre also nur zu retten, wenn man zeigen könnte, daß der vierte Fall tatsächlich nicht eintreten kann. Bei zwei endlichen Mengen ist dies in der Tat der Fall; denn ist, mit Numerierung der Elemente, $A = \{a_1, a_2, \ldots, a_m\}$, $B = \{b_1, b_2, \ldots, b_n\}$ und bildet man behufs eineindeutiger Zuordnung die Paare (a_1, b_1), (a_2, b_2) usw., so kommt man zu Ende, sobald eine der Mengen verbraucht ist. (Auch eine endliche und

```
 1  kleinere.) Auf eine "ahnliche, aber erst sp"ater exakt zu begr"undende Art
 2  (mittels Wohlordnung der Mengen, \S\ 13) werden wir k"unftig zeigen, da"s
 3  der Fall~(4) niemals eintreten kann, zwei Kardinalzahlen also stets
 4  vergleichbar sind.
 5
 6  Ist $A$ einer Teilmenge von $B$ "aquivalent (Fall~(1) oder~(2) tritt ein),
 7  so ist $\frak a = \frak b$ oder $\frak a < \frak b$, was wir in
 8  $\frak a \leqq \frak b$ zussammenfassen. Der "Aquivalenzsatz schreibt sich
 9  dann in der einleuchtenden Form: wenn $\frak a \leqq \frak b$ und
10  $\frak a \geqq \frak b$, so ist $\frak a = \frak b$.
11
12  \line{\rlap{\indent Ist}\hfil $\frak a = \frak b$, $\frak a < \frak b$,
13  $\frak a > \frak b$, $\frak a \parallel \frak b$,\hfil}
14  \smallskip
15  \line{\rlap{so ist}\hfil $\frak b = \frak a$, $\frak b < \frak a$,
16  $\frak b > \frak a$, $\frak b \parallel \frak a$.\hfil}
17
18  Ist $\frak a = \frak b$, $\frak b\, \rho\, \frak c$, so ist
19  $\frak a\, \rho\, \frak c$, wenn $\rho$ eine der vier Relationen bedeutet.
20
21  Ist $\frak a < \frak b$, $\frak b < \frak c$, so ist
22  $\frak a < \frak c$; die Relation $<$ ist transitiv.
23
24  Jede unendliche M"achtigkeit ist $\geqq\aleph_0$ (Satz I); $\aleph_0$ ist
25  die kleinste unendliche M"achtigkeit.
26
27  Jede unendliche Teilmenge einer abz"ahlbaren Menge (z.~B. die Menge der
28  Primzahlen) ist abz"ahlbar, da ihre M"achtigkeit $\leqq\aleph_0$ und
29  zugleich $\geqq \aleph_0$, also nach dem "Aquivalenzsatz $=\aleph_0$ ist.
30
31  \Section{Summe, Produkt, Potenz.}
32
33  Die Mengen $A$, $B$ seien fremd. Sind beide endlich und besteht $A$ aus
34  $m$, $B$ aus $n$ Elementen, so besteht $A+B$ aus $m+n$ Elementen. Ferner:
35  ist $A\sim A_1$, $B\sim B_1$ und auch $A_1$, $B_1$ fremd, so ist
36  $A+B\sim A_1+B_1$. Diese Bemerkungen berechtigen zu folgender Definition:
37
38  Die {\it Summe\/} $\frak a + \frak b$ zweier Kardinalzahlen ist die
39  M"achtigkeit der Mengensumme $A+B$, wenn $A$, $B$ irgend zwei fremde Mengen
40  mit den M"achtigkeit $\frak a$, $\frak b$ sind.
41
42  Und allgemein: sind den Elementen $m$ einer Menge
43  $$ M=\{m, n, p,\ldots\,\} $$
44  M"achtigkeiten $\frak a_m$  zugeordnet, so ist
45  $$ \mathop{\textstyle\sum}\limits_m^M \frak a_m = \frak a_m + \frak a_n +
46  \frak a_p + \cdots $$
47  die M"achtigkeit der Mengensumme
48  $$ \mathop{\textstyle\sum}\limits_m^M A_m = A_m + A_n + A_p + \cdots $$
49  wenn die $A_m$ disjunkte (paarweise fremde) Mengen von den M"achtigkeiten
50  $\frak a_m$ sind.
51  \EnoughAlready \bye
```

NOTES _____

1, 6, etc.: Observe the use of " to get an umlaut.
2: Similarly, note the use of the non-standard "s, to get ß.

7, 8, 10, etc.: \frak gives fraktur characters; they come from the AMS collection of fonts.
8–10: \leqq and \geqq give certain AMS symbols.

Example 6

Source:
Mengenlehre
F. HAUSDORFF
© 1927
Walter de
Gruyter & Co.

eine unendliche Mächtigkeit sind stets vergleichbar, jene ist die kleinere.) Auf eine ähnliche, aber erst später exakt zu begründende Art (mittels Wohlordnung der Mengen, § 13) werden wir künftig zeigen, daß der Fall (4) niemals eintreten kann, zwei Kardinalzahlen also stets vergleichbar sind.

Ist A einer Teilmenge von B äquivalent (Fall (1) oder (2) tritt ein), so ist $\mathfrak{a} = \mathfrak{b}$ oder $\mathfrak{a} < \mathfrak{b}$, was wir in $\mathfrak{a} \leqq \mathfrak{b}$ zusammmenfassen. Der Äquivalenzsatz schreibt sich dann in der einleuchtenden Form: wenn $\mathfrak{a} \leqq \mathfrak{b}$ und $\mathfrak{a} \geqq \mathfrak{b}$, so ist $\mathfrak{a} = \mathfrak{b}$.

Ist $\qquad \mathfrak{a} = \mathfrak{b},\ \mathfrak{a} < \mathfrak{b},\ \mathfrak{a} > \mathfrak{b},\ \mathfrak{a} \parallel \mathfrak{b}$,

so ist $\qquad \mathfrak{b} = \mathfrak{a},\ \mathfrak{b} < \mathfrak{a},\ \mathfrak{b} > \mathfrak{a},\ \mathfrak{b} \parallel \mathfrak{a}$.

Ist $\mathfrak{a} = \mathfrak{b}$, $\mathfrak{b} \rho \mathfrak{c}$, so ist $\mathfrak{a} \rho \mathfrak{c}$, wenn ρ eine der vier Relationen bedeutet.

Ist $\mathfrak{a} < \mathfrak{b}$, $\mathfrak{b} < \mathfrak{c}$, so ist $\mathfrak{a} < \mathfrak{c}$; die Relation $<$ ist transitiv.

Jede unendliche Mächtigkeit ist $\geqq \aleph_0$ (Satz I); $\aleph_0$ ist die kleinste unendliche Mächtigkeit.

Jede unendliche Teilmenge einer abzählbaren Menge (z. B. die Menge der Primzahlen) ist abzählbar, da ihre Mächtigkeit $\leqq \aleph_0$ und zugleich $\geqq \aleph_0$, also nach dem Äquivalenzsatz $= \aleph_0$ ist.

§ 6. Summe, Produkt, Potenz.

Die Mengen A, B seien fremd. Sind beide endlich und besteht A aus m, B aus n Elementen, so besteht $A + B$ aus $m + n$ Elementen. Ferner: ist $A \sim A_1$, $B \sim B_1$ und auch A_1, B_1 fremd, so ist $A + B \sim A_1 + B_1$. Diese Bemerkungen berechtigen zu folgender Definition:

Die *Summe* $\mathfrak{a} + \mathfrak{b}$ zweier Kardinalzahlen ist die Mächtigkeit der Mengensumme $A + B$, wenn A, B irgend zwei fremde Mengen mit den Mächtigkeit $\mathfrak{a}$, $\mathfrak{b}$ sind.

Und allgemein: sind den Elementen m einer Menge

$$M = \{m, n, p, \dots\}$$

Mächtigkeiten $\mathfrak{a}_m$ zugeordnet, so ist

$$\sum_{m}^{M} \mathfrak{a}_m = \mathfrak{a}_m + \mathfrak{a}_n + \mathfrak{a}_p + \cdots$$

die Mächtigkeit der Mengensumme

$$\sum_{m}^{M} A_m = A_m + A_n + A_p + \cdots$$

wenn die A_m disjunkte (paarweise fremde) Mengen von den Mächtigkeiten $\mathfrak{a}_m$ sind.

$$\vdots$$

```
1  % A new file begins; it will use the AMS Euler and Computer Concrete fonts.
2  \input OutForm.tex      % Brings in the 'output format' file.
3  \RealPno=93
4  \ExampleNumber=6        % The number of the previous example.
5
6  \BookTitle={Functional Integration and Quantum Physics}
7  \Author={Barry Simon}
8  \CopyrightDate={1979}
9  \PublishInfo={Academic Press}
10 \DontMarkPageCorner
11
12 \ExampleType={\TeX\ and New Typefaces}%
13 \NewExample
14 A relatively new typeface for mathematics, called {\sl AMS Euler}, is shown
15 below. The accompanying text typeface is even newer; it is called
16 {\sl Computer Concrete}. Some information about the design of these faces is
17 given in Chapter~5, along with tables of available characters.
18    ..................................................................
19 \input eulconc  % The file that will switch everything to Euler/Concrete.
20
21 \ShortExample <49>
22 \noindent{\it Remark}\ \ Since $V(\omega(s))$ is almost everywhere
23 (in $\omega$) continuous, $\int_0^t V(\omega(s))\,ds$ can be taken as a
24 Riemann integral and so is $\omega$-measurable as a limit of Riemann sums.
25 \medskip
26 \noindent{\it First Proof}\ \ By the Trotter product formula:
27 $$
28 (f, e^{-tH}g)=\lim_{n\to\infty}(f, (e^{-tH_0/n}e^{-tV/n})^n g)
29 $$
30 so by Theorem~4.8:
31 $$
32 (f, e^{-tH}g) = \lim_{n\to\infty}\int \overline{f(\omega(0))} g(\omega(t))
33 \exp \biggl[ -{t\over n}\sum_{j=0}^{n-1} V \biggl( \omega \biggl( {tj\over n}
34 \biggr) \biggr) \biggr] d\mu_0(\omega)\quad \eqno\equ(6.2)
35 $$
```

NOTES

Most of the commands here are the formatting commands that have been seen, and briefly discussed, at the start of earlier examples.

19: This file is a modification by the author of one originally written by Donald Knuth. It contains all the font commands necessary to make a smooth switch to the new typefaces. The inside back cover carries information on how to get the new faces and the associated file of commands.

27–40: A simple device is used here in order to make the input for displayed equations stand out more sharply. Actual blank lines are not a good idea because they can end up leaving too much space around the display (the interparagraph gap gets added to the normal space): this is the next best thing.

34: \equ is a new command (defined in the file eulconc.tex) that selects an appropriate typeface for equation numbers.

TEX and New Typefaces

——————————— **Example 7** ———————————

From *Functional Integration and Quantum Physics*
BARRY SIMON
Academic Press, 1979

A relatively new typeface for mathematics, called *AMS Euler*, is shown below. The accompanying text typeface is even newer; it is called *Computer Concrete*. Some information about the design of these faces is given in Chapter 5, along with tables of available characters.

The present paragraphs appear in *Computer Modern*—the typeface that has so far been used for the vast majority of documents produced with TEX. New typefaces designed to work with TEX are, however, increasingly becoming available. The example below shows that it is possible to make a complete face change—while still hanging on to the old commands \it, \bf, etc.—merely by calling upon a single file. Furthermore, the new typeface can be switched on in the middle of a document and, for text at least, it can be switched off again very easily later. Switching back successfully to the old style of mathematics takes a bit more work; but, in any case, switching typefaces a lot in the same document isn't a good idea.

Of course, somebody has to do the initial work of writing such typeface-changing files, and other users have to know where to get them. The inside back cover provides information on where such files may be obtained.

(Page 49)

Remark Since $V(\omega(s))$ is almost everywhere (in ω) continuous, $\int_0^t V(\omega(s))\,ds$ can be taken as a Riemann integral and so is ω-measurable as a limit of Riemann sums.

First Proof By the Trotter product formula:

$$(f, e^{-tH}\,g) = \lim_{n\to\infty} (f, (e^{-tH_0/n}\,e^{-tV/n}\,)^n g)$$

so by Theorem 4.8:

$$(f, e^{-tH}\,g) = \lim_{n\to\infty} \int \overline{f(\omega(0))}g(\omega(t)) \exp\left[-\frac{t}{n}\sum_{j=0}^{n-1} V\left(\omega\left(\frac{tj}{n}\right)\right)\right] d\mu_0(\omega) \quad (6.2)$$

```
 1 Since $\omega$ is almost everywhere continuous,
 2 $$
 3 {t\over n} \sum_{j=0}^{n-1} V \biggl( \omega \biggl( {jt\over n} \biggr)
 4 \biggr) \biggr) \to \int_0^t V(\omega(s))\,ds
 5 $$
 6 as $n\to\infty$ for almost every $\omega$. Moreover, the integrand in
 7 \equ(6.2) is dominated by $\vert f(\omega(0))\vert\,
 8 \vert g(\omega(t))\vert\,\exp(t\Vert V\Vert_{\infty}$ which is $L^1$ since
 9 $$
10 \int \vert f(\omega(0))\vert\,\vert g(\omega(t))\vert\,d\mu_0 =
11 (\vert f\vert, e^{-tH_0}\vert g\vert) < \infty
12 $$
13 Thus, by the dominated convergence theorem, \equ(6.1) holds.
14 \EndShortExample
15
16 \ShortExample <140>
17 $$
18 \eqalign{C_\lambda & = \lambda \vert C_1\vert^{1/n}\tilde C_1 +
19 (1-\lambda)\vert C_0\vert^{1/n}\tilde C_0 \cr
20 & = [\lambda \vert C_1\vert^{1/n} + (1-\lambda) \vert C_0\vert^{1/n}]
21 \tilde C_{\tilde\lambda} \cr}
22 $$
23 \EndShortExample
24
25 \ShortExample <233>
26 $$
27 \phi_i({\bf x}) = \cases{0,&$\vert{\bf x}-{\bf x}_i\vert \le r$\cr
28 \noalign{\smallskip}
29 \displaystyle{\vert{\bf x}-{\bf x}_i\vert-r\over r},&
30 $r\le \vert{\bf x}-{\bf x}_i\vert \le 2r$\cr
31 \noalign{\smallskip}
32 1,&$\vert{\bf x}-{\bf x}_i\vert \ge 2r$}
33 $$
34 \EndShortExample
35
36 \ShortExample<246>
37 $$
38 \displaylines{\quad \rho_\Lambda^{(n)}({\bf x}_1,\ldots,{\bf x}_n;
39 \varepsilon_1,\ldots,\varepsilon_n; z, \beta)\hfill\cr
40 \noalign{\smallskip}
41 \hfill = \Xi_\Lambda^{-1}z^n \sum_{N=0}^\infty {z^N\over N!} 2^{-N} \!\!\!
42 \sum_{\scriptstyle \varepsilon'_j=\pm1\atop \scriptstyle j=1,\ldots, N}
43 \int_{\Lambda^N} d^{N\nu}x' \exp(-\beta U_{N=n}({\bf x}, {\bf x}';
44 \varepsilon, \varepsilon'))\quad\cr
45 \hfill \equ(23.5)}
46 $$
47 \NegVspace % Command from `OutForm.tex': it subtracts a little vertical space.
48 \EndShortExample
49 \smallskip
50 \CMtext    % Back to Computer Modern text; the command is in `eulconc.tex'.
51 \noindent {\it Note\/}: As the input for these displays indicates, all of
52 Plain~\TeX's standard formula-formatting commands can continue to be used
53 exactly as before.
54 \EndPage  \bye
```

Example 7

Source:
*Functional
Integration
and Quantum
Physics*
BARRY SIMON
© 1979
Academic Press

Since ω is almost everywhere continuous,

$$\frac{t}{n} \sum_{j=0}^{n-1} V\left(\omega\left(\frac{jt}{n}\right)\right) \to \int_0^t V(\omega(s)) \, ds$$

as $n \to \infty$ for almost every ω. Moreover, the integrand in (6.2) is dominated by $|f(\omega(0))|\,|g(\omega(t))|\,\exp(t\|V\|_\infty$ which is L^1 since

$$\int |f(\omega(0))|\,|g(\omega(t))|\,d\mu_0 = (|f|, e^{-tH_0}\,|g|) < \infty$$

Thus, by the dominated convergence theorem, (6.1) holds.

(Page 140)

$$C_\lambda = \lambda|C_1|^{1/n}\,\tilde{C}_1 + (1-\lambda)|C_0|^{1/n}\,\tilde{C}_0$$
$$= [\lambda|C_1|^{1/n} + (1-\lambda)|C_0|^{1/n}]\tilde{C}_{\tilde{\lambda}}$$

(Page 233)

$$\phi_i(x) = \begin{cases} 0, & |x - x_i| \le r \\ \dfrac{|x - x_i| - r}{r}, & r \le |x - x_i| \le 2r \\ 1, & |x - x_i| \ge 2r \end{cases}$$

(Page 246)

$$\rho_\Lambda^{(n)}(x_1, \ldots, x_n; \varepsilon_1, \ldots, \varepsilon_n; z, \beta)$$
$$= \Xi_\Lambda^{-1} z^n \sum_{N=0}^{\infty} \frac{z^N}{N!} 2^{-N} \sum_{\substack{\varepsilon_j'=\pm 1 \\ j=1,\ldots,N}} \int_{\Lambda^N} d^{N\nu} x' \exp(-\beta U_{N=n}(x, x'; \varepsilon, \varepsilon'))$$

$$(23.5)$$

Note: As the input for these displays indicates, all of Plain TeX's standard formula-formatting commands can continue to be used exactly as before.

```
1  % A new file begins here:
2  \input OutForm.tex     % Brings in the 'output format' file.
3  ...............................................................
4  \medskip
5  \input Times  % The file that will switch everything to 'Times-Roman'.
6
7  % Load two extra fonts, in order to produce some special symbols:
8  \font\BigMTSY=MTSY at 12.25pt % 'MTSY' is the MathTime symbols font.
9  \font\Script=pzcmi % This is a standard PostScript Font ('Zapf Chancery').
10 % Construct a symbol (to be used as a binary operation):
11 \def\BgTrLeft{\mathbin{\raise.5pt\hbox{\BigMTSY\char'107}}}
12 % Allow access to the second font loaded above:
13 \def\Scr #1{\mathord{\hbox{\Script #1}\,}}
14
15 \ShortExample <102>
16 \smallskip
17 {\bf Theorem 4.10}\quad (Galois).\quad Suppose $F$ is a field of
18 characteristic $0$, $f(x)\in F[x]$, and the Galois group of $f(x)$ is
19 solvable. Then $f(x)$ is solvable by radicals over $F$.
20 \smallskip
21 {\it Proof}.\quad Let $K$ be a splitting field for $f(x)$ over $F$; set
22 $G=G(K\colon F)$, and say $[K\colon F]=n$. Let $L$ be a splitting field over
23 $K$ for $x^n-1$, and let $\zeta\in L$ be a primitive $n$th root of unity, so
24 $L=K(\zeta)$. Set $E=F(\zeta)$; then clearly $L$ is a splitting field for
25 $f(x)$ over $E$. If we set $H=G(L\colon E)$, then $H$ is isomorphic with a
26 subgroup of $G$ by Proposition~4.9, and hence $H$ is also solvable. By
27 Theorem~I.5.3, $H$ has a subnormal series $H=H_0\ge H_1\ge \cdots \ge H_k=1$
28 with abelian factors, and by refining to a composition series we may assume
29 that $H_{i-1}/H_i$ is cyclic of prime order $p_i$, $1\le i\le k$. In the
30 setting of $E\subseteq L$ we set $L_i=\Scr F H_i$, $0\le i\le k$, so
31 $E=L_0\subseteq L_1\subseteq \cdots \subseteq L_k=L$ and $[L_i\colon L_{i-1}]
32 =p_i$. Since $G(L\colon L_i)=H_i\BgTrLeft H_{i-1}=G(L\colon L_{i-1})$ we
33 have $L_i$ Galois over $L_{i-1}$, and $L_{i-1}$ contains a primitive $p_i$th
34 root of unity (it is a power of $\zeta$). By Proposition~4.8, $L_i$ is a
35 simple radical extension of $L_{i-1}$, $1\le i\le k$. Thus $L$ is an
36 extension of $E$ by radicals, and hence also of $F$ since $E=F(\zeta)$.
37 \EndShortExample
```

NOTES

The procedure at the start of every example has been shown so often that it will be omitted from now on.

5: As with the file called **eulconc** in the previous example, this one contains all the font-switching commands. See the inside back cover for more information on getting the commands that go into this file.

8: A font is loaded at a special size in order to construct a large symbol. Since the fonts used here are PostScript fonts, they may be asked for at essentially arbitrary sizes (or scaled essentially arbitrarily).

9, 13: Another font is loaded, to be used for producing script letters. Even more "scripty" (i.e., curly) letters can be found in specialized font packages. The *MathTime* manual discusses one such package. The definition on line 13 allows the use of the new font in mathematics; more systematic ways of adding new typefaces are discussed in § 5.5.

11, 32: The new symbol is constructed and used; the location of the required character (at position *'107* in the font) was found by looking at a font table in the manual for the *MathTime* fonts. See Chapter 5 for information on fonts and font tables.

TEX and New Typefaces

──────── **Example 8** ────────

From *Algebra*
LARRY C. GROVE
Academic Press, 1983

The typeface that is exhibited below belongs to a category called *Times-Roman*. The name comes from a text face originally designed for *The Times*, the English newspaper. One of the requirements imposed when the face was being designed was that it be somewhat horizontally compressed (presumably to squeeze in all the news fit to print).

The American Mathematical Society uses TEX as its typesetting system, with a variant of Times-Roman as its main typeface ("efficient space utilization" appears to have been an important reason behind the choice; see YOUNGEN *et al.* [1989]).

The Times-Roman fonts used below are PostScript fonts. The text face is standard, the math face—called *MathTime*—is new. The pages here have been processed both by TEX and by a .dvi to PostScript converter. Chapter 5 explains the procedure. The definitions and font summonses that were needed have once again been stored in a separate file. Information about the availability of these fonts, and the associated files, is on the inside back cover.

(Page 102)

Theorem 4.10 (Galois). Suppose F is a field of characteristic 0, $f(x) \in F[x]$, and the Galois group of $f(x)$ is solvable. Then $f(x)$ is solvable by radicals over F.

Proof. Let K be a splitting field for $f(x)$ over F; set $G = G(K:F)$, and say $[K:F] = n$. Let L be a splitting field over K for $x^n - 1$, and let $\zeta \in L$ be a primitive nth root of unity, so $L = K(\zeta)$. Set $E = F(\zeta)$; then clearly L is a splitting field for $f(x)$ over E. If we set $H = G(L:E)$, then H is isomorphic with a subgroup of G by Proposition 4.9, and hence H is also solvable. By Theorem I.5.3, H has a subnormal series $H = H_0 \geq H_1 \geq \cdots \geq H_k = 1$ with abelian factors, and by refining to a composition series we may assume that H_{i-1}/H_i is cyclic of prime order p_i, $1 \leq i \leq k$. In the setting of $E \subseteq L$ we set $L_i = \mathcal{F}H_i$, $0 \leq i \leq k$, so $E = L_0 \subseteq L_1 \subseteq \cdots \subseteq L_k = L$ and $[L_i:L_{i-1}] = p_i$. Since $G(L:L_i) = H_i \triangleleft H_{i-1} = G(L:L_{i-1})$ we have L_i Galois over L_{i-1}, and L_{i-1} contains a primitive p_ith root of unity (it is a power of ζ). By Proposition 4.8, L_i is a simple radical extension of L_{i-1}, $1 \leq i \leq k$. Thus L is an extension of E by radicals, and hence also of F since $E = F(\zeta)$.

```
1  \ShortExample <104>
2  $$
3  \vbox{\offinterlineskip
4  \halign{\hfil$#$\ &#&$#$\hfil\cr
5  &&\ \ x^{n-1} + (x_n-\sigma_1)x^{n-2} + (\sigma_2-(x_n-\sigma_1)x_n)x^{n-3}
6  + \cdots \cr
7  \noalign{\vskip3pt}
8  &&\hrulefill\cr
9  x-x_n&\vrule height11pt depth4pt&\ \
10 x^n - \sigma_1x^{n-1} + \sigma_2x^{n-2} - \sigma_3x^{n-3} + \cdots\cr}}
11 $$
12 \EndShortExample
13
14 \ShortExample <109>
15 \noindent Since $k\cdot h(0)\in{\bf Z}$ and $p\!\not\vert\;k\cdot h(0)$ we
16 conclude that
17 $$ 0 \ne \sum_{j=1}^r h(b_j)+k\cdot h(0) \in {\bf Z}. $$
18 But
19 $$ \int_0^1 e^{(1-s)b_j}g(sb_j)\,ds = \int_0^1 e^{(1-s)b_j}
20 {c_r^{rp-1}\over (p-1)!} b_j^{p-1} s^{p-1} f(sb_j)^p\,ds. \eqno(\ast\ast) $$
21 \EndShortExample
22
23 \def\StretchRtArr#1{{\count255=0 \loop \relbar\joinrel \advance\count255 by1
24                   \ifnum\count255<#1 \repeat \longrightarrow}}
25 \def\OverArrow#1#2{\;{\buildrel #1\over {\StretchRtArr#2}}\;}
26 \def\TOR#1{\OverArrow{R_{#1}}{1}}
27 \def\TOC#1#2{\OverArrow{C_{#1(#2)}}{1}}
28 \def\BM#1#2#3#4{\left[\matrix{\hfill#1&#2\hfill\cr \hfill#3&#4\hfill
29       \cr}\right]}
30
31 \ShortExample <149>
32 $$
33 \displaylines{\eqalign{A=&\BM46{-6}8 \TOR{211} \BM46{-2}{14} \TOR{121}
34 \BM2{20}{-2}{14} \TOR{211} \cr
35 \noalign{\medskip}
36 &\BM2{20}0{34} \TOC{21}{-10} \BM200{34} = D,\cr} \cr
37 \noalign{\medskip}
38 \BM1001 \TOR{211} \BM1011 \TOR{121} \BM2111 \TOR{211} \BM2132 = P,\cr
39 \noalign{\smallskip \noindent and \smallskip}
40 \BM1001 \TOC{21}{-10} \BM1{-10}0{\hphantom{-}1} = Q. \cr
41 $$
42 \EndShortExample
43 \EndPage
```

NOTES

23–29: These are utility commands that help in typesetting subsequent examples. Users may find the first of the lot useful in other contexts as well; it gives right arrows of length determined by the number that is given to it. For example,

\StretchRtArr4

gives an arrow stretched by 4 units.

Example 8

Source:

Algebra

LARRY C.
GROVE

© 1983
Academic Press

(Page 104)

$$x - x_n \overline{\big)\ x^n - \sigma_1 x^{n-1} + \sigma_2 x^{n-2} - \sigma_3 x^{n-3} + \cdots}$$

with quotient

$$x^{n-1} + (x_n - \sigma_1)x^{n-2} + (\sigma_2 - (x_n - \sigma_1)x_n)x^{n-3} + \cdots$$

(Page 109)

Since $k \cdot h(0) \in \mathbf{Z}$ and $p \nmid k \cdot h(0)$ we conclude that

$$0 \neq \sum_{j=1}^{r} h(b_j) + k \cdot h(0) \in \mathbf{Z}.$$

But

$$\int_0^1 e^{(1-s)b_j} g(sb_j)\, ds = \int_0^1 e^{(1-s)b_j} \frac{c_r^{rp-1}}{(p-1)!} b_j^{p-1} s^{p-1} f(sb_j)^p\, ds. \qquad (**)$$

(Page 149)

$$A = \begin{bmatrix} 4 & 6 \\ -6 & 8 \end{bmatrix} \xrightarrow{R_{211}} \begin{bmatrix} 4 & 6 \\ -2 & 14 \end{bmatrix} \xrightarrow{R_{121}} \begin{bmatrix} 2 & 20 \\ -2 & 14 \end{bmatrix} \xrightarrow{R_{211}}$$

$$\begin{bmatrix} 2 & 20 \\ 0 & 34 \end{bmatrix} \xrightarrow{C_{21(-10)}} \begin{bmatrix} 2 & 0 \\ 0 & 34 \end{bmatrix} = D,$$

$$\begin{bmatrix} 1 & 0 \\ 0 & 1 \end{bmatrix} \xrightarrow{R_{211}} \begin{bmatrix} 1 & 0 \\ 1 & 1 \end{bmatrix} \xrightarrow{R_{121}} \begin{bmatrix} 2 & 1 \\ 1 & 1 \end{bmatrix} \xrightarrow{R_{211}} \begin{bmatrix} 2 & 1 \\ 3 & 2 \end{bmatrix} = P,$$

and

$$\begin{bmatrix} 1 & 0 \\ 0 & 1 \end{bmatrix} \xrightarrow{C_{21(-10)}} \begin{bmatrix} 1 & -10 \\ 0 & 1 \end{bmatrix} = Q.$$

```
1  \def\SetupforCD {\ifmmode \def\quad{\hskip.5em\relax}
2    \def\normalbaselines{\baselineskip22pt \lineskip3pt \lineskiplimit3pt}\fi}
3  \def\LeftofArrow #1{\llap{$\vcenter{\hbox{$\scriptstyle#1$}}$}\bigg\downarrow}
4  \def\RightofArrow #1{\bigg\downarrow
5        \rlap{$\vcenter{\hbox{$\scriptstyle#1$}}$}}
6
7  \ShortExample <84>
8  $$\SetupforCD
9  \matrix{
10 F_1[x] & \hidewidth \OverArrow{\phi}{7} \hidewidth & F_2[x] \cr
11 \LeftofArrow{\eta_1} && \RightofArrow{\eta_2} \cr
12 F_1[x]/(m_1(x)) & \OverArrow{\theta}{1} & F_2[x]/(m_2(x)) }
13 $$
14 \EndShortExample
15
16 \def\Vsubseteq{\left.\smash{\lower.5pt\hbox{$\cap$}}\!\right\arrowvert}
17
18 \ShortExample <74>
19 $$
20 \matrix{
21 I_0(0) & \subseteq & I_0(1) & \subseteq & \cdots & \subseteq & I_0(s-1) &
22 \subseteq&\cdots \cr
23 \Vsubseteq && \Vsubseteq &&&& \Vsubseteq \cr
24 I_1(0) & \subseteq & I_1(1) & \subseteq & \cdots & \subseteq & I_1(s-1) &
25 \subseteq&\cdots \cr
26 \Vsubseteq && \Vsubseteq &&&& \Vsubseteq \cr
27 I_2(0) & \subseteq & I_2(1) & \subseteq & \cdots & \subseteq & I_2(s-1) &
28 \subseteq&\cdots \cr
29 \Vsubseteq && \Vsubseteq &&&& \Vsubseteq \cr
30 \vdots && \vdots &&&& \vdots }
31 $$
32 \EndShortExample
33
34 \ShortExample <279>
35 $$\eqalignno{y_1={}&u_1+v_1 = \bigl(-q/2+\sqrt{-D_g/108}\,\bigr)^{1/3}
36 + \bigl(-q/2-\sqrt{D_g/108}\,\bigr)^{1/3},\cr
37 y_2={}&\omega u_1+\omega^2 v_1 = \omega\bigl(-q/2+\sqrt{-D_g/108}
38 \,\bigr)^{1/3}\cr
39 &+ \omega^2\bigl(-q/2-\sqrt{-D_g/108}\,\bigr)^{1/3},\cr
40 y_3={}&\omega^2 u_1+\omega v_1 = \omega^2\bigl(-q/2+\sqrt{-D_g/108}
41 \,\bigr)^{1/3}\cr
42 &+ \omega\bigl(-q/2-\sqrt{-D_g/108}\,\bigr)^{1/3}.\cr} $$
43 \EndShortExample
44 \EndPage  \bye
```

NOTES

1–5: These commands help in the setting of simple commutative diagrams. The first is similar to \CompressMatrices from page 40, but it alters the vertical spacing as well. (Both vertical and horizontal spacing are deeply built into Plain TeX's \matrix command, making such workarounds necessary when the spacing has to be altered.)

10: \OverArrow is defined on the previous page; the second argument ('7') specifies how much the arrow is to stretch. \hidewidth allows the arrow to stretch into the adjoining column.

16: The construction of a vertical "subset or equals" symbol. The \smash is put in to hide the height of the symbol given by \cap. An alternative to this construction is the rotation by 90 degrees of the standard \subseteq symbol. (PostScript permits rotations of characters; see a PostScript manual for details.)

Example 8

Source:
Algebra
LARRY C.
GROVE
© 1983
Academic Press

(Page 84)

$$F_1[x] \xrightarrow{\phi} F_2[x]$$

$$\eta_1 \downarrow \qquad\qquad \downarrow \eta_2$$

$$F_1[x]/(m_1(x)) \xrightarrow{\theta} F_2[x]/(m_2(x))$$

(Page 74)

$$
\begin{array}{ccccccccc}
I_0(0) & \subseteq & I_0(1) & \subseteq & \cdots & \subseteq & I_0(s-1) & \subseteq & \cdots \\
\cap & & \cap & & & & \cap & & \\
I_1(0) & \subseteq & I_1(1) & \subseteq & \cdots & \subseteq & I_1(s-1) & \subseteq & \cdots \\
\cap & & \cap & & & & \cap & & \\
I_2(0) & \subseteq & I_2(1) & \subseteq & \cdots & \subseteq & I_2(s-1) & \subseteq & \cdots \\
\cap & & \cap & & & & \cap & & \\
\vdots & & \vdots & & & & \vdots & &
\end{array}
$$

(Page 279)

$$y_1 = u_1 + v_1 = \left(-q/2 + \sqrt{-D_g/108}\,\right)^{1/3} + \left(-q/2 - \sqrt{-D_g/108}\,\right)^{1/3},$$

$$y_2 = \omega u_1 + \omega^2 v_1 = \omega\left(-q/2 + \sqrt{-D_g/108}\,\right)^{1/3}$$
$$+ \omega^2\left(-q/2 - \sqrt{-D_g/108}\,\right)^{1/3},$$

$$y_3 = \omega^2 u_1 + \omega v_1 = \omega^2\left(-q/2 + \sqrt{-D_g/108}\,\right)^{1/3}$$
$$+ \omega\left(-q/2 - \sqrt{-D_g/108}\,\right)^{1/3}.$$

102

```
 1 | \input OutForm.tex    % Brings in the 'output format' file.
 2 | ......................................................
 3 | % New Fonts are added below:
 4 | \input lcdplain    % Brings in the file containing the new font definitions.
 5 | \input lcdfrak     % Brings in definitions that allow access to fraktur fonts.
 6 |
 7 | \def\LVdots#1{\smash{\lower#1pt\hbox{$\vdots$}}}
 8 |
 9 | \ShortExample<272>
10 | \NegVspace \NegVspace
11 | $$\displaylines{
12 | g(X)=\sum_{i=0}^m a_i X^{m-i}, \qquad h(X)=\sum_{i=0}^n b_i X^{n-i}, \qquad
13 | l(X)=\sum_{i=0}^{m+n-1} c_i X^{m+n-1-i}, \cr
14 | \phi(X)=\sum_{i=0}^{n-1} u_i X^{n-1-i}, \qquad
15 | \psi(X)=\sum_{i=0}^{m-1} v_i X^{m-1-i}. } $$
16 | \centerline{$\cdots$}
17 | $$ \def\quad{\hskip.7em\relax}
18 | \left\vert\matrix{
19 | a_0&&&&b_0\cr  a_1&a_0&&&b_1&b_0\cr
20 | \LVdots7&&\ddots&&&\LVdots7&\ddots\cr
21 | &&&a_0&&&&b_0\cr  a_m&a_{m-1}&&&a_1&b_n&b_{n-1}&&&b_1\cr
22 | &a_m&&\LVdots9&&&b_n&&\LVdots9\cr
23 | &&\ddots&&&&\ddots\cr  &&&a_m&&&&b_n \cr
24 | \multispan4\upbracefill&\multispan4\quad\upbracefill \cr
25 | \multispan4\hfil $n$\hfil&\multispan4\quad\hfil $m$\hfil}\right\vert
26 | $$
27 | (with zeros everywhere else); that is, it equals the resultant $\rho=R(g,h)$.
28 | Since $\rho\ne 0$, this system has a unique solution, and since all the
29 | constant terms $\rho c_i$ are divisible by $\rho$, then the values of $u_i$
30 | and $v_i$ will belong to the ring $\frak v$. The lemma is proved.
31 |
32 | Now let $k$ be complete under the valuation $\nu$, with $\frak v$ the ring of
33 | integral elements of $k$ and $\pi$ a prime element of $\frak v$. Two
34 | polynomials $f(X)$ and $f_1(X)$ of $\frak v[X]$ are called {\it congruent
35 | modulo\/} $\pi^k$, and we write $f(X)\equiv f_1(X)\ (\bmod\;\pi^k)$, if this
36 | congruence holds for the coefficient of each power of $X$.
37 |
38 | {\bf Theorem 1.} Let $f(X)\in \frak v[X]$ have degree $m+n$. \dots
39 | \EndShortExample
40 | \EndPage  \bye
```

TEX and New Typefaces

─────────── **Example 9** ───────────

From *Number Theory*
Z. I. BOREVICH & I. R. SHAFAREVICH
Academic Press, 1966

The mathematics below is in a new PostScript typeface, designed to match a previously existing text face (which readers may recognize as the one used in *Scientific American*). More information is given on the inside back cover.

(Page 272)

$$g(X) = \sum_{i=0}^{m} a_i X^{m-i}, \qquad h(X) = \sum_{i=0}^{n} b_i X^{n-i}, \qquad l(X) = \sum_{i=0}^{m+n-1} c_i X^{m+n-1-i},$$

$$\phi(X) = \sum_{i=0}^{n-1} u_i X^{n-1-i}, \qquad \psi(X) = \sum_{i=0}^{m-1} v_i X^{m-1-i}.$$

$$\cdots$$

$$\left|
\begin{array}{cccccccc}
a_0 & & & & b_0 & & & \\
a_1 & a_0 & & & b_1 & b_0 & & \\
\vdots & & \ddots & & \vdots & & \ddots & \\
& & & a_0 & & & & b_0 \\
a_m & a_{m-1} & & a_1 & b_n & b_{n-1} & & b_1 \\
& a_m & & \vdots & & b_n & & \vdots \\
& & \ddots & & & & \ddots & \\
& & & a_m & & & & b_n
\end{array}
\right|$$

$$\underbrace{\qquad}_{n} \qquad \underbrace{\qquad}_{m}$$

(with zeros everywhere else); that is, it equals the resultant $\rho = R(g, h)$. Since $\rho \neq 0$, this system has a unique solution, and since all the constant terms ρc_i are divisible by ρ, then the values of u_i and v_i will belong to the ring $\mathfrak{v}$. The lemma is proved.

   Now let k be complete under the valuation v, with $\mathfrak{v}$ the ring of integral elements of k and π a prime element of $\mathfrak{v}$. Two polynomials $f(X)$ and $f_1(X)$ of $\mathfrak{v}[X]$ are called *congruent modulo* π^k, and we write $f(X) \equiv f_1(X)$ (mod π^k), if this congruence holds for the coefficient of each power of X.

   Theorem 1. Let $f(X) \in \mathfrak{v}[X]$ have degree $m + n$. ...

```
1   % A new file begins here:
2   \input OutForm.tex     % Brings in the `output format' file.
3   ........................................................................
4
5   \ShortExample <68>
6   \leftline{\indent
7   \vbox to 96bp{\hsize160bp\vss                  % The first picture begins.
8      \special{" newpath 0 0 moveto
9               60 72 lineto 160 72 lineto 100 0 lineto
10              closepath
11              gsave 0.9 setgray fill grestore stroke
12              }                                  % \special is over.
13     \medskip
14     \line{\hfil The plane\hfil \hskip60bp}      % The first caption.
15     \line{\hfil}                                % To balance the next caption.
16  }                                              % \vbox is over.
17  \qquad\qquad
18  \vbox to 108bp{\hsize72bp\vss                  % The second picture begins.
19     \special{" newpath 0 72 moveto
20              24 62 48 62 72 72 curveto
21              48 82 24 82 0 72 curveto
22              0.95 setgray fill
23              newpath 0 72 moveto
24              21 41 21 31 0 5 curveto
25              24 -5 48 -5 72 5 curveto
26              51 31 51 41 72 72 curveto
27              48 62 24 62 0 72 curveto
28              0.75 setgray fill
29              }                                  % \special is over.
30     \medskip
31     \centerline{\vbox{\hbox{The catenoid}
32         \hbox{$\sqrt{x^2+y^2}=\cosh z$}}} % The second caption.
33  }                                              % \vbox is over.
34  }                                              % \leftline is over.
35  \EndShortExample
36  \EndPage
```

NOTES

7–33: The lines show how it is possible to give explicit PostScript instructions to the printing device from within a `\special` command. The exact statement at the start will depend on the particular `.dvi` to `.ps` converter being used, but the other instructions are pure PostScript. What they do is often obvious—once one is told that commands like `lineto` act on numbers to their left. The numbers are mostly coordinates in PostScript points (72 to an inch, exactly the same as the TeX unit `bp`); they represent positions relative to the bottom left-hand corner of the `\vbox`. Further information about PostScript commands can be found in the manuals and tutorials put out by the company that developed PostScript (Adobe Systems Incorporated).

Real pictures are not usually drawn by direct PostScript instructions like this; the method commonly used is shown on the next page. The pictures in this example are not particularly spectacular; more eye-catching pictures are shown in the Glossary under **diagrams**.

HOW THIS FILE WAS PROCESSED

The file was typed as shown in the input boxes. It was then run through TeX to make a `.dvi` file. That file was further translated into a PostScript file using a `.dvi` to `.ps` translation program. Finally, the PostScript file was printed on a printer that understands PostScript.

T_EX and Pictures

―――――――――― **Example 10** ――――――――――

From *Geometric Measure Theory*
FRANK MORGAN
Academic Press, 1988

The inclusion of pictures in T_EX documents is sometimes seen as problematic. In fact, it is rather easy to do, as long as one has access to the appropriate graphical tools. The problem—if there is one—lies not so much with T_EX itself, but with the absence of a universally agreed-upon approach to graphics. On the one hand, T_EX is specifically designed to run the same way (essentially) on all machines; on the other, the production of graphical material often depends on the exact nature of the machinery available. To accommodate this, T_EX has a command called \special that allows users to directly address their printing devices with special instructions. (T_EX itself will not normally do anything with these instructions, except to store them so that they can eventually be passed on to the printer.) The command is used by saying \special{*stuff*}, where *stuff* consists of instructions to the printer in a language that it understands.

PostScript is a widely used such language. The examples below offer glimpses of how it is used. They have been kept simple, since the main point is to illustrate the principles involved. The examples are loosely based on pictures in the book named above. The reproduction of diagrams is, however, an altogether different proposition from the reproduction of formulas: the author is no artist, and any resemblance to the original pictures is purely accidental.

After the examples, there follow some comments and a short discussion of potential problems.

(Page 68)

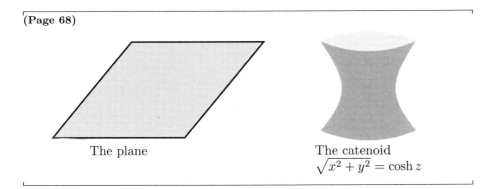

The plane

The catenoid
$$\sqrt{x^2 + y^2} = \cosh z$$

```
 1 | \input epsfmod              % A file of specialized 'dvips' commands.
 2 |
 3 | \ShortExample <43>
 4 | \NegVspace\NegVspace
 5 | $$\epsffile{remmid.ps}$$
 6 | \NegVspace\NegVspace
 7 | \EndShortExample
 8 |
 9 | The first two examples show how ''literal graphics''---explicit instructions
10 | to the printer---can be specified in PostScript. In general, however,
11 | this isn't the way to go, especially if the pictures are even slightly
12 | complex. The preferred approach is to create a picture (a drawing, a
13 | graph, a flowchart) separately with a specialized drawing program, and then
14 | save the image as a PostScript file. Many graphical programs now allow this
15 | option. Suppose that the file is called {\tt picture.ps} and that it
16 | represents an image 2 inches high; the image can, in principle,
17 | be summoned in the
18 | \TeX\ document by saying something~like
19 |
20 | {\tt\string\vbox\ to 2in\string{\string\vss\string\special\string{%
21 | psfile=picture.ps\string}\string}}
22 |
23 | \noindent at the place where it is to go. The exact commands
24 | within {\tt\string\special} will depend on the translation program
25 | (see below) being used.
26 | (In general, it is better to use the special \TeX\ commands that are
27 | supplied with translation programs rather than a 'raw' specification
28 | of a PostScript file. See, for example, the input for the figure above.)
29 |
30 | The size specification is necessary, since \TeX\ itself does not normally
31 | process the ''special'' material and so has no way of knowing how much room to
32 | leave for it when it is composing the {\tt.dvi} file; the {\tt\string\vss}
33 | merely asks for stretchable or shrinkable vertical white space to be left,
34 | as needed.
35 | \par
```

NOTES

1, 5: This is the broad way in which a PostScript file can be summoned by a TeX one. The commands on these lines are specific to the program that was used here: a public domain program called **dvips** (written by Tomas Rokicki). The method is, however, similar in all programs that translate .dvi to .ps.
1: A file of utility commands is brought in first.

The one used is a slightly modified version of the one that comes with **dvips**.
5: A graphical file called **remmid.ps** (for "removed middle") is embedded here. The **\epsffile** command (from the **dvips** package) will scan this file, extract information about the size of the picture and leave vertical space for it in the .dvi file.

Source:

*Geometric
Measure
Theory*
FRANK
MORGAN
© 1988
Academic Press

(Page 43)

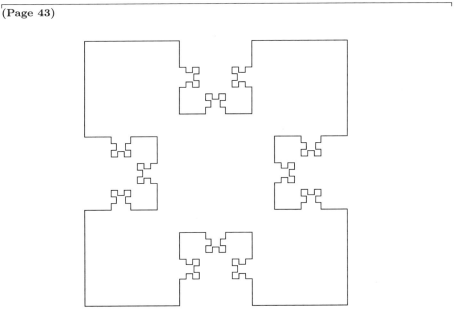

The first two examples show how "literal graphics"—explicit instructions to the printer—can be specified in PostScript. In general, however, this isn't the way to go, especially if the pictures are even slightly complex. The preferred approach is to create a picture (a drawing, a graph, a flowchart) separately with a specialized drawing program, and then save the image as a PostScript file. Many graphical programs now allow this option. Suppose that the file is called `picture.ps` and that it represents an image 2 inches high; the image can, in principle, be summoned in the TeX document by saying something like

    \vbox to 2in{\vss\special{psfile=picture.ps}}

at the place where it is to go. The exact commands within `\special` will depend on the translation program (see below) being used. (In general, it is better to use the special TeX commands that are supplied with translation programs rather than a 'raw' specification of a PostScript file. See, for example, the input for the figure above.)

The size specification is necessary, since TeX itself does not normally process the "special" material and so has no way of knowing how much room to leave for it when it is composing the `.dvi` file; the `\vss` merely asks for stretchable or shrinkable vertical white space to be left, as needed.

```
 1  In order to create a single integrated document, one needs to have access not
 2  only to a printing device that can understand PostScript, but also to a
 3  {\tt.dvi} to PostScript translation program. The program is used to process
 4  the {\tt.dvi} file in order to make a PostScript file (the usual extension is
 5  {\tt.ps}). The file will contain the summoned image, embedded at the right
 6  spot.
 7
 8  Most {\tt.dvi} to PostScript converters offer additional utilities that
 9  make the task somewhat easier, and PostScript itself offers a file format
10  called {\sl Encapsulated PostScript\/} that is specifically aimed at meeting
11  the needs of embedding one PostScript document inside another. Such files are
12  generated by a number of drawing and plotting programs. A principal
13  defining characteristic of an Encapsulated PostScript File (usually called
14  an EPS File, or just an EPSF) is that it contains a statement about
15  how big the image in it is. The statement, usually near the very top of
16  the file, looks something like this:
17
18  {\tt\%\%BoundingBox: 54 484 306 738}
19
20  \noindent (PostScript files are written in ordinary text; they can be
21  opened with an ordinary text editor and their contents examined to see if
22  such a ``bounding box comment'' is indeed present.)
23  The numbers represent the size and location of the image, and they will
24  therefore vary from file to file. They are in PostScript
25  points, 72 to an inch. The first two numbers are the coordinates of the
26  bottom left corner of the picture and the next two those of the top
27  right corner, both relative to the bottom left corner of the page.
28
29  Typical utilities offered with programs that translate {\tt.dvi} to {\tt.ps}
30  will cause \TeX\ to examine the contents of the PostScript file to be
31  embedded, extract the bounding box information, and use it in placing the
32  image---in principle. In practice, problems can arise in many ways (an
33  inaccurate, or non-existent, bounding box comment being one), so adjustments
34  are often necessary. Adjustments are sometimes also necessary when the
35  document is to be printed on devices with differing resolutions (say
36  page proofs on a laser printer and final copy on a phototypesetter).
37  The manuals that come with the translation program being used will usually
38  offer suggestions on dealing with such~problems.
39  \vfil
40  \line{\epsfxsize.25in \epsffile{remmid.ps}
41  \hfil
42  \epsfxsize.75in \epsffile{remmid.ps}
43  \hskip-.25in
44  \epsfxsize1in\hskip\epsfxsize\Rotangle=45 \epsffile{remmid.ps}
45  \hskip-.25in
46  \epsfxsize.75in \epsffile{remmid.ps}
47  \hfil
48  \epsfxsize.25in \epsffile{remmid.ps}}
49  \EndPage
50  \bye
```

NOTES

40–48: PostScript pictures can be scaled and rotated at will. These lines show how this was done for the pictures on the facing page. The actual commands will vary from system to system, but the main point being made here is that it can be done.

Example 10

Source:

*Geometric
Measure
Theory*
FRANK
MORGAN
© 1988
Academic Press

In order to create a single integrated document, one needs to have access not only to a printing device that can understand PostScript, but also to a `.dvi` to PostScript translation program. The program is used to process the `.dvi` file in order to make a PostScript file (the usual extension is `.ps`). The file will contain the summoned image, embedded at the right spot.

Most `.dvi` to PostScript converters offer additional utilities that make the task somewhat easier, and PostScript itself offers a file format called *Encapsulated PostScript* that is specifically aimed at meeting the needs of embedding one PostScript document inside another. Such files are generated by a number of drawing and plotting programs. A principal defining characteristic of an Encapsulated PostScript File (usually called an EPS File, or just an EPSF) is that it contains a statement about how big the image in it is. The statement, usually near the very top of the file, looks something like this:

    `%%BoundingBox:   54 484 306 738`

(PostScript files are written in ordinary text; they can be opened with an ordinary text editor and their contents examined to see if such a "bounding box comment" is indeed present.) The numbers represent the size and location of the image, and they will therefore vary from file to file. They are in PostScript points, 72 to an inch. The first two numbers are the coordinates of the bottom left corner of the picture and the next two those of the top right corner, both relative to the bottom left corner of the page.

Typical utilities offered with programs that translate `.dvi` to `.ps` will cause TeX to examine the contents of the PostScript file to be embedded, extract the bounding box information, and use it in placing the image—in principle. In practice, problems can arise in many ways (an inaccurate, or non-existent, bounding box comment being one), so adjustments are often necessary. Adjustments are sometimes also necessary when the document is to be printed on devices with differing resolutions (say page proofs on a laser printer and final copy on a phototypesetter). The manuals that come with the translation program being used will usually offer suggestions on dealing with such problems.

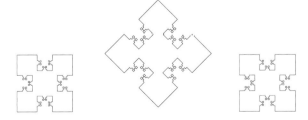

```
 1  %A new file begins here:
 2  \input OutForm.tex      % Brings in the 'output format' file.
 3  .........................................................................
 4  \input pictex           % Brings in the drawing package.
 5  \ShortExample <58>
 6  \NegVspace $$\beginpicture
 7  \setcoordinatesystem units <.5in,.5in>  % First picture.
 8  \multiput {{\bf.}} at 0 0 *100 .01 .00325 /      % Angled rules are built
 9  \multiput {{\bf.}} at 1 .325 *100 -.01 .01376 /  % out of periods placed
10  \multiput {{\bf.}} at 0 0 *100 -.01 .00325 /     % close to each other
11  \multiput {{\bf.}} at -1 .325 *100 .01 .01376 /  % (\bf for thick lines).
12  \linethickness=.3pt
13  \putrule from 0 0 to 0 1.701    % This is the vertical rule.
14  \setcoordinatesystem units <.5in,.5in> point at -2.5 0 % Second picture.
15  \multiput {.} at 0 0 *100 .01 .01376 /
16  \multiput {.} at 0 1.051 *100 .01 .00325 /
17  \multiput {.} at 0 0 *100 -.01 .01376 /
18  \multiput {.} at 0 1.051 *100 -.01 .00325 /
19  \endpicture $$ \NegVspace
20  \EndShortExample
21
22  \ShortExample <75>
23  \NegVspace $$\beginpicture
24  \setcoordinatesystem units <1.5cm,1.5cm>
25  \setplotarea x from 0 to 3.5, y from 0 to 2.5
26  \axis bottom ticks in length <6pt> width <.5pt> from 1 to 3 by 1 /
27  \axis left ticks in length <6pt> width <.5pt> from 1 to 2 by 1 /
28  \multiput {$\bullet$} at 0 0 *1 0 1 *1 1 0 *1 1 1 *1 1 0 /
29  \multiput {$\circ$} at 1.01 1.98 *1 .99 -.98 /
30  \linethickness1.4pt
31  \putrule from 0 0 to 0 1
32  \putrule from 0 1 to 1 1
33  \putrule from 2 2 to 3 2
34  \multiput {{\bf.}} at 1 1 *200 .005 .005 /
35  \setdashes
36  \putrule from 1 1 to 2 1
37  \putrule from 2 1 to 2 2
38  \putrule from 1 1 to 1 2
39  \putrule from 1 2 to 2 2
40  \multiput {1} at -.15 .5 *1 1 1 *1 1.3 0 /
41  \multiput {$\tau$} at .5 .85 *1 1 0 *1 0 1.3 *1 1 0 /
42  \put {\vbox{\hsize 12cm Figure 22. In case of doubt between $1\tau$ and
43  $\tau 1$ we proceed diagonally. This gives the quasicrystal ''the
44  benefit of the doubt.''}} [Bl] at -2.25 -.7
45  \endpicture $$ \NegVspace
46  \EndShortExample
47  \EndPage
```

NOTES

PₖCTₑX, developed by Michael Wichura, is a complex package; it is impossible to do it justice here (see WICHURA [1987]). The basic idea is that the program thinks of the drawing area as a coordinate grid, with units picked as in lines 7 and 24. (Only that portion of the area that is actually drawn on gets placed in the document.) The structure of commands is important in the program, especially the use of reserved characters like *.

TeX and Pictures

───────── **Example 11** ─────────

From *Introduction to the Mathematics of Quasicrystals*
MARKO V. JARIĆ
Academic Press, 1989

Simple pictures can be drawn directly in TeX itself by fitting together horizontal and vertical rules and other elements. The process is slow, and it is difficult to achieve really spectacular results with it; but it has the enormous advantage of being fully integrated with TeX, rather than being an outside element. There are packages that automate some of the drudgery involved; this example is based on one such package, called PiCTeX. (The package was also used to draw the figure on page 6 of this book.) Again, the pictures here are only loosely based on the originals.

(Page 58)

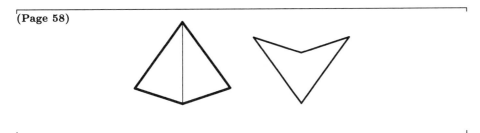

(Page 75)

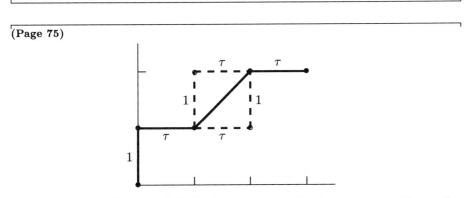

Figure 22. In case of doubt between 1τ and $\tau 1$ we proceed diagonally. This gives the quasicrystal "the benefit of the doubt."

```
1  \BookTitle={Computability, Complexity and Languages}
2  ................................................................
3  \def\Lbl#1{\put {\vbox{\hsize1.2cm\centerline{\fiverm#1}}} [l] at}
4  \def\Dbl#1#2{\vbox{\baselineskip6pt
5          \hbox{$\scriptscriptstyle#1$}\hbox{$\scriptscriptstyle#2$}}}
6
7  \ShortExample <79>
8  $$\beginpicture
9  \setcoordinatesystem units <1.2cm,1.1cm>
10 \multiput{\beginpicture \linethickness .02cm    % BOXES
11             \putrectangle corners at 0 .3 and 1 -.3
12         \endpicture} [l] at  0 0   2 0   2 1.6   2 -1.6   4 1.6   5 0
13         6.5 -1.6   6.5 -3.2   8.5 -1.6 /
14 \linethickness .04cm
15 \multiput{\beginpicture                       % HORIZONTAL RULES
16             \putrule from  0 0 to 1 0
17         \endpicture} [l] at  1 0   3 0   4 0   3 1.6   6 0   5.5 -1.6
18         5.5 -3.2   7.5 -1.6 /
19 \multiput{\beginpicture                       % VERTICAL RULES
20             \putrule from 0 0 to 0 1
21         \endpicture} [b] at  2.5 .3   2.25 -1.3   2.75 -1.3   7 -2.9 /
22 \putrule from 5.5 -3.22 to 5.5 -1.59
23 \putrule from 7 -1.3 to 7 0.02
24 \Lbl{BEGIN} 0 0
25 \Lbl{TEST X} 2 0  \Lbl{TEST X} 6.5 -1.6  \Lbl{END} 8.5 -1.6  \Lbl{END} 4 1.6
26 \Lbl{$\scriptscriptstyle {\rm Y}\leftarrow s_1{\rm Y}$} 2 1.6
27 \Lbl{\Dbl{{\rm X}\leftarrow{\rm X}^-}{{\rm Y}\leftarrow s_1{\rm Y}}} 2 -1.6
28 \Lbl{\Dbl{{\rm X}\leftarrow{\rm X}^-}{{\rm Y}\leftarrow s_i{\rm Y}}} 6.5 -3.2
29 \Lbl{\Dbl{{\rm X}\leftarrow{\rm X}^-}{{\rm Y}\leftarrow s_{i+1}\!{\rm Y}}} 5 0
30 \Lbl{\Dbl{\rm Carry}{\rm propagates}} 1.15 -.8
31 \Lbl{$\scriptscriptstyle x\ {\rm ends}\ s_n$} 2.8 -.8
32 \Lbl{$\scriptscriptstyle x=0$} 2.3 .8
33 \Lbl{$\scriptscriptstyle x\ {\rm ends}\ s_i$} 3.5 .15
34 \Lbl{$\scriptscriptstyle i<n$} 3.5 -.2
35 \Lbl{$\scriptscriptstyle x=0$} 7.5 -1.45
36 \Lbl{$\scriptscriptstyle x\ {\rm ends}\ s_i$} 7 -2.4
37 \multiput{\beginpicture                       % RIGHT ARROWS
38             \arrow <.18cm> [.15,.5] from 0 0 to .18 0
39         \endpicture} [l] at 1.46 0   3.3 0   3.46 1.6   6.46 0   7.96 -1.6
40         5.96 -1.6 /
41 \multiput{\beginpicture                       % UP ARROWS
42             \arrow <.18cm> [.15,.5] from 0 0 to 0 .18
43         \endpicture} at 2.25 -1.09   2.5 .8   5.5 -2.36 /
44 \multiput{\beginpicture                       % DOWN ARROWS
45             \arrow <.18cm> [.15,.5] from 0 0 to 0 -.18
46         \endpicture} at 2.75 -.91   7 -2.3 /
47 \arrow <.18cm> [.15,.5] from 6.14 -3.2 to 5.96 -3.2
48 \endpicture $$
49 \EndShortExample
50 \EndPage
```

NOTES

3–5: Commands that are meant to help in labelling the diagram.

10–13, etc.: The P$_I$CT$_E$X command \multiput

can be used to put multiple copies of entire pictures at specified spots, as shown. The [l] tells it to place the left edge of the picture at the given point.

TEX and Pictures

-------- **Example 12** --------

From *Computability, Complexity and Languages*
MARTIN D. DAVIS & ELAINE J. WEYUKER
Academic Press, 1983

Here is a slightly more complex application of PICTEX. It uses the program's ability to put pictures within pictures, an ability which allows the easy placement of a repeated element, without having to draw it from scratch each time. Still, the instructions that have to be issued are not simple, and it would be a formidable task to have to explicitly specify them for a really complicated diagram (even if it only involves horizontal and vertical lines). But, just as complicated PostScript instructions aren't usually programmed from scratch—they are generated instead behind the scenes by visually-oriented drawing programs—there are "front ends" for graphical instructions in TEX. For example, there is a public domain program called TEXCAD that automatically translates pictures drawn on the screen (using, say, a mouse) into TEX (more precisely, LATEX) instructions.

(Page 79)

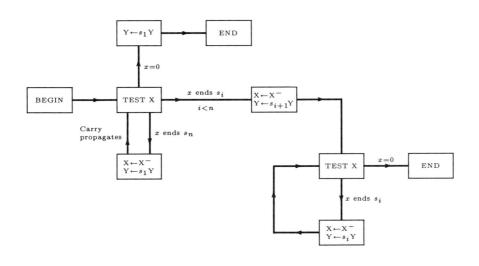

113

114

```
 1 | \font\Chss=chess10
 2 | \BookTitle={On Numbers and Games}
 3 |       ..............................................................
 4 | such a font is a drawing of a piece, and these drawings can
 5 | be placed in the document as easily as one can say {\Chss  N p Q r}.
 6 |       ..............................................................
 7 | \input latexpic.tex     % Brings in the LaTeX 'picture environment'.
 8 | \ShortExample <168>
 9 | $$
10 | \beginpicture
11 | \setcoordinatesystem units <1pt,1pt>
12 | \unitlength=1pt
13 | \put {\circle{3}}         [Bl] at -10 -1.5
14 | \put {\circle{3}}         [Bl] at 10 -1.5
15 | \put {\line(-1,2){10}}    [Bl] at -10 0
16 | \put {\circle*{3}}        [Bl] at -20 20
17 | \put {\line(1,2){10}}     [Bl] at 10 0
18 | \put {\circle*{3}}        [Bl] at 20 20
19 | \put {\line(1,1){20}}     [Bl] at -20 20
20 | \put {\line(-1,1){20}}    [Bl] at 20 20
21 | \put {\circle*{3}}        [Bl] at 0 40
22 | \put {\line(0,5){20}}     [Bl] at 0 40
23 | \put {{\sevenrm 6}} at 4 50
24 | \put {\circle*{3}}        [Bl] at 0 60
25 | \put {\circle{24}}        [Bl] at 0 72
26 | \put {\line(3,-1){24}}    [Bl] at 0 60
27 | \put {{\sevenrm 2}} at 14 60
28 | \put {\line(-5,-1){30}}   [Bl] at 0 60
29 | \put {{\sevenrm 7}} at -15 52
30 | \put {\circle*{3}}        [Bl] at 24 52
31 | \put {\circle*{3}}        [Bl] at -30 54
32 | \put {\line(1,1){20}}     [Bl] at 24 52
33 | \put {\line(0,5){20}}     [Bl] at -30 54
34 | \put {\circle*{3}}        [Bl] at 44 72
35 | \put {\circle*{3}}        [Bl] at -30 74
36 | \put {$\scriptstyle x\colon 6$} at -22 64
37 | \endpicture
38 | $$ \NegVspace \NegVspace
39 | \EndShortExample
40 | \EndPage \bye
```

NOTES

1: **chess10** is part of a collection of chess fonts developed by Piet Tutelaers; see TUTELAERS [1992].

The rest of the page contains a mixture of PiCTEX commands (e.g., \put) and LaTEX picture commands (e.g., \line). As with all the examples on pictures in the present book, the main point being made is that such pictures can be drawn if one has the right tools. Readers are urged to go out and get the tools (which usually come with detailed instructions of their own).

7: A file of LaTEX picture commands that PiCTEX makes available.

TEX and Pictures

──────── **Example 13** ────────

From *On Numbers and Games*
J. H. CONWAY
Academic Press, 1976

Of the two approaches to including pictures with TEX that have been discussed so far—importing drawings from the outside with the \special command, and drawing them with TEX itself—the first clearly offers far greater power. There is a third approach, comparable in power to the first, but as yet imperfectly explored. (See, for example, HOENIG [1991].) The approach involves the design and use of graphical fonts, out of whose "characters" pictures can be built (or whose individual characters might be entire pictures in themselves). TEX's companion program, METAFONT, is powerful enough to draw a variety of graphical images; these images can be invoked as an integral part of the document. For example, some work has been done on chess fonts: each character in such a font is a drawing of a piece, and these drawings can be placed in the document as easily as one can say ♘♟♕♜.

Most of this work has so far been confined to small groups of enthusiasts. In the mainstream, a very small step in this direction has been taken in LATEX, where a rudimentary "picture environment" is provided along with a set of commands that allow users to manipulate simple graphical elements: arcs of circles, angled straight lines, etc. It is possible to use this picture environment even from outside LATEX; programs like PICTEX, for example, make it available. The exceedingly simple picture shown below (taken from a book that contains many novelties, mathematical and typographical) has been drawn using these LATEX facilities. Though it is not obvious from looking at it, most of the elements of that picture are characters from a set of circle and line fonts.

(Page 168)

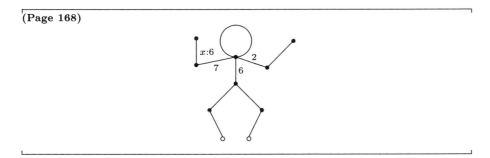

115

```
 1 | %A new file begins here:
 2 | \input OutForm.tex   % Brings in the 'master format'.
 3 | ..........................................................
 4 | The samples shown below demonstrate how formula alignments are done with
 5 | \AmSTeX, since that is one area where there are differences from the approach
 6 | used in Plain~\TeX.
 7 |
 8 | \input AmSTeX
 9 | \UseAMSsymbols      % This is an AmSTeX command.
10 |
11 | \ShortExample <4>
12 | To calculate the divergence, one certainly has $n^3$ functions
13 | $\Gamma_{ij}^k$, $i, j, k=1,\dots, n$, known as {\it Christoffel symbols},
14 | determined by
15 | $$\gather
16 | \nabla_{\partial_j}\partial_i = \sum \Gamma_{ij}^k \partial_k \tag 23 \\
17 | \intertext{on $U$. Thus for $\xi$ given by (19) and $X$ given by}
18 | X = \sum_j\eta^j \partial_j, \tag24 \\
19 | \intertext{we have, using (7) and (8),}
20 | \nabla_{\xi}X = \sum_{j,k} \xi^j\left\{ \partial_j\eta^k +
21 | \sum_l \eta^l \Gamma_{lj}^k \right\} \partial_k. \tag 25
22 | \endgather $$
23 | \NegVspace \EndShortExample
```

NOTES

$\mathcal{A}_{\mathcal{M}}S$-TeX commands are usually easy to use: in fact, one of the chief driving forces behind $\mathcal{A}_{\mathcal{M}}S$-TeX is that it simplifies mathematical input. The most efficient way to read the input here is to look up unfamiliar commands in the Glossary: they are all listed there.

$\mathcal{A}\mathcal{M}S$-TEX

—————— Example 14 ——————

From *Eigenvalues in Riemannian Geometry*
ISAAC CHAVEL
Academic Press, 1984

A few features of $\mathcal{A}\mathcal{M}S$-TEX are shown in this and the next two examples. The exact relationship of $\mathcal{A}\mathcal{M}S$-TEX to other flavors of TEX has been discussed in Chapter 1, but one fact about it bears repeating here: for most purposes, it envelopes Plain TEX. Thus the effects illustrated in the preceding examples should work even for those who use $\mathcal{A}\mathcal{M}S$-TEX. The present examples will concentrate on features not available in other places.

A further aspect of $\mathcal{A}\mathcal{M}S$-TEX must also be kept in mind if the examples shown here are to be used as models elsewhere: the importance in $\mathcal{A}\mathcal{M}S$-TEX of the *document style*. A lot of what $\mathcal{A}\mathcal{M}S$-TEX does depends on the style that has been chosen for the document; different choices of style can result in identical commands having drastically different effects. The notion is discussed in Chapter 3. This feature makes it difficult to demonstrate efficiently the uses of $\mathcal{A}\mathcal{M}S$-TEX solely through examples. Those shown here are supplemented by a complete discussion in the Glossary of all of $\mathcal{A}\mathcal{M}S$-TEX's mathematical commands.

The samples shown below demonstrate how formula alignments are done with $\mathcal{A}\mathcal{M}S$-TEX, since that is one area where there are differences from the approach used in Plain TEX.

(Page 4)

To calculate the divergence, one certainly has n^3 functions Γ_{ij}^k, $i, j, k = 1, \ldots, n$, known as *Christoffel symbols*, determined by

$$(23) \qquad \nabla_{\partial_j} \partial_i = \sum \Gamma_{ij}^k \partial_k$$

on U. Thus for ξ given by (19) and X given by

$$(24) \qquad X = \sum_j \eta^j \partial_j,$$

we have, using (7) and (8),

$$(25) \qquad \nabla_\xi X = \sum_{j,k} \xi^j \left\{ \partial_j \eta^k + \sum_l \eta^l \Gamma_{lj}^k \right\} \partial_k.$$

117

```
1   \define\DIV{\operatorname{div}}
2   \define\DIX{\operatorname{dix}}
3   \define\TR{\operatorname{tr}}
4   \ShortExample <5>
5   In particular, we have by (26) and (31),
6   $$\align
7   \DIV X&=\sum_j \Biggl\{ \partial_j\eta^j + \eta^j \sum_{k,l} \tfrac12
8   g^{kl} \partial_j g_{lk} \Biggr\} \\
9   &=\sum_j \left\{ \partial_j\eta^j + \tfrac12 \eta^j \TR (G^{-1}
10  \partial_j G ) \right\} \\
11  &=\sum_j \{ \partial_j \eta^j + \tfrac12 \eta^j \partial_j (\ln g) \} \\
12  &=(1/\sqrt g) \sum_j \partial_j (\eta^j \sqrt g), \\
13  \intertext{that is,}
14  \DIX X&=(1/\sqrt g) \sum_j \partial_j (\eta^j \sqrt g). \tag 32
15  \endalign $$
16  \EndShortExample
17
18  \define\GRAD{\operatorname{grad}}
19  \ShortExample <49>
20  \NegVspace \NegVspace
21  $$\align
22  \int_{\bold B(\delta)} \vert\GRAD\phi\vert^2 dV
23  & = \int_{\Bbb B^n(\tau)} \frac{2^{n-2} dx^1 \dotsb dx^n}
24  {\{1-\vert x\vert^2\}^{n-2}} \\
25  & = \roman{const}\,\int_0^\tau \frac{r^{n-1}}{\{1-r^2\}^{n-2}}\,dr \\
26  & \le \roman{const}\,
27  \cases
28  1, & n=2, \\ \vert\ln(1-\tau)\vert, & n=3, \\ (1-\tau)^{3-n}, & n>3.
29  \endcases
30  \endalign $$
31  \EndShortExample
32
33  \ShortExample <177>
34  \NegVspace \NegVspace
35  $$\multline
36  \left\vert \int_{\Gamma_1(w_0;\delta)} -\frac{\partial G}{\partial \nu_w}
37  (x,w)\{\varphi(w) - \varphi(w_0)\}\, dA(w) \right\vert \\
38  \aligned
39  &\le \{\sup_{\Gamma_1(w_0;\delta)} \vert\varphi(w)-\varphi(w_0)\vert\}
40  \int_{\Gamma_1(w_0;\delta)}-\frac{\partial G}{\partial \nu_w}(x,w)\,dA(w) \\
41  & \le \sup_{\Gamma_1(w_0;\delta)} \vert\varphi(w)-\varphi(w_0)\vert.
42  \endaligned
43  \endmultline $$
44  \EndShortExample
```

Source:

*Eigenvalues in
Riemannian
Geometry*
ISAAC CHAVEL
© 1984
Academic Press

(Page 5)

In particular, we have by (26) and (31),

$$\operatorname{div} X = \sum_j \left\{ \partial_j \eta^j + \eta^j \sum_{k,l} \tfrac{1}{2} g^{kl} \partial_j g_{lk} \right\}$$

$$= \sum_j \left\{ \partial_j \eta^j + \tfrac{1}{2} \eta^j \operatorname{tr}(G^{-1} \partial_j G) \right\}$$

$$= \sum_j \{ \partial_j \eta^j + \tfrac{1}{2} \eta^j \partial_j (\ln g) \}$$

$$= (1/\sqrt{g}) \sum_j \partial_j (\eta^j \sqrt{g}),$$

that is,

$$(32) \qquad \operatorname{dix} X = (1/\sqrt{g}) \sum_j \partial_j (\eta^j \sqrt{g}).$$

(Page 49)

$$\int_{\mathbf{B}(\delta)} |\operatorname{grad} \phi|^2 \, dV = \int_{\mathbb{B}^n(\tau)} \frac{2^{n-2} dx^1 \cdots dx^n}{\{1 - |x|^2\}^{n-2}}$$

$$= \operatorname{const} \int_0^\tau \frac{r^{n-1}}{\{1 - r^2\}^{n-2}} \, dr$$

$$\leq \operatorname{const} \begin{cases} 1, & n = 2, \\ |\ln(1 - \tau)|, & n = 3, \\ (1 - \tau)^{3-n}, & n > 3. \end{cases}$$

(Page 177)

$$\left| \int_{\Gamma_1(w_0;\delta)} -\frac{\partial G}{\partial \nu_w}(x, w) \{\varphi(w) - \varphi(w_0)\} \, dA(w) \right|$$

$$\leq \left\{ \sup_{\Gamma_1(w_0;\delta)} |\varphi(w) - \varphi(w_0)| \right\} \int_{\Gamma_1(w_0;\delta)} -\frac{\partial G}{\partial \nu_w}(x, w) \, dA(w)$$

$$\leq \sup_{\Gamma_1(w_0;\delta)} |\varphi(w) - \varphi(w_0)|.$$

```
1  \ShortExample <187>
2  $$\align
3  v(x,t)& = \int_\Omega v(y,t_1) h(y) q_\Omega (x,y,t-t_1)\, dV(y) \\
4  & \hphantom{=} - \int_{t_1}^t d\tau \int_\Omega
5  \{2\langle \GRAD_y v, \GRAD h\rangle + v\Delta h\} (y, \tau)
6  q_\Omega (x,y,t-\tau)\, dV(y) \\
7  & = \int_\Omega v(y,t_1) h(y) q_\Omega (x,y,t-t_1)\, dV(y) \\
8  & \hphantom{=} + \int_{t_1}^t d\tau \int_\Omega v(y,\tau)
9  \{2\langle(\GRAD h)(y), (\GRAD_y q_\Omega)(x,y,t-\tau)\rangle \\
10 & \hphantom{= + \int_{t_1}^t d\tau \int_\Omega }
11 + (\Delta h)(y)q_\Omega(x,y,t-\tau)\}\, dV(y)
12 \endalign $$
13 \EndShortExample
14
15 \ShortExample <304--305>
16 $$\topaligned
17 \vert \hat f(\xi)\vert \le (2\pi)^{-n/2}\Vert f\Vert_{L^1}
18 \quad\text{for all $\xi\in \Bbb R^n$,\quad and}
19 \endtopaligned
20 \quad
21 \topaligned
22 \widehat{\tau_yf}(\xi)&=e^{-i\langle y,\xi\rangle}\hat f(\xi), \\
23 \widehat{e^{i\langle y,\;\rangle}f}(\xi)&=(\tau_y \hat f)(\xi), \\
24 \widehat{f(\lambda x)}(\xi)&=\lambda^{-n}\hat f(\xi/\lambda).
25 \endtopaligned $$
26 \EndShortExample
27
28 \loadeufm  % To use fraktur
29 \ShortExample <128>
30 $$\align
31 \{V(\Omega)\}^2&=\int_\Omega dV(p) \int_{\frak S_p} d\mu_p(\xi)
32 \int_0^{l(\xi)}\sqrt{\bold g}(t;\xi)dt \\
33 &\;\vdots \\
34 &\ge \frac{\bold c_n}{2\pi^n c_{n-1}}
35 \frac{\bigl\{\smallint_{\frak S^+\partial\Omega} l(\xi)
36 \langle \xi, \bar\nu_{\pi(\xi)} \rangle d\sigma(\xi) \bigr\}^{n+1}} % NUM
37 {\bigl\{\smallint_{\frak S^+\partial\Omega}\langle\xi,\bar\nu_{\pi(\xi)}
38 \rangle d\sigma(\xi) \bigr\}^n} % DEN
39 \endalign $$
40 \EndShortExample
41 \EndPage
```

Example 14

Source:

*Eigenvalues in
Riemannian
Geometry*
ISAAC CHAVEL
© 1984
Academic Press

(Page 187)

$$v(x,t) = \int_\Omega v(y,t_1)h(y)q_\Omega(x,y,t-t_1)\,dV(y)$$

$$- \int_{t_1}^t d\tau \int_\Omega \{2\langle \mathrm{grad}_y\, v, \mathrm{grad}\, h\rangle + v\Delta h\}(y,\tau)q_\Omega(x,y,t-\tau)\,dV(y)$$

$$= \int_\Omega v(y,t_1)h(y)q_\Omega(x,y,t-t_1)\,dV(y)$$

$$+ \int_{t_1}^t d\tau \int_\Omega v(y,\tau)\{2\langle(\mathrm{grad}\, h)(y),(\mathrm{grad}_y\, q_\Omega)(x,y,t-\tau)\rangle$$

$$+ (\Delta h)(y)q_\Omega(x,y,t-\tau)\}\,dV(y)$$

(Page 304–305)

$$|\hat{f}(\xi)| \le (2\pi)^{-n/2}\|f\|_{L^1} \quad \text{for all } \xi \in \mathbb{R}^n, \quad \text{and} \qquad \widehat{\tau_y f}(\xi) = e^{-i\langle y,\xi\rangle}\hat{f}(\xi),$$

$$\widehat{e^{i\langle y, \rangle}f}(\xi) = (\tau_y\hat{f})(\xi),$$

$$\widehat{f(\lambda x)}(\xi) = \lambda^{-n}\hat{f}(\xi/\lambda).$$

(Page 128)

$$\{V(\Omega)\}^2 = \int_\Omega dV(p) \int_{\mathfrak{S}_p} d\mu_p(\xi) \int_0^{l(\xi)} \sqrt{\mathbf{g}}(t;\xi)dt$$

$$\vdots$$

$$\ge \frac{\mathbf{c}_n}{2\pi^n c_{n-1}} \frac{\left\{\int_{\mathfrak{S}^+\partial\Omega} l(\xi)\langle\xi,\bar{\nu}_{\pi(\xi)}\rangle d\sigma(\xi)\right\}^{n+1}}{\left\{\int_{\mathfrak{S}^+\partial\Omega} \langle\xi,\bar{\nu}_{\pi(\xi)}\rangle d\sigma(\xi)\right\}^n}$$

```
122    1  \BookTitle={Differential Geometry, Lie Groups and Symmetric Spaces}
       2  \Author={Sigurdur Helgason}
       3  \CopyrightDate={1978}
       4  \PublishInfo={Academic Press}
       5  \NewExample
       6  \AmSTeX's {\tt\string\matrix} commands have a slightly different structure
       7  from those of Plain~\TeX. Along with the closely related
       8  {\tt\string\cases} command, these are the only major points at which
       9  there are discrepancies between commands that bear the same name in the two
      10  packages. A full discussion of matrices occurs in the Glossary, but
      11  here are some real-life examples:
      12
      13  \ShortExample <444>
      14  $$ K_{p,q}=
      15  \pmatrix -I_p & 0   & 0    & 0 \\
      16           0 & I_q & 0    & 0 \\
      17           0 & 0   & -I_p & 0 \\
      18           0 & 0   & 0    & I_q \endpmatrix $$
      19  \EndShortExample
      20
      21  \input EightPt
      22  \LoadEight   % Private command to load 8-point typefaces.
      23
      24  \ShortExample <446>
      25  $$\align
      26  \frak{sp}(n, \bold R)&:\left\{\pmatrix X_1 & X_2 \\ X_3 & -^t\!X_1 \endpmatrix
      27  \Bigm\vert
      28  \matrix\format\l\\
      29    \text{\Eightpoint $X_1$, $X_2$, $X_3$ real $n\times n$ matrices,} \\
      30    \text{\Eightpoint $X_2$, $X_3$ symmetric}
      31  \endmatrix \right\}, \\ \vspace{5pt}
      32  % The next line of the overall alignment begins here.
      33  \frak{sp}(p,q)&:\left\{
      34  \pmatrix\format&\r\ \\
      35  Z_{11}&Z_{12}&Z_{13}&Z_{14} \\ ^t\bar Z_{12}&Z_{22}&^tZ_{14}&Z_{24} \\
      36  -\bar Z_{13}&\bar Z_{14}&\bar Z_{11}&-\bar Z_{12} \\
      37  ^t\bar Z_{14}&-\bar Z_{24}&-^tZ_{12}&\bar Z_{22} \endpmatrix
      38  \!\left\vert \;
      39  \foldedtext\foldedwidth{1.8in}{\Eightpoint
      40  $Z_{ij}$ complex matrix; $Z_{11}$ and $Z_{13}$ of
      41  order $p$, $Z_{12}$ and $Z_{14}$ $p\times q$ matrices, $Z_{11}$ and $Z_{22}$
      42  are skew Hermitian, $Z_{13}$ and $Z_{24}$ are symmetric.}
      43  \right. \right\}.
      44  \endalign $$
      45  \EndShortExample
      46  \EndPage
```

$\mathcal{A}\mathcal{M}\mathcal{S}$-TEX

──────────── **Example 15** ────────────

From *Differential Geometry, Lie Groups and Symmetric Spaces*
SIGURDUR HELGASON
Academic Press, 1978

$\mathcal{A}\mathcal{M}\mathcal{S}$-TEX's `\matrix` commands have a slightly different structure from those of Plain TEX. Along with the closely related `\cases` command, these are the only major points at which there are discrepancies between commands that bear the same name in the two packages. A full discussion of matrices occurs in the Glossary, but here are some real-life examples:

(Page 444)

$$K_{p,q} = \begin{pmatrix} -I_p & 0 & 0 & 0 \\ 0 & I_q & 0 & 0 \\ 0 & 0 & -I_p & 0 \\ 0 & 0 & 0 & I_q \end{pmatrix}$$

(Page 446)

$$\mathfrak{sp}(n, \mathbf{R}) : \left\{ \begin{pmatrix} X_1 & X_2 \\ X_3 & -{}^t X_1 \end{pmatrix} \ \middle| \ \begin{matrix} X_1, X_2, X_3 \text{ real } n \times n \text{ matrices,} \\ X_2, X_3 \text{ symmetric} \end{matrix} \right\},$$

$$\mathfrak{sp}(p, q) : \left\{ \begin{pmatrix} Z_{11} & Z_{12} & Z_{13} & Z_{14} \\ {}^t \bar{Z}_{12} & Z_{22} & {}^t Z_{14} & Z_{24} \\ -\bar{Z}_{13} & \bar{Z}_{14} & \bar{Z}_{11} & -\bar{Z}_{12} \\ {}^t \bar{Z}_{14} & -\bar{Z}_{24} & -{}^t Z_{12} & \bar{Z}_{22} \end{pmatrix} \ \middle| \ \begin{matrix} Z_{ij} \text{ complex matrix; } Z_{11} \text{ and } Z_{13} \text{ of} \\ \text{order } p, Z_{12} \text{ and } Z_{14} \ p \times q \text{ matrices,} \\ Z_{11} \text{ and } Z_{22} \text{ are skew Hermitian,} \\ Z_{13} \text{ and } Z_{24} \text{ are symmetric.} \end{matrix} \right\}.$$

```
1  \BookTitle={An Introduction to Mathematical Logic and Type Theory}
2  \Author={Peter B.~Andrews}
3  \CopyrightDate={1986}
4  \PublishInfo={Academic Press}
5
6  \NewExample
7  Here is a further assortment of delicacies made with \AmSTeX.
8
9  \mathchardef\Neg="0218  % See Chapter 5 (section 5) for a discussion.
10 \define\BM #1#2{\bmatrix\format\r\\ #1 \\ #2 \endbmatrix}
11 \define\CoverAll{\delimitershortfall-2pt}
12 \ShortExample <32>
13 $$\CoverAll
14 \bmatrix\format\c&\ \c\ &\c\\
15 \bmatrix\format\c&\ \c\ &\c\\  \BM pq &\vee& \BM{\Neg p}{\Neg q} \\
16 \vspace{1.5pt}&r \endbmatrix
17 &\vee&
18 \bmatrix\format\c&\ \c\ &\c\\  \BM p{\Neg q} &\vee& \BM{\Neg p}q \\
19 \vspace{1.5pt}&\ \llap{$\Neg r$}& \endbmatrix \\ \vspace{1\jot}
20 \bmatrix\format\c&\ \c\ &\c\\  &\ \llap{$\Neg p$}&\\ \vspace{1.5pt}
21 \BM qr &\vee& \BM{\Neg q}{\Neg r} \endbmatrix
22 &\vee&
23 \bmatrix\format\c&\ \c\ &\c\\  &p \\ \vspace{1.5pt}
24 \BM q{\Neg r} &\vee& \BM{\Neg q}r \endbmatrix \endbmatrix$$
25 \EndShortExample
26
27 \define\BS{\mathord{\sssize\blacksquare}}
28 \ShortExample <35>
29 \NegVspace  \NegVspace
30 $$\alignat2
31 p \supset q & \equiv \BS \Neg q \supset \Neg p &&\qquad
32 \text{(Contrapositive law)} \\
33 p \wedge q \supset r & \equiv \BS p \supset \BS q \supset r \\
34 p \supset q \supset p & \supset p &&\qquad \text{(Peirce's law)} \\
35 p & \supset \BS p \vee q \\
36 p \wedge q & \supset p
37 \endalignat $$
38 \NegVspace
39 \EndShortExample
40
41 \font\ScaledSy=cmsy5 scaled\magstep5
42 \define\TinT{\setbox0=\hbox{\ScaledSy\char53}\raise1.7pt\hbox to\wd0
43    {\hfil$\sssize\bigtriangledown$\hfil}\llap{\box0}}
44
45 \ShortExample <143>
46 \medskip
47 \flushpar The value of $(\boxplus \bold x)\bold A(\bold x)$ is
48 $\TinT$ iff there is an $n$ such that the value of $\bold A(n)$ is~$\TinT$.
49 \medskip
50 \EndShortExample
51 \EndPage
```

$\mathcal{AMS}$-T_EX

—————— **Example 16** ——————

From *An Introduction to Mathematical Logic and Type Theory*
PETER B. ANDREWS
Academic Press, 1986

Here is a further assortment of delicacies made with $\mathcal{AMS}$-T_EX.

(Page 32)

$$\left[\left[\left[\begin{matrix}p\\q\end{matrix}\right]\vee\left[\begin{matrix}\sim p\\\sim q\end{matrix}\right]\right]_r\vee\left[\left[\begin{matrix}p\\\sim q\end{matrix}\right]\vee\left[\begin{matrix}\sim p\\q\end{matrix}\right]\right]_{\sim r}\right]$$

$$\left[\left[\begin{matrix}\sim p\\\left[\begin{matrix}q\\r\end{matrix}\right]\vee\left[\begin{matrix}\sim q\\\sim r\end{matrix}\right]\end{matrix}\right]\vee\left[\begin{matrix}p\\\left[\begin{matrix}q\\\sim r\end{matrix}\right]\vee\left[\begin{matrix}\sim q\\r\end{matrix}\right]\end{matrix}\right]\right]$$

(Page 35)

$$p \supset q \equiv \blacksquare \sim q \supset \sim p \qquad \text{(Contrapositive law)}$$
$$p \wedge q \supset r \equiv \blacksquare p \supset \blacksquare q \supset r$$
$$p \supset q \supset p \supset p \qquad \text{(Peirce's law)}$$
$$p \supset \blacksquare p \vee q$$
$$p \wedge q \supset p$$

(Page 143)

The value of $(\boxplus\mathbf{x})\mathbf{A}(\mathbf{x})$ is $\bigtriangledown$ iff there is an n such that the value of $\mathbf{A}(n)$ is $\bigtriangledown$.

```
1  \define\EndWith#1{\hfill {#1}\par}
2  \define\R{\hbox{''}}
3  \define\BBS{\bigstar\bigstar}
4  \ShortExample <249>
5  \medskip
6  {\baselineskip=14pt
7  \flushpar {\bf 7101\quad G\"odel's Incompleteness Theorem} [G\"odel 1931].
8  Let $\Cal A$ be any recursively axiomatized pure extension of
9  $\Cal Q_0^\infty$. Let $\bigstar_{o\sigma}$ be the wff
10 $$
11 [\lambda x_\sigma \dot\forall y_\sigma \sim
12 \italic{Proof}_{o\sigma\sigma}^{\Cal A} [\overline{128} \ast x_\sigma \ast
13 [\italic{Num}_{\sigma\sigma}x_{\sigma}] \ast \overline{2048}] y_\sigma]
14 $$
15 Let $m=``\bigstar_{o\sigma}\R$ and let $\BBS_o$ be
16 $[\bigstar_{o\sigma}\bar m]$.
17
18 \itemitem{(1)} If $\Cal A$ is consistent, then not $\vdash_{\Cal A}\sim \BBS$.
19 \itemitem{(2)} If ${\Cal A}$ is $\omega$-consistent, then not
20 $\vdash_{\Cal A}\sim\BBS$.
21
22 \smallpagebreak\flushpar
23 {\it Proof}. We shall write $\vdash_{\Cal A}\bold B$ simply as
24 $\vdash\bold B$.
25 \item{(.1)} $\vdash\overline{128}\ast\bar m\ast[\italic{Num}_{\sigma\sigma}
26 \bar m]\ast \overline{2048}=\overline{``\BBS\R}$ since the
27 wffs $\ast$ and {\it Num\/} represent the corresponding functions, and since
28 $128\ast m\ast \roman{Num}(m)\ast 2048=``[\R\ast
29 ``\bigstar_{o\sigma}\R\ast ``\bar m\R \ast ``]\R =
30 ``[\bigstar_{o\sigma}\bar m]\R=``\BBS\R$.
31 \item{(.2)} $\vdash\BBS=\dot\forall y_{\sigma}\sim \italic{Proof}^{\Cal A}
32 \overline{``\BBS\R}y_\sigma$
33 \EndWith{$\lambda$, Def of $\BBS$; R: .1}
34
35 Now suppose $\vdash\BBS$. Then
36 \item{(.3)} $\vdash\dot\forall y_\sigma \sim \italic{Proof}^{\Cal A}
37 \overline{``\BBS\R}y_\sigma$
38
39 \flushpar Let $n$ be the number of a proof $\BBS$.
40 \item{(.4)} $\vdash\italic{Proof}^{\Cal A}\overline{``\BBS\R}\bar n$\newline
41 since the wff {\it Proof\/} represents the numerical relation Proof.
42 \item{(.5)} $\vdash\sim\italic{Proof}^{\Cal A}\overline{``\BBS\R}\bar n$
43 \EndWith{$\forall$I: .3}
44 \item{(.6)} $\vdash F_0$ \EndWith{.4, .5}
45 \flushpar Thus if $\vdash\BBS$, then $\Cal A$ is inconsistent.
46
47 Next suppose $\Cal A$ is $\omega$-consistent and $\vdash\sim\BBS$.
48 \item{(.7)} $\vdash\dot\exists y_\sigma\italic{Proof}^{\Cal A}
49 \overline{``\BBS\R}y_\sigma$ \EndWith{.2, 6000}
50 \flushpar $\Cal A$ is consistent by 7100, so not $\vdash\BBS$; so for each
51 natural number $k$,
52 \item{(.8)} $\vdash\sim\italic{Proof\/}\overline{``\BBS\R}\bar k$
53
54 \flushpar since $k$ is not the number of a proof of $\BBS$, and the wff
55 {\it Proof\/} represents the relation Proof. But this contradicts the
56 $\omega$-consistency of $\Cal A$. Hence if $\Cal A$ is $\omega$-consistent,
57 not $\vdash\sim\BBS$. \EndWith{$\blacksquare$}
58 \medskip } \EndShortExample \bye
```

Source:

*An
Introduction
to
Mathematical
Logic and
Type Theory*
PETER
B. ANDREWS
© 1986
Academic Press

(Page 249)

7101 Gödel's Incompleteness Theorem [Gödel 1931]. Let $\mathcal{A}$ be any recursively axiomatized pure extension of $\mathcal{Q}_0^\infty$. Let $\bigstar_{o\sigma}$ be the wff

$$[\lambda x_\sigma \dot{\forall} y_\sigma \sim Proof_{o\sigma\sigma}^{\mathcal{A}} [\overline{128} * x_\sigma * [Num_{\sigma\sigma} x_\sigma] * \overline{2048}] y_\sigma]$$

Let $m = \text{``}\bigstar_{o\sigma}\text{''}$ and let $\bigstar\bigstar_o$ be $[\bigstar_{o\sigma} \bar{m}]$.

    (1) If $\mathcal{A}$ is consistent, then not $\vdash_\mathcal{A} \sim \bigstar\bigstar$.

    (2) If $\mathcal{A}$ is ω-consistent, then not $\vdash_\mathcal{A} \sim \bigstar\bigstar$.

Proof. We shall write $\vdash_\mathcal{A} \mathbf{B}$ simply as $\vdash \mathbf{B}$.

(.1) $\vdash \overline{128} * \bar{m} * [Num_{\sigma\sigma} \bar{m}] * \overline{2048} = \overline{\text{``}\bigstar\bigstar\text{''}}$ since the wffs $*$ and *Num* represent the corresponding functions, and since $128 * m * \text{Num}(m) * 2048 = \text{``[''} * \text{``}\bigstar_{o\sigma}\text{''} * \text{``}\bar{m}\text{''} * \text{``]''} = \text{``}[\bigstar_{o\sigma}\bar{m}]\text{''} = \text{``}\bigstar\bigstar\text{''}$.

(.2) $\vdash \bigstar\bigstar = \dot{\forall} y_\sigma \sim Proof^{\mathcal{A}}\overline{\text{``}\bigstar\bigstar\text{''}} y_\sigma$ λ, Def of $\bigstar\bigstar$; R: .1

Now suppose $\vdash \bigstar\bigstar$. Then

(.3) $\vdash \dot{\forall} y_\sigma \sim Proof^{\mathcal{A}}\overline{\text{``}\bigstar\bigstar\text{''}} y_\sigma$

Let n be the number of a proof $\bigstar\bigstar$.

(.4) $\vdash Proof^{\mathcal{A}}\overline{\text{``}\bigstar\bigstar\text{''}} \bar{n}$

since the wff *Proof* represents the numerical relation Proof.

(.5) $\vdash \sim Proof^{\mathcal{A}}\overline{\text{``}\bigstar\bigstar\text{''}} \bar{n}$ $\forall$I: .3

(.6) $\vdash F_0$ .4, .5

Thus if $\vdash \bigstar\bigstar$, then $\mathcal{A}$ is inconsistent.

    Next suppose $\mathcal{A}$ is ω-consistent and $\vdash \sim \bigstar\bigstar$.

(.7) $\vdash \dot{\exists} y_\sigma Proof^{\mathcal{A}}\overline{\text{``}\bigstar\bigstar\text{''}} y_\sigma$ .2, 6000

$\mathcal{A}$ is consistent by 7100, so not $\vdash \bigstar\bigstar$; so for each natural number k,

(.8) $\vdash \sim Proof\overline{\text{``}\bigstar\bigstar\text{''}} \bar{k}$

since k is not the number of a proof of $\bigstar\bigstar$, and the wff *Proof* represents the relation Proof. But this contradicts the ω-consistency of $\mathcal{A}$. Hence if $\mathcal{A}$ is ω-consistent, not $\vdash \sim \bigstar\bigstar$.   ■

```
 1 | \input amstexl         % A modified version of AmSTeX.
 2 | \input lamstex         % The LamSTeX package.
 3 | \input OutForm.tex      % The 'output format' file.
 4 | \RealPno=129           % The real page number.
 5 | \ExampleNumber=16      % The number of the previous example.
 6 | .....................................................................
 7 | \input ptmatrix.tex    % The partitioned matrices package.
 8 | \font\Bigrm=cmr12 scaled1200
 9 | \define\Bg#1[#2]{\smash{\lower#2pt\hbox{\Bigrm#1}}}\mathstrut}
10 |
11 | \ShortExample <143>
12 | $$B=
13 | \left( \partition
14 | \matrix
15 |   b_1^1&\dotsb&b_1^p& \\ \vdots&&\vdots&\Bg0[0] \\
16 |   b_1^p&\dotsb&b_p^p& \\
17 |   \vspace{6pt}
18 |   b_1^{p+1}&\dotsb&b_p^{p+1}& \\ \vdots&&\vdots&\quad\Bg I[0]\quad \\
19 |   b_1^m&\dotsb&b_p^m& \\
20 | \endmatrix
21 | \vdashed3:06  % Gives a dashed vertical line after column 3, from rows 0 to 6.
22 | \hdashed3:04  % Gives a dashed horizontal line after row 3, from cols 0 to 4.
23 | \endpartition \right) $$
24 | \EndShortExample
25 |
26 | \ShortExample <222>
27 | \NegVspace
28 | $$A = \left( \partition
29 | \matrix \format\quad\c&\quad\c&\quad\c&\quad\l\\
30 |   0&\dotsb&0&\alpha_m \\ \vspace{6pt}
31 |   1&&\Bg0[4]&\alpha_{m-1} \\
32 |   &\ddots&&\vdots \\ \Bg0[0]&&1&\alpha_1
33 | \endmatrix
34 | \vsolid3:03  \hsolid3:03 % These give solid lines; see \vdashed, etc., above.
35 | \endpartition \right) $$
36 | \EndShortExample
37 |
38 | \input CDsty    % Definitions for commutative diagrams.
39 |
40 | \ShortExample <98>
41 | \NegVspace
42 | $$\CD
43 | Y\times \{0\} \Est \Sth & E \LBL \l\pi \Sth \\
44 | Y\times [0,1] \LBL \l{f} \NE \Est & X
45 | \endCD $$
46 | \EndShortExample
47 | \EndPage
```

NOTES

It is impossible in a few notes below the input boxes to do justice to a large package like LA$_{\mathcal{M}}$S-TeX. These examples are meant primarily to illustrate the possibilities of the package; readers who want more information about it will have to get the actual files and associated manuals. The inside back cover carries information about sources (also see § 4.1).

8–9: The first line introduces a font of slightly large characters, needed for occasional entries in parti-tioned matrices. The second line introduces a command that helps place the large characters where they are needed.

15, 18, 31, 32: The lines show uses of the command \Bg in placing big characters.

38: LA$_{\mathcal{M}}$S-TeX makes available a set of special arrow fonts. This file introduces private names for some of them, following suggestions in the official manual. The names are listed on the next page.

$\mathcal{A}_{\mathcal{M}}\mathcal{S}$-$T_{E}X$

──────── **Example 17** ────────

From *Vector Bundles*
HOWARD OSBORN
Academic Press, 1982

$\mathcal{A}_{\mathcal{M}}\mathcal{S}$-$T_{E}X$ is a package of commands that adds several capabilities to $\mathcal{A}_{\mathcal{M}}\mathcal{S}$-$T_{E}X$. The general features of the package are briefly discussed in Chapter 4. What will be illustrated here are the facilities that it provides for the easy typesetting of complicated commutative diagrams and partitioned matrices.

(Page 143)

$$
B = \left(
\begin{array}{ccc|c}
b_1^1 & \cdots & b_1^p & \\
\vdots & & \vdots & 0 \\
b_1^p & \cdots & b_p^p & \\
\hline
b_1^{p+1} & \cdots & b_p^{p+1} & \\
\vdots & & \vdots & I \\
b_1^m & \cdots & b_p^m &
\end{array}
\right)
$$

(Page 222)

$$
A = \left(
\begin{array}{ccc|c}
0 & \cdots & 0 & \alpha_m \\
\hline
1 & & 0 & \alpha_{m-1} \\
 & \ddots & & \vdots \\
0 & & 1 & \alpha_1
\end{array}
\right)
$$

(Page 98)

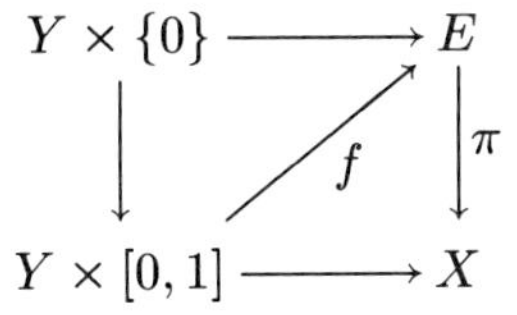

$$
\begin{CD}
Y \times \{0\} @>>> E \\
@VVV \nearrow^{f} @VV{\pi}V \\
Y \times [0,1] @>>> X
\end{CD}
$$

129

```
1  \BookTitle={Theory of Categories}
2  \Author={Barry Mitchell}
3  \CopyrightDate={1965}
4  \PublishInfo={Academic Press}
5
6  \NewExample
7  Here are some further commutative diagrams. \LamSTeX\ is actually capable of
8  handling much more complicated diagrams, but these should
9  provide some glimpses of what the package can do.
10
11 \ssizeCDlabels  % Gives script-size (s-size) labels.
12
13 \ShortExample <144>
14 $$\rgaps{.5; .5} \cgaps{.5} % Reduces the gaps between rows and columns.
15 \CD
16 &M\otimes_{\bold R}C \LBL \L{\eta(f\mu)} \SW
17  \LBL \l{\mu\otimes_{\bold R}C} \SS \\
18 A \\ &M'\otimes_{\bold R}C \LBL \l{\eta(f)} \NW
19 \endCD $$
20 \EndShortExample
21
22 \define\PPsp {\hphantom{{}''}}  % To give a space the width of two 'primes'.
23 \ShortExample <202>
24 \NegVspace $$ \CD
25 O \Est & A' \LBL\L\mu \Est \LBL\L\kappa \Sth & A \LBL\L\lambda \Est
26 \LBL\l\rho \Sth & \;A'' \\
27 O \Est & Q \LBL\L\alpha \p2 \Est \LBL\l\delta \p{-2}\0h\1e \Est &
28 \PPsp Q\oplus A'' \LBL\L\beta \p2 \Est \LBL\l\gamma \p{-2}\0h\1e \Est &
29 \PPsp A'' \LBL \p{-2}\a= \Nth
30 \endCD $$
31 \EndShortExample
32
33 \ShortExample <250>
34 \NegVspace
35 $$\CD
36 P \LBL \l\alpha \SSE \LBL \L{\mu_P} \ESE \\
37 & E \Sth \Est & \bar R_XP \LBL \l{\bar R_X(\alpha)} \Sth \\
38 & F \LBL \L{\mu_F} \Est & \bar R_X F
39 \endCD $$
40 \EndShortExample
41 \EndPage  \bye
```

NOTES

Here are the privately defined names for the arrows used in the examples. The official LAMS-TeX manual carries information on defining such names. The four basic arrows are called \Est, \Sth, \Wst and \Nth (after east, south, west and north, respectively). The four basic diagonal arrows are called \NE, \SE, \SW and \NW. Three other arrows are used here: \SS (for a double-length southward arrow), \SSE (after south-southeast) and \ESE (east-southeast). Other names can be defined as needed. **16, 17, etc.**: An arrow that is to be labelled is marked with \LBL (a private command for a longer LAMS-TeX command); \L puts the label on one side, \l on the other. **27, 28**: \p is used to translate the arrows, and the combination \0h\1e to switch the head and the end.

$\mathcal{A}_{\mathcal{M}}S$-T$_{\!E\!}$X

—————————— **Example 18** ——————————

From *Theory of Categories*
BARRY MITCHELL
Academic Press, 1965

Here are some further commutative diagrams. $\mathcal{A}_{\mathcal{M}}S$-T$_{\!E\!}$X is actually capable of handling much more complicated diagrams, but these should provide some glimpses of what the package can do.

(Page 144)

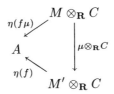

(Page 202)

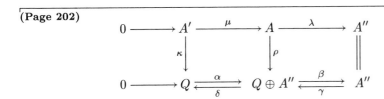

(Page 250)

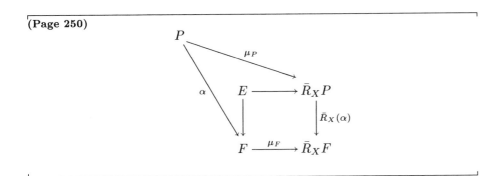

3

A Summary of
$\mathcal{AMS}$-TEX

3.1 Introduction

$\mathcal{AMS}$-TEX is the official TEX of the American Mathematical Society. It was written principally by Michael Spivak. His book, *The Joy of TEX* (see SPIVAK [1982, 1990]), is the official guide to the package.

One of the aims of $\mathcal{AMS}$-TEX is the simplification of the input of mathematical expressions. The package also supplies a wide range of special symbols, everything from $\hbar$ to ¥ (with, among other things, ✓, ®, ⤳, ○ and ≲ along the way). It should therefore be of interest to a wide range of scientists, not just straight mathematicians. The Glossary contains discussions of many $\mathcal{AMS}$-TEX commands as well as lists of all its special symbols.

The package comes as a set of files, the main one being `amstex.tex`. It is used simply by typing

    \input amstex

at the top of the document. It is also possible to make a format file for $\mathcal{AMS}$-TEX and to invoke it—possibly automatically—when the `tex` command is given; see §1.2. Once this is done, several features (special equation alignment commands, for example) are at one's disposal. Calling `amstex.tex` does not, however, automatically give access to the special symbols of $\mathcal{AMS}$-TEX. To get those, the command `\UseAMSsymbols` must also be issued (see page 116 and §3.3). It is possible to use these special symbols without using all the other mechanisms of $\mathcal{AMS}$-TEX; this is discussed in §3.3.

3.2 Style

The notion of a *document style* is important in A$\mathcal{M}$S-TEX. Though the package does not insist on a style declaration, some commands do not work unless a style is chosen (e.g., \footnote and \proclaim). Furthermore, many of its formatting decisions are based on the document style—identical commands can have different effects in different styles. An example is given by the equation numbering command, \tag (see page 116 and the Glossary). The command itself is consistently used in the same way for all documents, but effects like the placement of equation numbers to the left or to the right depend on the style.

Styles are defined in style files (their names usually have extension .sty). The most widely disseminated of the AMS styles is the AMS Preprint Style, defined in the file amsppt.sty. The dictates of this style are invoked by typing the following at the top of the document:

```
\input amstex
\documentstyle{amsppt}
```

The same approach is used for other styles. What is to be typed after such a style declaration depends on the style being used.

In the AMS preprint style there typically follows a preamble in which new commands may be defined, specifications may be made about whether running heads are to be used or not, etc. Then come some optional \topmatter commands; different ones are required in different contexts. Here is a partial list:

```
\topmatter
\title ... \endtitle
\author ... \endauthor
\address ... \endaddress
\email ... \endemail
\thanks ... \endthanks
\abstract ... \endabstract
\endtopmatter
\document
```

The names of these commands are self-explanatory. The AMS (or any other organization that requires the use of this style) will specify which of these optional pieces of information are required, and in what order.

Each pair of commands will impose some automatic sub-style. For instance, the \abstract ... \endabstract pair that sandwiches the abstract of the document will set it (the abstract) in 8-point type, indented away from both margins. If these automatic choices are not desired, there is a command called \nofrills that may be used to override these default style choices.

3.3 Fonts

$\mathcal{AMS}$-T$_E$X currently comes with eight special mathematical fonts of its own, plus some additions to T$_E$X's standard fonts, and a set of Cyrillic fonts. Six of the mathematics fonts belong to the specially designed **AMS Euler typefaces**. Tables that show the characters in essentially all of these are displayed in Chapter 5. The font files of the eight mathematical fonts have these names: `msam` and `msbm` (special symbols); `eufm` and `eufb` (Euler fraktur, medium and bold); `eusm` and `eusb` (Euler script, medium and bold); `eurm` and `eurb` (Euler "roman"— more accurately cursive—medium and bold). In each case, the point size must be attached to the name; for example, the 10-point medium fraktur font has `eufm10` as its filename.

Adding new fonts eats up memory. $\mathcal{AMS}$-T$_E$X makes several levels of font loading available, so that users can add only what they need. The `amstex.tex` file defines eight separate `\load` commands: `\loadmsam`, ... , `\loadeurb`. It also defines the command `\loadbold`; the command may be used to add bold versions of some of T$_E$X's standard symbols. The file does not go beyond these definitions and load any of the fonts itself.

Users may load on their own any combination that they need (and that the available memory permits) of the extra fonts. The two most important ones for regular use are the symbol fonts. For example, a set of `msam` fonts can be added by saying `\loadmsam` (the command will add 5-, 7- and 10-point sizes). The loading of such a font does not by itself allow direct access to the symbols in it; for that, names also have to be assigned to each special symbol. For the `msam` and `msbm` fonts, this can be done symbol by symbol, through a command called `\newsymbol` (see the Glossary); this is useful if memory is particularly tight and only a few of the special symbols are needed. If this approach is found cumbersome, $\mathcal{AMS}$-T$_E$X makes available a further command, defined as follows:

`\def\UseAMSsymbols{\loadmsam\loadmsbm \input amssym.tex }`

The file `amssym.tex` contains names (over 200 of them) for all the special symbols. The command `\UseAMSsymbols` (see, for example, page 116) loads the symbol fonts and gives a name to each symbol.

The other commonly used fonts are the ones with bold versions of standard symbols (summoned by `\loadbold`) and the medium fraktur fonts (summoned by `\loadeufm`). Once these fonts are loaded, $\mathcal{AMS}$-T$_E$X provides ways to get at the characters in them. See `\boldkey`, `\boldsymbol` and `\frak` in the Glossary.

The file `amsppt.sty`, which defines the AMS preprint style, contains the statements `\loadeufm` and `\UseAMSsymbols`. Thus, users of that style have access to all the special symbols and to the fraktur characters.

Finally, users who wish to use the special fonts, but not any of the other $\mathcal{AMS}$-T$_E$X commands, can call upon the file `amssym.def` instead of `amstex.tex` (see pages 74 and 84). This file takes the minimal steps necessary for loading fonts.

The other fonts in the AMS collection have special uses. See Example 7 in Chapter 2, and the individual discussions of the fonts in Chapter 5.

3.4 Other Features

There are many additional aspects to A$\mathcal{M}$S-T$_E$X beyond the simplifying of mathematical input and the supplying of special symbols. Style files associated with it provide a flexible set of list-making commands (under a general command called \roster) and a comprehensive collection of commands that help in the typesetting of bibliographies and lists of references. The package does not provide special commands for making tables, for drawing pictures, or for things like automatic cross-referencing. The relatively new package called L$\mathcal{A}$$\mathcal{M}$S-T$_E$X (see Chapter 4) provides some of these additional features.

3.5 A$\mathcal{M}$S-T$_E$X versus Plain T$_E$X

Plain T$_E$X is almost wholly a subset of A$\mathcal{M}$S-T$_E$X. Here is a list of discrepancies between the two packages (each of the commands mentioned here is discussed in greater detail in the Glossary):

- *Plain T$_E$X commands that are undefined in A$\mathcal{M}$S-T$_E$X.*
  \=, \>, \cal, \mit, \oldstyle, \footnote and \proclaim (the last two are defined in style files like amsppt.sty and are thus undefined only in "raw" A$\mathcal{M}$S-T$_E$X).
- *Plain T$_E$X commands that have different meanings in A$\mathcal{M}$S-T$_E$X.*
  \., \bf, \it, \rm, \sl (the typeface-switching commands in the two packages work in the same way for text, but they do not work inside formulas in A$\mathcal{M}$S-T$_E$X). Another difference in this general category is the use of '' to get a superscript "prime": A$\mathcal{M}$S-T$_E$X does not permit things like x''^2, whereas Plain T$_E$X does.
- *Commands with the same names, but used differently.*
  \matrix, \pmatrix and \cases.
- *Other differences.*
  The spacing around the dots given by things like \ldots is different, as is that used in \pmod in certain contexts.

The list isn't, of course, meant to cover those features of A$\mathcal{M}$S-T$_E$X that are not present at all in Plain T$_E$X.

4

Other Packages

4.1 L&A_M_S-T_EX

L&A_M_S-T_EX is also from Michael Spivak (the author of A_M_S-T_EX). It adds several LATEX-like functions to A_M_S-T_EX. These include automatic equation numbering, the easy formatting of tables, cross-referencing, the making of indexes and the generation of tables of contents. The official guide to the package is *L&A_M_S-T_EX: The Synthesis* (SPIVAK [1989]).

In addition to these functions, L&A_M_S-T_EX also brings with it some new mathematical features. The principal ones are the handling of partitioned matrices and of complicated commutative diagrams. The features are illustrated in Examples 17 and 18 in Chapter 2. L&A_M_S-T_EX handles commutative diagrams by extracting arrows from a new set of arrow fonts that come with the package; the fonts allow it to draw arrows of a variety of slopes and lengths.

L&A_M_S-T_EX sits "on top of" a modified version of A_M_S-T_EX, defined in a file called `amstex1.tex`. It is used by typing

```
\input amstex1
\input lamstex
```

at the top of the file. (Or format files can be made with these built in; see §1.2). Most A_M_S-T_EX constructions continue to work unchanged in L&A_M_S-T_EX: for the most part, all that it does is to add new features.

The L&A_M_S-T_EX package is large enough that using all of it at the same time can strain the available memory (the computer's, if not the user's). For example, a supplement to its manual says that it uses around 2,700 command names on its own. To avoid problems, it is built around modules; only the ones that are needed are brought in. For instance, Example 17 (Chapter 2) brought in the file `ptmatrix.tex` (see page 128); the file allowed the construction of partitioned matrices. Modules may be purged from memory after they are used: a command called `\purge` does just that. They may also be brought back later, if needed, using a command called `\unpurge`.

4.2 $\mathcal{AMS}$-LaTeX

$\mathcal{AMS}$-LaTeX is another attempt to marry $\mathcal{AMS}$-TeX and LaTeX, but this time from the LaTeX direction. It is a project of the American Mathematical Society, to which many people have contributed: among others, Romesh Kumar, Michael Downes, Frank Mittelbach and Rainer Schöpf.

The package comes as a set of files from the American Mathematical Society, with a User's Guide included. The guide is in a file called `amslatex.tex`: users must "latex" and print it themselves. (They may have to edit it first—the instructions are in the file—and switch on the table of contents and index generator, run it through LaTeX, edit the file once more to switch off the auxiliary file generator, then run it through LaTeX a second time. The procedure will be familiar to LaTeX users.)

For the most part, $\mathcal{AMS}$-LaTeX follows $\mathcal{AMS}$-TeX faithfully in what it makes available to users. The commands, however, now all have the standard LaTeX syntax. For example, `\matrix ... \endmatrix` becomes `\begin{matrix}` ... `\end{matrix}`. In a few minor places, $\mathcal{AMS}$-LaTeX deliberately disables features of $\mathcal{AMS}$-TeX. A complete list of such differences is given in the User's Guide.

$\mathcal{AMS}$-LaTeX incorporates the New Font Selection Scheme of Frank Mittelbach and Rainer Schöpf. The scheme is discussed in § 5.7.

5

Typefaces

ᵀᴬᴼᴬᴱᴨᴱᵞ OW DOES TEX choose typefaces? What can users do to influence
H its choices? The questions often arise; this chapter gives detailed
answers. The chapter is largely self-contained: the discussions in
it should be accessible even to new users of TEX. There may be
occasional references to the examples (Chapter 2) or to the Glossary, but all the
essential information is given fully here. None of this information is necessary
for routine uses of TEX—all that is needed for such tasks is a knowledge of how
and when to use commands like \it, \bf, \sl, and so on. The information is
needed, however, for an understanding of what happens behind the scenes.

In the examples and discussions given below it is assumed that "text"
refers to text in English; despite that assumption, the discussions are often
broadly applicable to other languages as well.

5.1 Basic Ideas

It is not always appreciated that what a program like TEX does when it typesets
a document is that it selects characters (also called *type*) from some preexisting
set, and then places them in precise positions on the page. Beginners sometimes
tend to think of the characters as being made up on the spot as the page is put
together, but that is far from the truth. The only things that TEX constructs
on the spot, from scratch, are vertical and horizontal rules (i.e., straight lines)
and regions of white space. Everything else comes from type that was created
beforehand. It helps, therefore, to understand the meanings of some of the
terms related to the notion of type.

The terms are all historical, coming from days when a piece of type was
a material object, say a piece of metal, carved or molded into the required
shape. The same terms continue to be used today in discussions of computer
typesetting, but they now refer to more abstract entities: a "piece of type," or
a "character," is a computer program that tells the printing device where to

put ink. Thus a character that was "created beforehand" is simply a computer program that was written beforehand.

A *typeface* is a complete assortment of type in one style. The notion of a *complete assortment* is itself a little difficult to define precisely. In English text typefaces, for example, the assortment should contain at least the upper- and lowercase letters, and it should also contain suitable punctuation marks, and so forth. A *font* of type is a typeface at one size. For example, it is clear that ℜ𝔇, RD, *RD* and ᴿᴰ all have different styles or different sizes. They therefore belong to different fonts.

A font that is used for text usually has variations associated with it. For example, the text in this book is in the commonly used roman style. Here are some of its variations: **bold**, **bold extended**, *slanted*, *italic*, ***bold extended slanted***, ***bold extended italic*** and CAPITALS AND SMALL CAPITALS.

Terms like *style*, *variation* and *font* are not always used by everybody in exactly the same way. Variations in a font are often considered separate fonts in their own right. Even the notion of the *size* of a font is vaguer than it may appear. Font sizes are typically stated in points; there are approximately 72 points to the inch, but the exact size of a point is slightly different in different typesetting systems. Worse, even after an exact value is chosen, the precision seemingly implied by a statement like "10-point font" is fake: there isn't a universally agreed-upon font characteristic (like, say, the size of a particular character) that can be used to quantify the font size exactly.

Despite this looseness in their meanings, these typographical terms are useful. For example, it is mostly true that letters in a 10-point font will not have a vertical size that is sharply greater than 10 points and that characters like parentheses will often have a size close to the stated (or "design") size of a font. Slight differences in the use of typographical terms aren't usually a problem, and it is mostly possible to tell from the context how they are being used.

• *Fonts in TEX*

The term *font* has a technical meaning in TEX. When typesetting a document, the program expects to find files that contain information about the characters (their heights, widths, etc.) that it is to typeset. Each such file represents a font. The file may indeed contain information about a complete assortment of characters in one style and size—and many of the commonly used ones do—but it does not necessarily have to. In fact, as the tables at the end of the chapter show, the placement of characters in TEX's standard fonts can be a little unusual at times: fonts called *roman* contain, in addition to all the necessary letters and punctuation marks, capital Greek letters (but not lowercase ones); fonts called *math italic* contain, in addition to the special italic letters that are used in formulas, a few miscellaneous symbols as well. The reason for these placements is historical (see the end of §5.4); new font layouts (for the "256-character fonts") are organized differently.

There are two stages at which external font files are sought. The first stage occurs when TEX processes the `.tex` file to create the `.dvi` file. It looks at this stage for a file with the extension `.tfm` (for *TEX Font Metrics*), in which it expects to find information about the sizes of characters and the spacings it should use. Appendix H of *The* METAFONT*book* (KNUTH [1986b]) gives a precise description of the information contained in `.tfm` files. Interestingly, the program is quite disinterested at this stage in the actual shapes of the individual characters: all that it wants to know is where to place them and how much room they take up. Information about character shapes is sought only when a command is given to print the `.dvi` file, or to preview it on a monitor screen. The particular printing or previewing programs that are being used look at this stage for a separate file (not the `.tfm` one) that contains information about digitized character shapes at a suitable resolution. Common extensions for the names of these character shape files are `.gf` (for *generic font*) and `.pk` (for *packed*). The `.gf` files are produced when METAFONT (§5.3 briefly discusses this program) is used to create fonts. The `.pk` files are compact representations of the contents of `.gf` files; the format was designed by Tomas Rokicki in 1985 (ROKICKI [1985]; KNUTH, ROKICKI and SAMUEL [1990]) and is now widely used. There is also an old format for character shape files, labelled by the extension `.pxl` (for *pixel*); these files tend to be large and unwieldy, and their use is declining.

It is also possible to process a `.dvi` file further and translate it into a page description language that is understood by the printing device being used. Even if this is the method chosen, information about character shapes of the fonts is still needed—but now in a format that is compatible with the page description language. An approach that is popular these days is to convert a `.dvi` file into a PostScript file; the process is discussed in §5.8.

- *Loading A New Font*

Suppose that it has been decided to use ***boldface italics***. Suppose, further, that it is known that an appropriate font is indeed available. What this means in TEX is that a `.tfm` file and, say, a `.pk` file exist for the desired style. The `.tfm` file is needed by TEX in order to produce the `.dvi` file; the `.pk` file, or an equivalent one, is needed in order to print or view the document. Suppose that the files are called `cmbxti10.tfm` and `cmbxti10.pk` (the system used to name such files is discussed in §5.3). A suitable name is chosen for the new font, say `\BoldItal`, and the line

```
\font\BoldItal=cmbxti10
```

is placed in the input file. The name `\BoldItal` is arbitrary: any legitimate command name may be chosen. The `\font` command links this name to the external `cmbxti10` file. Once this is done, the font `cmbxti10` is considered *loaded.* One can now type '`{\BoldItal Listen up now!}`' to get '***Listen up now!***'. This is a 10-point font. If a larger size is needed—say, 15 points—but the font does not exist at that size (i.e., `cmbxti15` files don't exist), then a smaller font can often be magnified to the required size. There are three ways of asking for magnified sizes in general; here are examples:

```
\font\BigBoldItal=cmbxti10 scaled 1500
\font\BigBoldItal=cmbxti10 scaled\magstep2
\font\BigBoldItal=cmbxti10 at 15pt
```

In the conventions used in specifying scale factors, 1000 is considered the normal size; thus, a font scaled 2000 is twice as big, and one scaled 1500 is one and one-half times as big. As far as the `.tex` to `.dvi` conversion goes, arbitrary scale factors are admissible (the size information in the `.tfm` file is just scaled accordingly). But when it comes to the actual printing, there's a catch: character shape files like the `.pk` ones are available only for specific sizes. (If something like the "`.dvi` to PostScript"' method is being used, then essentially arbitrary magnifications can in fact be handled; see § 5.8.) So, Plain TeX makes available the commands `\magstephalf`, `\magstep1`, ..., `\magstep5` to represent sizes that are likely to be available. Each `\magstep` n stands for a scale factor of $1000 \times (1.2)^n$ (`half` corresponds to $n = 1/2$). Thus in the second statement above, '`\magstep2`' stands for 1440. Finally, as the third statement illustrates, it is also possible to ask for the font directly at a specific size, if that size is known to be available. A few words of caution against indiscriminate magnification are probably in order here: fonts of high quality are meant to be used at their design sizes. Thus, it is better to use, say, a true 12-point size rather than a 10-point size scaled 1200, unless one deliberately wants the heavier appearance of the magnified font.

Plain TeX automatically loads 16 basic fonts (a list is given towards the end of § 5.3) and assigns them names like `\rm`, `\bf`, etc. (Actually, the statement about names is not totally precise: the whole truth will emerge in § 5.5.) These fonts are sufficient for straightforward uses of TeX, relieving many users from ever having to load fonts themselves. If a style is needed that does not correspond to one of these fonts (bold italics, bold math symbols and sans serif faces—the term *sans serif* is explained later—are examples of styles that are not included in the basic 16), then the appropriate font must be loaded by the user.

In order to load new fonts, one needs to know their external filenames. These names can usually be obtained from whoever is involved with maintaining TeX on the system being used. The fonts most commonly used with TeX come from the Computer Modern family (see § 5.3), but it must be stressed that any font at all can be used, as long as there are external files that contain the information about it that TEX seeks.

It must also be stressed that the discussion in this section of using a new font glosses over some complexities. If the new font is to be used for limited portions of text, then what was shown is indeed all that one has to do. If, on the other hand, one is planning to switch to the new typeface for an extended part of the document, or if one wishes to use new fonts in math formulas, then there are other issues to be considered. For example, if the new font has a size that is different from the one currently being used, then the line spacing must also be appropriately altered by resetting `\baselineskip`. All of these issues will be dealt with systematically in the sections that follow.

5.2 Character and Category Codes

A two-step translation is carried out when TEX typesets material. In the first step, input characters are translated into a numerical code that identifies them internally to TEX. The code for an input character is called its *character code*; for a standard keyboard character this code is its ASCII (American Standard Code for Information Interchange) number. The details of the coding are discussed below. In the next step, character code numbers are translated into output characters. The second translation is done differently for text and for mathematics: the precise details are given in §5.4 and §5.5.

Throughout the rest of this chapter, the expression "input character" will refer to a character that is not part of the name of a command when it occurs. Command names do not get broken down into their individual characters when TEX processes a file. This is discussed under **tokens** in the Glossary. For example, if the input is 'A \quad B', then only A and B are translated into character codes; \quad makes its way through TEX's processing system as a single entity, not as five separate character codes. Further, not all input characters are eventually translated into output characters: this is discussed under *Category Codes* at the end of this section.

As an example of the first step in the translation, consider the opening words of an early paper on infinity[†] as they might have been typed into a computer file:

```
That all Magnitudes infinitely great,
```

If the paper were being typeset in TEX, these characters from the input would be represented internally as:

[†] E. HALLEY [1692]. *An account of several species of Infinite Quantity, and of the Proportions they bear one to the other,* Philosophical Transactions of the Royal Society of London, **16**, 556.

84 104 97 116   97 108 108   77 97 103 110 105 116 117 100 101 115

105 110 102 105 110 105 116 101 108 121   103 114 101 97 116 44

This sequence of numbers is in no way meant to represent the full stream of typesetting instructions that flows through TEX. A lot of other information goes along with these code numbers: what roles the characters are playing, whether there are spaces in the input between characters, and so on. Though an appreciation of these issues is needed for a proper understanding of exactly how TEX processes an input file, the issues are not directly relevant to the question of typeface choices, so they will not be discussed here.

The numbers shown above are decimal code numbers; octal or hexadecimal representations are more commonly used. Table I lists all three versions of the code number to the right of each character. Octal representations are labelled by using ' as the leading character, and hexadecimal by using ". It has also become conventional to use italics for octal representations, and typewriter type for hexadecimal, thus revealing quickly what the underlying radix is: 110, *'156* and "6E all represent the same number.

- *The Character Code Table*

Computer systems in use today employ 94 standard visible characters. They are assigned ASCII codes *'41–'176*. A "single space" is assigned code *'40*. These code numbers are also precisely the TEX character codes. The remaining keys on a keyboard will either give special characters or perform unprintable functions (i.e., functions not expressible as single characters), such as a carriage return or a backspace. TEX may not accept exotic input characters, so it is often best to stick to the standard ones (unless the system being used has been correctly configured).

Though individual characters are usually individually translated into separate code numbers, there is one situation that receives special treatment. TEX has a convention built into it that translates certain short sequences of input characters into single numbers instead of into separate numbers, one for each member of the sequence. These sequences consist of two superscript characters next to each other immediately followed either by a character with code $c < 128$, or by two of the hexadecimal digits 0, ..., 9, a, ..., f (where a, ..., f are in lower case). In the first case, TEX interprets the sequence as the code number that differs from c by 64 (or, equivalently, *'100* or "40); in the second (this is a new feature added in Version 3.0 of TEX), it interprets it as the code number given by the hexadecimal digits. For example, the sequence ^^B in the input is not translated into the three separate code numbers "5E, "5E and "42 (for ^, ^ and B, respectively), but instead into the single number "2. And ^^5a is translated into the single number "5a. In all of this, ^ stands for whatever character plays the role of a superscript command, usually literally

Table I: Character Codes (internal codes for input characters)

^^@ 0 '0 "0	^^A 1 '1 "1	^^B 2 '2 "2	^^C 3 '3 "3	^^D 4 '4 "4	^^E 5 '5 "5	^^F 6 '6 "6	^^G 7 '7 "7
^^H 8 '10 "8	^^I 9 '11 "9	^^J 10 '12 "A	^^K 11 '13 "B	^^L 12 '14 "C	^^M 13 '15 "D	^^N 14 '16 "E	^^O 15 '17 "F
^^P 16 '20 "10	^^Q 17 '21 "11	^^R 18 '22 "12	^^S 19 '23 "13	^^T 20 '24 "14	^^U 21 '25 "15	^^V 22 '26 "16	^^W 23 '27 "17
^^X 24 '30 "18	^^Y 25 '31 "19	^^Z 26 '32 "1A	^^[27 '33 "1B	^^\ 28 '34 "1C	^^] 29 '35 "1D	^^^ 30 '36 "1E	^^_ 31 '37 "1F
␣ 32 '40 "20	! 33 '41 "21	" 34 '42 "22	# 35 '43 "23	$ 36 '44 "24	% 37 '45 "25	& 38 '46 "26	' 39 '47 "27
(40 '50 "28	) 41 '51 "29	* 42 '52 "2A	+ 43 '53 "2B	, 44 '54 "2C	- 45 '55 "2D	. 46 '56 "2E	/ 47 '57 "2F
0 48 '60 "30	1 49 '61 "31	2 50 '62 "32	3 51 '63 "33	4 52 '64 "34	5 53 '65 "35	6 54 '66 "36	7 55 '67 "37
8 56 '70 "38	9 57 '71 "39	: 58 '72 "3A	; 59 '73 "3B	< 60 '74 "3C	= 61 '75 "3D	> 62 '76 "3E	? 63 '77 "3F
@ 64 '100 "40	A 65 '101 "41	B 66 '102 "42	C 67 '103 "43	D 68 '104 "44	E 69 '105 "45	F 70 '106 "46	G 71 '107 "47
H 72 '110 "48	I 73 '111 "49	J 74 '112 "4A	K 75 '113 "4B	L 76 '114 "4C	M 77 '115 "4D	N 78 '116 "4E	O 79 '117 "4F
P 80 '120 "50	Q 81 '121 "51	R 82 '122 "52	S 83 '123 "53	T 84 '124 "54	U 85 '125 "55	V 86 '126 "56	W 87 '127 "57
X 88 '130 "58	Y 89 '131 "59	Z 90 '132 "5A	[91 '133 "5B	\ 92 '134 "5C	] 93 '135 "5D	^ 94 '136 "5E	_ 95 '137 "5F
` 96 '140 "60	a 97 '141 "61	b 98 '142 "62	c 99 '143 "63	d 100 '144 "64	e 101 '145 "65	f 102 '146 "66	g 103 '147 "67
h 104 '150 "68	i 105 '151 "69	j 106 '152 "6A	k 107 '153 "6B	l 108 '154 "6C	m 109 '155 "6D	n 110 '156 "6E	o 111 '157 "6F
p 112 '160 "70	q 113 '161 "71	r 114 '162 "72	s 115 '163 "73	t 116 '164 "74	u 117 '165 "75	v 118 '166 "76	w 119 '167 "77
x 120 '170 "78	y 121 '171 "79	z 122 '172 "7A	{ 123 '173 "7B	| 124 '174 "7C	} 125 '175 "7D	~ 126 '176 "7E	^^? 127 '177 "7F

the character ^ itself. (The roles that characters play in TEX are discussed at the end of this section under category codes.) If there is any ambiguity about interpretation, the second rule supersedes the first; for example, ^^b5 will be translated into the code number "b5 and not into the code for ^^b followed by the code for 5. Table I shows how input characters (or input sequences that start with ^^) translate into internal character codes; only the first 128 code numbers are shown; numbers 128–255 are produced from standard keyboards through the second part of the mechanism just explained.

The ^^ mechanism makes possible a visible representation of otherwise invisible or unprintable input. (i.e., input not representable as a single visible character.) For example, it is sometimes necessary to instruct TEX to do something special at the end of every input line. Since the ASCII code for a carriage return is '15, such instructions can be given by using ^^M (which also translates into code number '15) to represent a carriage return. Plain TEX uses this approach, for instance, when defining the command \obeylines: every occurrence of ^^M (and, therefore, every carriage return) is defined to equal the paragraph-ending command \par.

Since, as far as TEX is aware, such ^^ sequences represent single characters, the expression "input character" will from now on encompass these sequences.

TEX also provides a device that allows code numbers to be represented symbolically, thus saving users from having to deal directly with character codes. If, for example, it is necessary to refer to the character code of h, the notation `h may be used in place of the actual code number. TEX then extracts the value of the character code on its own, without the user ever having to know what it was. The method also works with commands that have single character names: `\% may be used, for instance, to represent the character code of %. For some characters, the two methods may be used interchangeably; for instance, `\o and `o are equivalent. But for characters like % that play special roles when used as single characters, the second method must be used.

This device has obvious uses in situations where the character code of an input character is needed but memory has failed to serve and a code table is not on hand. It has even more valuable uses when setting up frameworks like the one provided by Plain TEX. In such frameworks it is necessary to instruct the bare core of TEX in a variety of rules and conventions: it has to be told which characters are to be used for grouping portions of input text ({ and } in Plain TEX), which character will switch the math mode on and off ($), which fonts are to be used for the digits 0 ... 9 when typesetting formulas (the normal roman font), etc. TEX requires that these instructions be given via character codes, and it makes the instructions enormously easier to give (and to later read) when they are given using this device. For example, it makes everything much clearer to say that character number `\$ will act as the "math shift" than to say that character number "24 will.

● *Category Codes*

In addition to character codes, input characters are each assigned a further code
number in TEX, called the *category code*. There are 16 categories of characters
in TEX's world—numbered from 0 to 15. The category code determines the role
that the character plays. For example, in Plain TEX an 'A' in the input is taken
as an instruction to print an output character, usually 'A', but a '$' is taken
as an instruction to switch on or switch off the math mode. It is the difference
in the category codes of the two characters that determines the difference in
TEX's response. *Only characters from categories 11 and 12 are automatically
interpreted as instructions to print output characters; all others are taken as
instructions to do other things.*

The 16 categories, along with the standard assignments of characters to
each category, are

0	*Escape*	`\`		8	*Subscript*	`_`, `^^A`
1	*Begin group*	`{`		9	*Ignored*	`null`, `^^@`
2	*End group*	`}`		10	*Space*	`space`, `tab`, `^^I`
3	*Math mode on/off*	`$`		11	*Letter*	A, ..., Z, a, ..., z
4	*Alignment tab*	`&`		12	*Other*	none of the others
5	*End of line*	`return`, `^^M`		13	*Active*	`~`, `^^L`
6	*Parameter*	`#`		14	*Comment*	`%`
7	*Superscript*	`^`, `^^K`		15	*Invalid*	`delete`, `^^?`

These assignments of category codes are the ones used in Plain TEX. Category
codes are assigned in that format file by statements like this:

```
\catcode`\{=1
\catcode`\$=3
```

The first of these makes { the "begin group" character, and the second makes
$ the "math shift."

Plain TEX does not have to assign all the category codes that it uses; in
some cases it uses assignments that are already built in. The initial assignments
that TEX begins with are as follows: the 52 letters A ... z are each assigned
category code 11; all other characters are assigned code 12 except \ (code 0),
^^M (code 5), ^^@ (code 9), ␣, i.e., an input space (code 10), % (code 14) and
^^? (code 15).

5.3 Computer Modern Typefaces

This is a class of typefaces created by Donald Knuth in collaboration with several celebrated typographers (see KNUTH [1986d]). Many of the faces in this class are based on a traditional typeface (called Monotype Modern 8A) that was often used for technical texts. *Modern* is a technical term in typography, referring to a particular class of typefaces. An important feature of the modern style is the very sharp contrast between the thick and the thin lines in its characters; look, for example, at this letter: M.

Computer Modern typefaces were created with the help of yet another program composed by Donald Knuth. This program, METAFONT, is graphical in nature, and it is used—as its name suggests—to create fonts. METAFONT automatically makes the .tfm files that TEX needs in order to use a font. It also makes the character shape files (.gf) that are needed in order to print or view a document.

METAFONT, the program, makes it possible to design not only fonts, but also meta-fonts. A *meta-font* is a parameterized class of fonts that will yield individual fonts when its parameters are set to specific values.

The set of Computer Modern typefaces roughly corresponds to such a class; the class has 62 parameters that determine things like the height of individual characters, whether they are slanted or not, the thickness of the lines in them, whether they have serifs or not, etc. By suitable adjustments of the parameters, one can get different styles. The following interpretations of the letters "rd" are, for example, obtained from different settings of the parameters:

rd *rd* **rd** **rd** rd **rd** **rd** rd rd

Standard TEX distributions come with around 75 Computer Modern typefaces, of which 16 are automatically loaded by Plain TEX. More will be said about these 16 at the end of this section. $\mathcal{AMS}$-TEX distributions add a few extra Computer Modern faces (just different sizes of some of the standard styles) and several new math fonts. The pages displaying font tables at the end of the chapter explicitly state which fonts come with $\mathcal{AMS}$-TEX. LATEX distributions commonly add several new fonts of their own. The font files that contain digitized shape information tend to be large, so all Computer Modern fonts are not always available at every institution. Users will have to check with local system experts, or experiment on their own (by trying to load a font and seeing if an error message ensues), to find out what they have available. As discussed in Chapter 3, $\mathcal{AMS}$-TEX allows users to control which extra fonts are to be loaded.

Table II gives a complete list of the standard Computer Modern distribution. (A further font, logo10, that contains just the characters needed in

Table II: External Filenames for Computer Modern Typefaces							

Typefaces distributed at more than one design size

cmr5*	cmr6	cmr7*	cmr8	cmr9	cmr10*	cmr12	cmr17
cmbx5*	cmbx6	cmbx7*	cmbx8	cmbx9	cmbx10*	cmbx12	
cmmi5*	cmmi6	cmmi7*	cmmi8	cmmi9	cmmi10*	cmmi12	
cmsy5*	cmsy6	cmsy7*	cmsy8	cmsy9	cmsy10*		
		cmti7	cmti8	cmti9	cmti10*	cmti12	
			cmss8	cmss9	cmss10	cmss12	cmss17
			cmssi8	cmssi9	cmssi10	cmssi12	cmssi17
			cmsl8	cmsl9	cmsl10*	cmsl12	
			cmtt8	cmtt9	cmtt10*	cmtt12	
			cmtex8	cmtex9	cmtex10		
					cmex10*		

Typefaces distributed at only one design size

cmb10	cmbxsl10	cmbxti10	cmbsy10	cmcsc10	cmdunh10	cmff10	cmfi10
cmfib8	cmitt10	cminch	cmmib10	cmsltt10	cmssbx10	cmssdc10	cmssq8
cmssqi8	cmtcsc10	cmu10	cmvtt10				

order to say METAFONT is often available as well.) The cm at the start of each name stands for *Computer Modern*; the number at the end is the design size in points (1 inch equals 72.27 pt in TEX)—a rough measure of the size of individual characters. A b stands for *bold*; bx for *bold extended*; ex for *extension*; csc for *capitals and small capitals*; dc for *demibold condensed*; dunh for *dunhill* (a style that gives elongated letters); ff for *funny* (an experimental style); fi for *funny italic* (another experiment); fib for *fibonacci* (an experimental font based on the Fibonacci sequence); i or ti for *italic* (text); inch for a special inch-high font; mi for *math italic*; q for *quotation style* (a special style used by Donald Knuth for quotations in his books); r for *roman*; sl for *slanted*; ss for *sans serif*; sy for *symbols*; t or tt for *typewriter*; u for *unslanted*; and v for *variable width*.

- *Fonts loaded by Plain TEX*

The fonts in Table II that are marked by an asterisk are the ones loaded by Plain TEX. They are assigned the following names: \fiverm, \sevenrm, \tenrm (cmr fonts); \fivebf, \sevenbf, \tenbf (cmbx fonts); \fivei, \seveni, \teni (cmmi fonts); \fivesy, \sevensy, \tensy (cmsy fonts); and \tenit, \tensl, \tentt and \tenex (cmti10, cmsl10, cmtt10 and cmex10, respectively). In other words, the Plain TEX file contains 16 statements of this type:

    \font\tenrm=cmr10

It also contains statements that, roughly speaking, define \rm, \it, \sl, \bf and \tt to be abbreviations for \tenrm, \tenit, \tensl, \tenbf and \tentt, respectively. The statements partially look like this: '\def\rm{\tenrm}'. But they also involve substatements that ensure that everything works correctly in formulas as well as in text. The actual statements are shown in §5.5. Finally, almost at the very end of the long file that defines the Plain TeX format, there occurs the simple statement '\rm'. This declaration is what makes that package automatically use the 10-point roman font as its starting font.

These 16 fonts satisfy the minimum needs of technical typing, but they provide few frills. The 10-point roman, italic, slanted, boldface and typewriter fonts cover many basic text needs. They also provide uppercase Greek letters— regular weight as well as bold—for use in formulas. The math italic, symbols and extension fonts provide the lowercase Greek alphabet, specially drawn italic letters for use in formulas, and a wide selection of special symbols. They come (as far as these 16 fonts are concerned) just in regular weight, and not in bold versions. The font tables at the end of the chapter show what is available in each font. The smaller sizes, 7- and 5-point, of these faces are used in indices. With these fonts at one's disposal, one can type {\it italics} to get *italics*, {\sl slanted type} to get *slanted type*, or {\fivebf small boldface} to get small boldface.

- ### *Some Font Suggestions*

If greater variety is desired, more fonts can be loaded—though the **temptation** to go overboard with too much variety should be strenuously **avoided**. For a piece of straight text, it is usually advisable to stick to one text typeface and its associated variations (bold, italic, slanted and small capitals), with possibly another style for titles. Some commonly loaded fonts in TeX, beyond Plain TeX's 16, include cmcsc10 (Capitals and small capitals) and cmss10 (sans serif—a style without the little protrusions, or *serifs*, at the tips of the strokes that make up characters). Since these fonts are 10-point, they may be loaded and their names used—just as \bf, \it, etc., are—without necessarily having to make any extra line-spacing adjustments. (See the bold italic example in §5.1.) The font cmssdc10 (**sans serif demibold condensed**) is good for titles, especially in its magnified versions. Bold lowercase Greek letters and bold math italic (α, β, x, y, etc.) can be obtained by loading cmmib10 and smaller sizes, and bold math symbols ($\oplus$, $\forall$, etc.) by loading cmbsy10. Using new fonts loaded for mathematical purposes isn't as straightforward as using new fonts in text, however. The details are given in §5.5.

5.4 Typefaces in Text

TₑX chooses typefaces for text (formulas are discussed in the next section) in a simple way: it just uses the *current font*—the last encountered typeface declaration. If the declaration occurred inside a group (roughly, enclosed by braces), then it is in force only inside; once the group is over (i.e., once one is past the closing right brace), the typeface that was being used when the group began becomes current again.

Suppose, for example, that a computer file begins in this way:

`DEFINITION. Consider three distinct sets of objects.`

How will this appear in the output? A large number of TₑX formats—Plain TₑX is one example—declare 10-point roman to be the starting typeface. This will cause the output to appear as:

DEFINITION. Consider three distinct sets of objects.

If a different typeface is preferred, a different declaration must be made in place of the statement `\rm` in the format file, or `\rm` itself must be redefined (before it is used), or a new declaration must be made after the `\rm` statement. For example, one way to get sans serif type is to put the lines

`\font\TenSans=cmss10`
`\TenSans`

either at the start of the computer file or in a separate format file that is used as input (via the `\input` command) at the start. The two lines do not necessarily have to appear adjacent to each other as shown here—other lines may intervene—but they must come in the order shown (the font must be loaded before it is used). The opening line shown above will then appear in the output in this style:

DEFINITION. Consider three distinct sets of objects.

So will everything else until the font is explicitly changed. The current font may be changed at any stage; if the change is required to hold all the way until the end of the file (or until the next explicit change), the new font declaration must not be made within a group (roughly, in text sandwiched by braces). Conversely, if the change is required to be local, it must be confined to the appropriate group. Suppose that the file under discussion continues as follows:

`Let the objects of the {\bf first} set be called`
`{\it points\/} and be denoted by $A$, $B$, $C$, \dots;`

The commands `\bf` and `\it` are the standard names for the bold and italic faces; the effects of the commands are confined by the surrounding braces. The output from these lines will be

Let the objects of the **first** set be called *points* and be denoted by A, B, C, ...;

In addition to the typeface changes made in response to explicit instructions, there is also an automatic one that is made every time a character appears between $ signs. This is discussed in the next section.

Many beginning users of TeX try at some point or another to combine typeface changes by combining commands: for example, they try to get boldface italics by saying \bf\it or \it\bf. The commands \bf and \it are, however, usually each defined just to switch the current typeface. Therefore, in either case, only the second declaration takes effect: \bf\it gives italics and \it\bf gives boldface. To get a combination effect, a separate font must be summoned, with its own special command. Boldface italic characters, for example, belong to a separate font from the regular bold or italic fonts; therefore a separate command, unconnected to the standard \bf and \it commands, must be issued (as shown in § 5.1) to get characters in this style.

Having said these discouraging things, it must also be added that there do exist more flexible approaches to the problem of typeface switching. In these approaches (see § 5.7), commands like \bf and \it have been redefined in sophisticated ways that allow combinations like \bf\it to give multiple typeface switches.

- *Common Errors*

When confining the effect of a typeface command, it is easy to forget to type the closing brace. A missing brace in such a situation will often not trigger an error message, merely a warning at the end of the processing of the file that will say something like '\end occurred inside a group'. It is only when the output is seen that it becomes clear that several pages of text have ended up italicized, say, whereas italics had been intended for only one word. A useful device to prevent this from happening is to type both braces right away instead of just the opening one—'{}' instead of '{'—and then to type text in between. Some systems allow keys to be programmed so that both braces appear automatically, with the cursor positioned between them, when the opening one is typed.

Errors are also commonly made when typeface commands are included in definitions. Suppose, for example, that there is a document where it is frequently necessary to enclose bold text inside bold brackets. It may be tempting to define a new command as follows:

```
\def\Bld #1{\bf [#1]}
```

Typing \Bld{stuff} will, however, cause all the subsequent text to appear in boldface (in addition to producing [**stuff**]), because the replacement text for \Bld is what lies between the braces in the definition and does not include the braces themselves. Using

```
\def\Bld #1{{\bf [#1]}}
```

instead will do the trick.

● *Font Switching*

Plain TEX makes 10-point roman its starting typeface, and it provides the convenient abbreviations \it, \bf, \sl, etc., to allow the easy switching of typefaces. When appropriate additional definitions are given, \bf may be made to summon the bold variant—if one exists—of whatever basic text typeface is currently being used, not just the roman typeface. Similar arrangements can be made for \it and \sl as well.

In order to allow \bf and its companions to play these flexible roles, comprehensive font-switching commands need to be defined. Such font switches also often involve readjusting line-spacing parameters. §6.2 displays an example of an elaborate typeface change to 8-point type, along with a change of the line spacing to a value more suited to smaller type. That example also takes into account the typefaces to be used in formulas. Here is a simpler example that illustrates the issues for straight text.

Suppose that a 12-point sans serif look has been chosen. The following lines allow a smooth switch to the new look:

```
\font\TwelveSans=cmss12
\font\TwelveSansI=cmssi12
\font\TwelveSansBf=cmssbx10 scaled \magstep1
% NOTE: A true 12-point bold sans serif face isn't part of
% standard distributions of Computer Modern faces.
\def\BigSSf {\let\ssf=\TwelveSans  \let\it=\TwelveSansI
          \let\bf=\TwelveSansBf \normalbaselineskip14pt
          \normalbaselines\ssf}
```

After this, all that has to be done is to type \BigSSf. Such a statement at the start of the file discussed above would produce this output:

> DEFINITION. Consider three distinct sets of objects. Let the objects of the **first** set be called *points* and be denoted by A, B, C, ...;

This works satisfactorily for pure text, but as is seen above, the typefaces that are used between $ signs (those used for A, etc.) continue to follow old rules. It takes a little more work to change the typeface used in formulas. The method is discussed in the next section.

The setting of the line spacing here may need explanation. The baseline-to-baseline gap between successive lines is determined by the current value of \baselineskip. If that value brings a pair of lines too close together—where "too close" is determined by the value of a parameter called \lineskiplimit—then extra space is added between the two: i.e., between the bottom of the top line and the top of the bottom one. The extra space to be added is given by the value of a third parameter called \lineskip. Instead of setting these values directly, Plain TEX introduces the auxiliary quantities \normalbaselineskip, etc., and assigns them appropriate values. The command \normalbaselines then sets the initial values of the primitive line-spacing parameters to be the "normal" values:

```
\def\normalbaselines{\lineskip=\normallineskip
                     \baselineskip=\normalbaselineskip
                     \lineskiplimit=\normallineskiplimit}
```

This may seem a tortured approach to a simple issue, but it has many advantages: it allows, for example, the temporary resetting of \baselineskip (directly, by typing \baselineskip=*new value*), etc., in certain situations—say, to display a few lines of spaced-out text—with the possibility of later setting everything back to their normal values just by typing \normalbaselines. It is therefore useful to have appropriate normal values tailored to different font sizes built right into the font-switching command.

- *How Input Characters Are Translated into Output Characters*

As discussed in §5.2, TEX internally represents input characters by their character codes. When typesetting text, these numerical representations are translated in turn (as long as they are from categories 11 or 12) into output characters chosen from the current font. Fonts used for text (English) in TEX have tended to have similar *layouts* (this means that a given code number has usually corresponded to the same output character in all of them) and text fonts that are to be used in future (the "256-character fonts"—see §5.28) are to have identical layouts. Thus character number 103 is usually 'g' in fonts used for text, though it can appear in the output in different styles—as **g**, or *g*, or ﬤ, for example, depending on what the current font is. The last part of this chapter displays tables that show the first 128 characters from several fonts.

As a concrete illustration, consider the first two words of the example on the previous page:

Input: DEFINITION. Consider
Translation into TEX's internal code:
68 69 70 73 78 73 84 73 79 78 46   67 111 110 115 105 100 101 114

The resulting output from several choices of current font (10-point roman and 10- and 12-point sans serif) has been shown above. If, for some unfortunate reason, a font that is not normally used for text has been declared to be the current font, the results can be unexpected. For example, if cmsy10, the 10-point math symbols font, were declared to be the current font (by typing \tensy), then the very same input would produce

$$\mathcal{DEFINITION}\diagup\mathcal{C}\iota\backslash\flat\rangle\sqcap\nabla$$

in the output, since these are the characters that occur in those positions of the symbols font. (The space between the words vanishes because of the particular values of certain font parameters for `\tensy`; see § 5.6 and § 5.16.)

- *Other Output Characters*

Each font in TeX contains up to 256 characters, numbered from 0 to 255. As discussed in § 5.2, TeX can also accept 256 input character code numbers. The total number of input character codes does not, however, have to automatically equal the maximum number of characters in a font. Indeed, till Version 3.0 of TeX became available, the two numbers were not equal (the program previously accepted only 128 input character codes).

Not all "places" have to be filled in a font—it is permissible to have some numbers that do not correspond to any character at all. The tables at the end of the chapter contain several such empty places. The first 128 places in the standard Computer Modern fonts are, however, all filled. This was a conscious design decision—to keep people from filling empty slots in these fonts in incompatible ways—even though it meant including stray characters in places where they did not strictly belong. This idiosyncrasy was mentioned at the start of this chapter; the explanation promised there is the one just given.

This chapter ends with several pages of font tables; the tables display the first 128 characters of the commonly used fonts. Positions 128–255 in the standard fonts are not filled in any special way. In 1990, the TeX Users Group decided to adopt a standard layout for text fonts, different in several ways from that of the Computer Modern fonts. All 256 positions are filled in the new layout. Work has been progressing since then on complementary sets of mathematics fonts. The new font layouts are eventually to replace the current ones entirely. This work is discussed briefly in § 5.28.

How does one gain access to all 256 characters in a font? The standard printable keyboard characters are obtained directly: the 'A' key, for instance, corresponds to code number 65, which in turn yields the character 'A' in text fonts. For other characters the answer is slightly more complicated. One way to invoke them is to use an extended keyboard and a special implementation of TeX that provides the correct matching of input and output characters; such a keyboard might, for example, have a key labelled ø, and TeX can then be set up to make the key produce ø in its roman fonts. Another way is to edit `.tfm` files of the appropriate fonts in order to define new ligatures. (A *ligature* is a combination of input characters that is represented in the output by a single character; e.g., an input 'fi' usually appears in the output as 'fi'.) Information can be added to the `.tfm` file so that the combination o/, for instance, is viewed as a ligature that produces ø.

There is yet another way. TeX provides a command called `\char` that allows users to obtain characters that match specified position numbers. For example, \char"1D gives character number "1D from the current font—in the roman fonts, this gives Æ. In practice, it is far more convenient to define names for such characters and then to use the names when the characters are needed. This is done through a command called \chardef. For example, the Plain TEX file contains several declarations of this type:

```
\chardef\ss="19
\chardef\AE="1D
```

and so on, allowing users to type \ss to get ß (the character that occurs in position "19 in standard text fonts), \AE to get Æ, and so on. Of course, the exact output character will depend, as always, on the current font.

Similar methods may be used to invoke any character at all from any font, not just the current one. A frequent need to say "β-testing," for example, can be satisfied by using the following definition:

```
\def\Tbeta {{\teni\char'14\/}}
```

The command \teni calls upon the already-loaded (in Plain TEX) math italic font, cmmi10; if a character from an unloaded font is needed, the font must first be loaded. (The command '\/' gives an "italic correction": it prevents the β from leaning too closely against the following character.) Then, typing '\Tbeta-testing' will have the desired effect. The character β used in this example was actually already available through a math mode command (but summoning it by typing β will give slightly different spacing). The method shown here works in general, however, even for characters with no pre-assigned names, as long as the desired character is indeed found in some font.

As another example, the METAFONT logo in this book is obtained through a command called \MF, which pulls the appropriate characters from a font called logo10. Here is how it is done:

```
\font\Logo=logo10
\def\MF{{\Logo META}\-{\Logo FONT}}
```

(The \- represents a discretionary hyphen; it is used, in general, to suggest possible hyphenation points to TEX at places that it might not choose on its own.)

- *Accents*

Accents are mostly obtained in a manner similar to the way characters are summoned by the command \char. The corresponding command here is \accent. For example, '\accent"17 B' gives 'B̊' (assuming that the standard roman font is being used). Again, it is more convenient to define names; here are samples from the Plain TEX file:

```
\def\'#1{{\accent"13 #1}}
\def\H#1{{\accent"7D #1}}
```

The second of these, for example, is what makes '\H A' produce 'Å'. It should be noted that some of Plain TEX's standard accents are not obtained through the use of \accent; they are constructed instead as carefully adjusted alignments.

There is a major drawback to constructing accented characters through the use of \accent: hyphenation incidentally gets suppressed. That was one major motivation behind the new 256-character font layout (see §5.28).

5.5 Typefaces in Mathematics

Consider these formulas: $\int x^2\,dx = \frac{x^3}{3} + C$, and

$$\frac{1}{2\pi}\left(\int_{\mathcal{M}} K\,d\mathcal{M}\right) = \chi(\mathcal{M}).$$

The input for the first of these is

    $\int x^2\,dx={x^3\over 3}+C$

and that for the second,

    $${1\over 2\pi}\left(\int_{\cal M} K\,d{\cal M}\right) =
       \chi({\cal M}).$$

The typesetting of these two simple formulas leads to several typeface questions: from what fonts does TEX take the numerals 1, 2, 3, ...? The symbols representing variables and constants: x, C, etc.? The lowercase Greek letters? How does it pick the different sizes of these characters, or of symbols like $\int$, or of things like parentheses? How does it pick the spacing around symbols like = and +?

The answers are slightly involved—as might be expected, given TEX's ability to automatically handle many varieties of formulas—but not overwhelmingly so. The next few subsections will give a systematic treatment of all the relevant issues. But, to reveal the ending right away, here is a very short summary: Every input character is assigned a *math code* number and a *delimiter code* number. The math code tells TEX from which font to take the matching output character (remember that input characters are translated into character codes, which in turn are then translated into output characters) and how to arrange spacing around it in a formula. The delimiter code tells TEX how the character should behave—for example, whether and how it should automatically grow in size—if it is used to delimit, or bound, a formula (parentheses and brackets are examples of delimiters). Special characters, like ℵ, are invoked by special commands (much as \AE produces Æ in text); these commands have math code or delimiter code information built in.

- *Font Families*

A key notion in TeX's handling of formulas is that of a font *family*. A family is a set of three fonts, one to be used at normal (or "text") size, one at "script" size (typically in sub- and superscripts, or in some fractions), and one at "scriptscript" size (typically in indices on indices, or in fractions within fractions). TeX can handle up to 16 different families; all the characters that are typeset in formulas come from one of these families.

The Plain TeX package assigns the normal roman fonts (`cmr`) to family 0, the math italic fonts (`cmmi`) to family 1, the symbol fonts (`cmsy`) to family 2 and the math extension font (`cmex`) to family 3. In the first three cases, the 10-, 7-, and 5-point sizes are chosen, respectively, for the text, script and scriptscript members of each family. For example, the declarations for family 1 are

```
\textfont1=\teni
\scriptfont1=\seveni
\scriptscriptfont1=\fivei
```

Similar declarations are made for families 0 and 2. There is no requirement that the sizes chosen for the three families necessarily be 10, 7 and 5 points, or even that the sizes decrease. The declarations for family 3, for example, are

```
\textfont3=\tenex
\scriptfont3=\tenex
\scriptscriptfont3=\tenex
```

All three members of family 3 have the same size; thus, characters from this family will often not change size when they appear in different places in a formula as ones from other families do (see, however, the discussion under *Size Selection* later in this section). For example, the integral signs used by TeX come from family 3; as a result the two occurrences of that sign in $\int_\alpha \int^{\alpha\,d\alpha} d\alpha$ are the same size, whereas the other characters have automatically shrunk in the superscript.

Although there is considerable freedom permitted in what fonts are placed in what family, there are a few restrictions. The delicate positioning rules that TeX follows when it typesets formulas assume that it has access to information about things like the height and depth at which to place super- and subscripts, the thickness of the rule that it uses in fractions, and so on. TeX expects to get this information from the `.tfm` files for the fonts in families 2 and 3, so these two families cannot be chosen entirely arbitrarily. The exact nature of the information supplied by font files is discussed in §5.6. Further, TeX is set up to take the 52 letters A ... z from family 1, and the digits 0 ... 9 from family 0. Therefore, it is usually necessary for the fonts assigned to these families to have regular interpretations of these characters.

As an illustration of the freedom that one has when assigning font families, consider this example:

```
\textfont1=\fivesy
\scriptfont1=\seveni
\scriptscriptfont1=\tenex
Okay, time for a formula: $q^{q^q}+r^2=1$.
```

The output from this is

Okay, time for a formula: $\sqrt[q]{q} + \mathtt{v}^2 = 1$.

This is a decidedly unusual interpretation of straightforward input. The three input qs give the output character that occupies the normal position of q in the chosen text, script and scriptscript fonts, respectively; the other effects arise similarly.

To ensure that families that have already been assigned fonts are not mistakenly reassigned new ones, TₑX provides the `\newfam` command that makes assignments and allows the use of symbolic names in place of explicit family numbers. Here, for example, are further family assignments that are made in Plain TₑX:

```
\newfam\itfam    \textfont\itfam=\tenit
\newfam\slfam    \textfont\slfam=\tensl
\newfam\bffam    \textfont\bffam=\tenbf
   \scriptfont\bffam=\sevenbf   \scriptscriptfont\bffam=\fivebf
\newfam\ttfam    \textfont\ttfam=\tentt
```

The names `\itfam`, etc., are arbitrarily chosen symbolic names for the family numbers assigned by `\newfam`. Since families 0–3 have already had fonts assigned to them, `\newfam` assigns to `\itfam` the value 4; to `\slfam` the value 5; to `\bffam` the value 6; and to `\ttfam` the value 7. So, `\textfont\bffam=\tenbf` is precisely equivalent here to `\textfont6=\tenbf`.

Only family 6 has been assigned the full set of three fonts; the absence of an assignment to the script and scriptscript members of families 4, 5 and 7 means that it is not possible to use the text italic, slanted or typewriter faces in superscripts, etc., in the normal way. For example, though it is possible to use `$X^{\rm Y}$` and `$X^{\bf Y}$` (the outputs are, respectively, X^{Y} and $X^{\mathbf{Y}}$), the occurrence of `$X^{\sl Y}$` will trigger an error message saying that nothing has been assigned to `\scriptfont5`.

- *A Summary of Plain TₑX's Families*

Family 0 is the roman family; family 1, math italic; family 2, symbols; family 3, math extension; family 4, text italic; family 5, slanted; family 6, boldface; family 7, typewriter. Families 4, 5 and 7 have only been assigned a text font. Families 8–15 are empty. $\mathcal{A}_{\mathcal{M}}\mathcal{S}$-TₑX fills some of these open families, as discussed later in this section.

When assigning a font to a family, the font must first be loaded (if it has not already been). The assignments of fonts to families have absolutely no effect on the typesetting of text; that process is carried out independently.

- *Math Codes*

An output character in a formula is labelled by several numbers: its position in the font from whence it came, the family that the font has been assigned to, and its *class*. The class determines the grammatical role that the character plays in a formula; there are eight possible roles in TEX:

0	*Ordinary* (e.g., '/')	4	*Opening* (e.g., '{', '[')
1	*Large operator* (e.g., '∫')	5	*Closing* (e.g., '}', ']')
2	*Binary operation* (e.g., '+', '−')	6	*Punctuation* (e.g., ';')
3	*Relation* (e.g., '>', '<')	7	*Variable family* (e.g., 'x')

The meanings of these classes should be fairly clear from their names (and the examples shown). The main visible effect that a character's class has on how it is typeset is the spacing used around it in a formula: each class follows different spacing rules. Class 7 is special, however. The spacings used for its members are the same as those used for the members of class 0. The real use of class 7 is explained later (under *Font Switching*).

Let the three numbers that label a character—its position number, its family and its class—be represented by p, f and c, respectively. Then its math character code, m, is given by $m = p + 16^2 f + 16^3 c$ $(= p + 256f + 4096c)$. If m is expressed in hexadecimal form, it will have four digits: the first will precisely equal c, the second precisely f, and the last two precisely p. Thus a character with math character code "321C will be a relation (class 3), and it will be taken from position "1C in family 2.

- *How Input Characters Are Translated into Output Characters*

Every input character is assigned a math character code; this is done through TEX's \mathcode command:

    \mathcode"49="7149

This means that the input character with character code "49 is assigned math code "7149. In such assignments it is preferable to use the symbolic names for character codes that TEX permits (see § 5.2); character number "49 happens to be 'A', so the statement above may be re-expressed in the preferred form

    \mathcode'A="7149

The assignments tell TEX how to translate input characters into output ones: the position and family numbers tell it where to find the output character, and the class tells it what spacing is appropriate.

TEX begins—before formats like Plain TEX have added anything to it—
with the following initial math code assignments: for the 10 digits 0 ... 9, the
math code is the character code plus "7000; for the 52 letters A ... z, the math
code is the character code plus "7100; for all other characters, the math code
is initially just the character code. Thus the digits and letters are in class 7,
with the digits coming from family 0 and the letters from family 1. All other
characters are initially in class 0 and family 0.

Plain TEX alters some of these basic math code assignments. The new
assignments are

```
\mathcode`\ ="8000    \mathcode`\'="8000    \mathcode`\_="8000
\mathcode`\;="603B    \mathcode`\,="613B    \mathcode`\!="5021
\mathcode`\)="5029    \mathcode`\?="503F    \mathcode`\]="505D
\mathcode`\}="5267    \mathcode`\(="4028    \mathcode`\{="4266
\mathcode`\[="405B    \mathcode`\>="313E    \mathcode`\<="313C
\mathcode`\:="303A    \mathcode`\==="303D   \mathcode`\+="202B
\mathcode`\*="2203    \mathcode`\-="2200    \mathcode`\.="013A
\mathcode`\\="026E    \mathcode`\/="031D    \mathcode`\|="026A
```

Thus, for example, the appearance of + in the input for a formula will lead
TEX to take character number "2B from family 0 (it will come as no surprise
that the character happens to be +) and to set it with the spacing appropriate
to class 2 (binary operation). A math code of 8000 is special: it causes the
character to become *active* (roughly, to function as a single character command;
this role transcends any that the character might play outside mathematics).
Observe that there is no necessary link in all of this between the input and
output characters. For example, it is possible (but, for the most part, pointless)
to declare

```
\mathcode`\+="2200
```

in one's file. This will make '$1+1=2$' produce '$1 - 1 = 2$' in the output.
Incidentally, this new interpretation of the plus sign is confined to math mode;
in text '+' still gives '+'. (As was said earlier, input characters are translated
into output characters in different ways in text and math.)

● *Other Output Characters*

As in text (but even more so here), there are many symbols that cannot be
obtained from single keystrokes—there are just too few keys. Therefore, special
commands have to be issued to produce special characters. Analogous to `\char`
and `\chardef` in text, there are `\mathchar` and `\mathchardef` available for use
in typesetting math. For example, $\mathchar"023B$ will produce character
number "3B from family 2 (the symbols font, `\tensy`, in Plain TEX) and place it
with spacing appropriate to class 0; i.e., it will produce ∅. The actual definitions
used in Plain TEX look like this:

```
\mathchardef\alpha="010B
\mathchardef\Gamma="7000
```

and so on. It takes many, many such definitions to allow users to invoke the rich variety of symbols in TeX and to have them placed in formulas with the correct spacing.

Thus, when TeX encounters something like `$\Gamma(1/2)=\pi^{1/2}$`, it uses the math code values for each input character (represented internally, as always, by the character code) and for each character-producing command in order to produce the correct output: $\Gamma(1/2) = \pi^{1/2}$. The current font, so important in text, is irrelevant in formulas—except when text is placed in an `\hbox` in a formula (and in two other situations). To illustrate, consider these:

`{\sl $1+x^2>0, \hbox{ for all } x$}`   gives $1 + x^2 > 0$, *for all* x,

whereas

`{\bf $1+x^2>0, \hbox{ for all } x$}`   gives $1 + x^2 > 0$, **for all** x.

The only thing in this example that is affected by the typeface in force when the formula begins is the text in the `\hbox`. In general, two other things inside a formula are also affected by the current font: the size of a `\quad` (see § 5.6) and that of the single space given by `\`␣.

- *Changing Typefaces in Formulas*

Text in formulas should be placed in an `\hbox`; it thus undergoes typeface changes as usual. But the situation is different for the formulas themselves: math code values are firmly attached both to characters and to character-producing commands like Γ. Is it possible, then, to change typefaces in formulas without going through an elaborate resetting of math codes? Consider this input:

`$\bf \Gamma(1/2)=\pi^{1/2}$`.

It produces $\mathbf{\Gamma(1/2)} = \boldsymbol{\pi^{1/2}}$. If the math codes assigned to digits force TeX to find them in family 0, and if the very definition of `\Gamma` requires that it be character number 0 in family 0 (i.e., the roman font in Plain TeX), how does this typeface change come about? And why, if typefaces can indeed change in a formula in this way, is the change in the formula only partial?

The answer lies in the meaning of class 7, the *variable family* class (to which the digits, all the letters and all the uppercase Greek letters are defined to belong). Characters in this class are allowed to change family allegiances temporarily and to switch to a font family other than the one originally specified. The new family is given by the current value of a parameter called `\fam` (if this value lies between 0 and 15—otherwise, `\fam` is ignored). The value of `\fam` is reset to -1 every time math mode is switched on, so TeX will initially be poised to obtain characters from the families specified in the original `\mathcode` declarations (if the input is a single character) or in the original `\mathchardef` definitions. Commands like \rm, \bf, etc., happen also to assign a new value to \fam (that they do more than just change the text typeface had been alluded to in §5.4), and they therefore cause some characters to be taken from new fonts. For example, \bf effectively sets \fam=6, thus yielding some boldface characters even in math. Lowercase Greek letters are not in class 7 (though, as will be seen below, this can be changed), nor are (, /, etc., and so these do not change appearance.

The full Plain TeX definitions of typeface commands like \rm and \bf, as well as ones like \cal, are

```
\def\rm{\fam0 \tenrm}
\def\it{\fam\itfam \tenit}
\def\sl{\fam\slfam \tensl}
\def\bf{\fam\bffam \tenbf}
\def\tt{\fam\ttfam \tentt}
\def\mit{\fam1 }
\def\cal{\fam2 }
```

Thus, the commands \rm ... \tt do two things: they change the current font and they specify from which family the characters of class 7 should temporarily be taken in a formula (\itfam, etc., are just the symbolic names for family numbers that were introduced earlier). The current font part of the command has essentially no effect on formulas; the \fam part has no effect on text. For example, though \bf and \tenbf have the same effect on text in Plain TeX, \tenbf (being a pure font command and making no statement about math families) has absolutely no typeface-changing effect in a formula. The remaining two commands merely switch the value of \fam, since they are meant for use in formulas.

- *Size Selection*

There remains the issue of size. In the sample formulas at the start of the section, different sizes of numerals and letters were automatically used in different contexts, two different sizes of $\int$ were employed, and the parentheses used in the displayed formula magically grew to the right sizes through the use of \left and \right. How did these size variations arise?

When TeX typesets a formula it keeps track of the current style: display, text, script or scriptscript. It picks the correct sizes of characters by using the matching font member of the appropriate family. For instance, when setting a superscript it is usually in script style, and it thus selects characters from \scriptfont. It will use \textfont for both the display and text styles.

Some symbols ($\int$ and $\sum$, for example) nevertheless still appear in different sizes in text and display styles. These are symbols that are classified as *large operators* (class 2); they usually come in two sizes in the appropriate \textfont, one for use in text-style formulas (typeset in the line of text where they occur) and one for use in displays (typeset on a separate line). The definitions of the names for these symbols refer initially just to the smaller size. For example, \sum is defined in Plain TeX through

    \mathchardef\sum="1350

The font `cmex10` is the Plain TeX choice for family 3, and character number "50 in this font is the smaller of the two \sum symbols. The .tfm file entry for the character says, however, that it has a *successor*—another version—in the font at location "58. When TeX comes across an instruction like \sum, it will look first at the position and family specified in the original \mathchardef; if the display style is currently being used, it will redirect itself from there to the successor (if there is one).

The typesetting of delimiters (parentheses, brackets, etc.) of variable size is more complicated, and it involves assigning yet another numerical attribute to characters—their delimiter code. All characters, except the period, start off initially with a delimiter code of −1; the period is given delimiter code 0. Very few characters are usually used as delimiters (none of the 52 letters A, ..., z is, for example), and ones that are not retain their negative delimiter codes. Positive delimiter codes have six hexadecimal digits, the first three of which specify where a small version of the character can be found, and the last three a large version. Here is how delimiter codes are assigned in Plain TeX:

    \delcode'\(="028300   \delcode'\/="02F30E   \delcode'\)="029301
    \delcode'\[="05B302   \delcode'\|="26A30C   \delcode'\]="05D303
    \delcode'\<="26830A   \delcode'\\="26E30F   \delcode'\>="26930B

The first of these says, for instance, that the character corresponding to character code '\((the character is just '('), when used as a delimiter, may be found in a small version at location "28 in family 0, and in a large version at location "00 in family 3. If the location of either the small or the large version is given as "000, that version is ignored. That is why, for example, \left. and \right. produce invisible—or "null"—delimiters. Every other input character is left with −1 as its delimiter code (this value is preset, even before the Plain TeX format file swings into action).

It is natural to invoke various sizes of parentheses through the single input characters (and), but many delimiters do not correspond to such natural keyboard input; e.g., \lmoustache and \rmoustache. There is a \delimiter command for such cases, to be used in a manner analogous to \mathchar. \delimiter requires the specification of seven hexadecimal digits: the first one specifies the class, the remaining six perform the functions of the six digits in a \delcode declaration. For example, \delimiter"5267309 invokes a character that is assigned to class 5 (the leading digit); the small version of the character is found in position "67 of family 2 (the next three digits), the large version

in position "09 of family 3 (the last three digits). The Plain TeX file contains
several definitions that make use of this:

```
\def\langle{\delimiter"426830A }
\def\rangle{\delimiter"526930B }
\def\lmoustache{\delimiter"437A340 }
\def\rmoustache{\delimiter"537B341 }
```

and so on. In fact every command listed under delimiters in the Glossary of the
present book is defined in this way.

One of two approaches is followed when a delimiter is typeset. If the
delimiter is not preceded by \left or \right in the input, then its size will not
grow. Only the small version of the delimiter is used in this case, and the last
three digits in the delimiter specification are ignored. The class specification is
used to determine the spacing around the output character.

If the command is preceded by \left or \right, the spacings implied
by those commands take over (for example, '\left)' is spaced as an opening
delimiter, not a closing one). The delimiter grows in this case, by using the
specifications of "small" and "large" sizes as follows: a search is conducted
starting with an appropriate size in the family of the small version. "Appropriate size" means that the search begins with the \scriptscriptfont if the
current style is scriptscript, with the \scriptfont if the current style is script,
or with the \textfont otherwise. The search is then continued, if necessary,
in the family of the large version. Fonts in either family may contain several
sizes of the delimiter or may contain pieces out of which arbitrarily large sizes
can be built (though this is more common in the large-version family). Information about the locations of larger sizes within a font, or of extensible pieces,
is contained in its .tfm file. The search goes on until the sought-after size is
found. The calculations that allow TeX to figure out what size it is looking for
are explained in the Glossary under \delimiterfactor.

- *Accents and Radicals*

Accents are chosen rather as characters are in math formulas, but using a
command called \mathaccent instead of \mathchar. Here are some sample
Plain TeX definitions:

```
\def\acute{\mathaccent"7013 }
\def\vec{\mathaccent"717E }
```

When a font is to be used as a source of characters in math formulas, a special
value is usually assigned to a parameter called \skewchar. The value is used in
placing accents—TeX looks for positioning information in the .tfm file entry at
the location given by \skewchar. Here is a typical statement assigning a value
to \skewchar, taken from the Plain TeX file:

```
\skewchar\teni="7F
```

This chooses position "7F in the .tfm file for the font \teni as the source of accent-positioning information for characters from that font. Plain TeX makes the same choice for the fonts \seveni and \fivei as well, and it chooses "30 for the symbols fonts \tensy, \sevensy and \fivesy.

Radicals are chosen as delimiters are, but this time using a command called \radical. This is from the Plain TeX file:

```
\def\sqrt{\radical"270370 }
```

The interpretation of the number "270370 is exactly the same as if it had occurred in a \delimiter command. In this case, it assigns class 0 to the radical (since there are only six digits here), and it asserts that the small and large versions are found in position "70 of families 2 and 3, respectively.

• *A Summary of the Codes Attached to Input Characters*

Each character is identified by its character code. It has an associated category code, assigned by \catcode, which specifies the role that it plays in TeX; it has a \mathcode that tells TeX how to treat it in a math formula; and it has a \delcode that tells TeX what to do if it is used as a delimiter.

• *Adding a New Typeface: An Example*

To tie together the discussion of the last few pages, here is an example of how to add bold lowercase Greek letters to Plain TeX's arsenal. It is assumed that one has access to the bold math italic font at 10, 7 and 5 points. Only the 10-point size is, in fact, included with the standard Plain TeX distribution, but the smaller sizes come with *AMS*-TeX.

```
% First load the needed fonts and fix values for '\skewchar':
\font\tenbi=cmmib10      \skewchar\tenbi="7F
\font\sevenbi=cmmib7     \skewchar\sevenbi="7F
\font\fivebi=cmmib5      \skewchar\fivebi="7F
% Next, start a new family:
\newfam\bmifam \textfont\bmifam=\tenbi
 \scriptfont\bmifam=\sevenbi \scriptscriptfont\bmifam=\fivebi
% Then, define a family-switching command:
\def\Bold{\fam\bmifam}
% Finally, redefine '\alpha'...'\omega' so that they are now
% in class 7 (only two sample redefinitions are shown):
\mathchardef\alpha="710B      \mathchardef\omega="7121
```

To know what numbers to use in the \mathchardef declarations, characters have to be looked up in font tables. The table for cmmi10 is shown later in this chapter; the layout of the bold math italic table is the same. To see the effects of these additions to Plain TeX, consider these examples:

$b\alpha = 2\sqrt{\omega^2}$ gives $b\alpha = 2\sqrt{\omega^2}$

$\Bold b\alpha = 2\sqrt{\omega^2}$ gives $b\alpha = 2\sqrt{\omega^2}$

$b\alpha = \Bold 2\sqrt{\omega^2}$ gives $b\alpha = 2\sqrt{\omega^2}$

and

$b{\Bold\alpha} = {\bf 2}\sqrt{\omega^2}$ gives $b\alpha = 2\sqrt{\omega^2}$

Thus, `\Bold` is used in a straightforward way. Observe that letters and digits are also affected by it since they are already placed in class 7 by Plain TeX. The effect of `\Bold` on the digits is to switch their source from family 0 (roman) to the family assigned to bold math italics: this results in "old style" digits (see the math italic font table in § 5.15). Using `\bf` for the digits, as in the last equation, maintains their normal style. Accents are affected by `\Bold` as well, since they are also placed in class 7. This can lead to problems because the new family does not have characters corresponding to most of the needed accents. For example, `$\Bold \hat\alpha$` gives α, since `\hat` is defined as `\mathaccent"705E`, and when families are switched it is the character $\smile$ that now occurs in position "5E. To get the correct accent, one must type `$\hat{\Bold\alpha}$`; this gives $\hat\alpha$. If a bold accent is needed, `${\bf\hat{\Bold\alpha}}$` will give $\hat\alpha$.

One way of systematically avoiding such problems with accents is to follow the lead of $\mathcal{A}_{\mathcal{M}}\mathcal{S}$-TeX in its definition of a similar command (called `\bold`), and define `\Bold` to be a command that acts on an argument:

`\def\Bold #1{{\fam\bmifam#1}}`

This allows one to type `$\Bold\alpha\hat\omega$` and get $\alpha\hat\omega$, without any misinterpretation of `\hat`. If this path is followed, then one has to use statements like

`$\Bold{\alpha\omega}$` or `$\Bold\alpha\Bold\omega$`

to get $\alpha\omega$.

It may seem like a lot of work to set up this command, but the effort has to be made just once—all the definitions and `\font` commands can be placed in a separate file, which can be used as input when needed—and the benefits last for years.

Another instructive example, and one that ties together the preceding discussions of both text and mathematics, is the `EightPt` file reproduced in § 6.2. That example shows how a new set of typefaces can be selected from the ground up, to give a new look (in this case, a smaller size) to text as well as mathematics.

- *Adding a New Typeface: An $\mathcal{A}_{\mathcal{M}}\mathcal{S}$-TeX Example*

$\mathcal{A}_{\mathcal{M}}\mathcal{S}$-TeX comes with the commands `\loadmsam` and `\loadmsbm`, which allow the loading of two extra fonts of symbols when needed. Character tables for both fonts are displayed later in this chapter. Here is a rundown of how these commands are defined; the actual input has been cleaned up a little by removing the @ signs from command names.

The first order of business in this part of the $\mathcal{A}_{\mathcal{M}}\mathcal{S}$-TEX file is the definition of a command that converts its argument into a hexadecimal number. The command is used later to insert the family number into things like `\mathchardef` declarations:

```
\def\hexnumber#1{\ifcase#1 0\or 1\or 2\or 3\or 4\or 5\or 6\or
    7\or 8\or 9\or A\or B\or C\or D\or E\or F\fi}
```

The main definition of `\loadmsam` begins next. The definition will be broken here by lines of commentary, but in the original file it stretches out in one long sequence of lines:

```
\def\loadmsam{%
    \font\tenmsa=msam10
    \font\sevenmsa=msam7
    \font\fivemsa=msam5
    \newfam\msafam
    \textfont\msafam=\tenmsa
    \scriptfont\msafam=\sevenmsa
    \scriptscriptfont\msafam=\fivemsa
```

All of the preceding lines are straightforward: fonts are loaded, and a new family declared. (The original file didn't contain `\newfam` but instead its breakdown in terms of more basic commands.) One can now use `\msafam` as a symbolic representation of the family number assigned to the msam fonts. The next step defines a command that converts `\msafam` into a hexadecimal number:

```
\edef\next{\hexnumber\msafam}%
```

Remember that this definition is given within the definition of `\loadmsam`. (`\edef` is a variation of the command `\def`; it expands its replacement text right away.) What follows is a string of definitions of special names for special characters. In each of these, `\next` is used to inject the hexadecimal representation of the family number into the right place. Here are two sample lines:

```
\edef\ulcorner{\delimiter"4\next70\next70 }%
\edef\urcorner{\delimiter"5\next71\next71 }%
```

Finally, the definition ends with redefining its own name, so that the command will have no effect if called a second time. This is an important precaution; without it subsequent uses of `\loadmsam` will create unnecessary new families. TEX can only accommodate 16 families, so this is something to watch out for.

```
\global\let\loadmsam\empty}%
```

This is the end of the definition of `\loadmsam`.

The definition of `\loadmsbm` is similar, but shorter (since it doesn't define special names for any of its symbols). Here is the full definition:

```
\def\loadmsbm{%
   \font\tenmsb=msbm10 \font\sevenmsb=msbm7
     \font\fivemsb=msbm5
   \newfam\msbfam
   \textfont\msbfam=\tenmsb
   \scriptfont\msbfam=\sevenmsb
   \scriptscriptfont\msbfam=\fivemsb
   \global\let\loadmsbm\empty}
```

Once these fonts are loaded, it still remains for the symbols they introduce to be named (even the `\loadmsam` command named only a small number of symbols, compared to the 127 actually included in the `msam` font). There is a separate file, `amssym.tex`, that names them all. It can be used after the fonts are loaded by typing

```
\input amssym.tex
```

In fact, there is a single standard $\mathcal{AMS}$-TeX command that does all the necessary work:

```
\def\UseAMSsymbols{\loadmsam\loadmsbm \input amssym.tex }
```

Users who wish to name only those symbols that they need may use `\newsymbol` (as discussed in the Glossary). They must, of course, first load the necessary fonts by using `\loadmsam` or `\loadmsbm`.

5.6 Font Parameters

Some attributes of fonts—the intercharacter spacing within a word in text, for example—are hardwired in the sense that changing them involves editing `.tfm` files. (Such files contain information not only about individual character sizes, but also about spacing.) There are, however, several other attributes to which users of TeX do have direct access. These attributes are called the `\fontdimen` parameters, and they are numbered 1, 2, 3, ... Values for these parameters, too, are given in the `.tfm` files, but users may call them up and examine them— and even ask TeX to use different values if they are not happy with the spacings that result from the ones in the `.tfm` files.

All `.tfm` files must contain the values of at least the first 7 parameters. Fonts assigned to family 2 must supply values for 15 additional parameters, and those assigned to family 3 the values for 6 additional ones. The examples of individual fonts at the end of this chapter list the actual values of the `\fontdimen` parameters for each. The rest of this section discusses what each value represents. First, here are the seven parameters common to all fonts:

- `\fontdimen1`: The slant per horizontal point of characters. The value of this is zero in upright fonts (like roman) and nonzero in slanted and italic faces. The parameter is used to position accents above characters.

- \fontdimen2: The interword space. In a math font like cmmi10, this parameter has value zero, which is why input spaces are ignored in formulas.
- \fontdimen3: The interword stretch. This specifies how much stretchability the space between words has.
- \fontdimen4: The interword shrink. This specifies how much the space between words can shrink.
- \fontdimen5: The x-height. This is a traditional font-dependent printer's unit, normally defined to equal the height of the lowercase 'x' in the font. In TEX, a more-or-less arbitrary value is assigned to this parameter; it fixes the size of the font-dependent unit called **ex**. Even though the unit doesn't have a direct interpretation (especially in fonts that do not contain letters), it is still very useful. The vertical position of an accent above a character, for instance, is determined internally by TEX through a comparison of the height of the character to the size of **ex**. The unit can also be used to set spacings that automatically adjust themselves if the font is changed. For example, the value of \baselineskip can be chosen to be a multiple of **ex**, thus guaranteeing automatic adjustment if the size of the font is changed.
- \fontdimen6: The quad width. This is another traditional printer's unit, normally defined to equal the width of the letter 'M'. In TEX, this value defines the size of the unit called **em**, which in turn defines the size of a \quad (1em) and a \qquad (2em). The discussion under \fontdimen5 broadly applies here as well, except that the **em** is useful in arranging horizontal rather than vertical spacings.
- \fontdimen7: The extra space. This determines how much space is to be added to the interword space after certain punctuation marks (like a period) under certain circumstances. See \spacefactor in the Glossary.

Though the values of these parameters are given in the .tfm files, it is possible for users to specify different ones. Given a font named (say) \FontName, the expression \fontdimen1\FontName gives access to the parameter \fontdimen1 of that font. The expression can be used to print the value of the parameter (by preceding it with \the) or to change its effective value, as shown below. Changing the value must be done carefully: changes to \fontdimen parameters are global, even if they are made within a group (roughly, within braces), so shifting back to the old values is difficult. Here is an example which, though frivolous, illustrates the procedure:

```
\dimen0=\fontdimen2\tenrm   % Store the old value.
\dimen1=\fontdimen3\tenrm   % Store the old value.
\fontdimen2\tenrm=-2pt      % The new interword space.
\fontdimen3\tenrm=0pt       % The new interword stretch.
Testing, testing, just testing: a b c.
\fontdimen2\tenrm=\dimen0   % Restore the old value.
\fontdimen3\tenrm=\dimen1   % Restore the old value.
```

This input will produce

        Testing,testing,justtesting:abc.

Other, more serious adjustments to spacing can be made in a similar way if TEX's normal spacing is considered unsuitable.

- ### *Mathematics Font Parameters*

Extra \fontdimen parameters are supplied by the fonts in families 2 and 3; they are listed next. The parameters are used, often in conjunction with each other, in setting spacings in formulas. The details of how the spacings are chosen will not be given here, since they are not directly relevant to typeface choices. Appendix G of *The TEXbook* (D. E. KNUTH [1984, 1986]) gives all the details.

Extra Parameters Supplied by Family 2:

- \fontdimen8: Num 1. The upward shift of the numerator of a fraction in display styles.
- \fontdimen9: Num 2. The upward shift of the numerator of a fraction in styles other than display, when the fraction bar has nonzero thickness.
- \fontdimen10: Num 3. The upward shift of the numerator of a fraction in styles other than display, when the fraction bar has zero thickness.
- \fontdimen11: Denom 1. The downward shift of the denominator of a fraction in display styles.
- \fontdimen12: Denom 2. The downward shift of the denominator of a fraction in styles other than display.
- \fontdimen13: Sup 1. Used in positioning superscripts in display style.
- \fontdimen14: Sup 2. Used in positioning superscripts in any of the cramped styles (i.e., used in places like denominators, where the superscripts are not to be raised as high as normal).
- \fontdimen15: Sup 3. Used in positioning superscripts in styles other than the ones listed in the previous two cases.
- \fontdimen16: Sub 1. Used in positioning subscripts when no superscripts are present.
- \fontdimen17: Sub 2. Used in positioning subscripts when superscripts are present.
- \fontdimen18: Sup drop. Used in a preliminary calculation of superscript positioning.
- \fontdimen19: Sub drop. Used in a preliminary calculation of subscript positioning.
- \fontdimen20: Delim 1. Determines the minimum sizes of delimiters used around certain fraction-like constructions (like the ones given by \overwithdelims) in display styles.

- \fontdimen21: Delim 2. Determines the minimum sizes of delimiters used around certain fraction-like constructions (like the ones given by \overwithdelims) in styles other than display.
- \fontdimen22: Axis height. Determines the height of the *axis*: the invisible horizontal line with respect to which formulas are centered.

Knowing what these parameters control has more than just academic interest. For example, as the descriptions of \fontdimen16 and \fontdimen17 indicate, the positioning of subscripts depends on whether superscripts are present: consider, for instance, $A_1^2 A_1$. It is, however, often necessary to ensure that all subscripts are set at the same height regardless of whether superscripts are present. The typesetting of chemical formulas is one situation where this is needed. Setting parameters 16 and 17 of the text font of family 2 to have the same value will ensure this. Here is how it can be done in Plain TeX:

```
\fontdimen16\textfont2=2.8pt
\fontdimen17\textfont2=2.8pt
```

Once these new values are assigned, all subscripts will be placed at the same vertical position: $A_1^2 A_1$.

Extra Parameters Supplied by Family 3:

- \fontdimen8: Default rule thickness. The thickness of fraction bars, etc.
- \fontdimen9: Big op spacing 1. Determines the minimum space between the top of an operator and an upper limit.
- \fontdimen10: Big op spacing 2. Used in determining the actual space to be left between the top of an operator and an upper limit.
- \fontdimen11: Big op spacing 3. Determines the minimum space between the bottom of an operator and a lower limit.
- \fontdimen12: Big op spacing 4. Used in determining the actual space to be left between the bottom of an operator and a lower limit.
- \fontdimen13: Big op spacing 5. Determines the extra space to be left above an upper limit and below a lower limit. TeX will then think of the entire construction as having this augmented size. (The point is also discussed in the Glossary, under \buffer.)

5.7 Font Selection Schemes

Readers who have persisted up to this point—and even more likely, those who have not—will agree that font selections and changes in TeX involve a certain degree of dirty work. Some packages do offer a few conveniences. For instance, LaTeX has a command called `\boldmath` that switches most characters in formulas to boldface. It also has convenient size selection commands called `\small`, `\large`, `\huge`, etc., and a wider range of typeface styles than Plain TeX. But, styles and sizes do not always work together harmoniously. For example, `\it\huge` has only the effect of `\huge` (though `\huge\it` does give a simultaneous switch both of style and size). Thus, even LaTeX is still far from offering its users anything like the effortless ability to switch typefaces at will.

Because of this, relatively few documents done with TeX have anything other than the standard look. There are, however, two recent developments that suggest that changes might be coming. The first is that new sets of typefaces, especially complete sets of faces for mathematics, have now become available from several sources. Samples of some of these are shown in Examples 7–9 in Chapter 2. But, though these new typefaces usually come with files that contain all the necessary background commands, they still leave open the question of selecting between them in a consistent and standard way, and even of easily making certain types of internal changes (for example, a switch to boldface italics, or a simultaneous switch of style and size).

The second development is the emergence of schemes aimed at allowing precisely such typeface selections in a systematic way. One of these, the New Font Selection Scheme (NFSS), has been widely circulated within the TeX community, and it is now included with distributions of AMS-LaTeX. The new version of LaTeX (version 3.0) is to also include it.

The NFSS is a new way of organizing and selecting fonts for use with TeX. It was proposed by Frank Mittelbach and Rainer Schöpf (F. MITTELBACH and R. SCHÖPF [1989, 1990]). The scheme is still evolving, so only the basic ideas will be given here.

The main purpose of the NFSS is to move the work of loading new fonts, arranging spacings, setting up new math families, etc., to the background by providing a set of high-level commands that effectively automate much of the process. The scheme is based on the observation that, in general, fonts can be conveniently classified by the following characteristics: their sizes (10-point, 8-point, and so on), their shapes (normal, slanted, italic, and so on), their weights—called *series* by the proposers (medium, bold, bold extended, and so on) and their style—called *family* by the proposers (roman, sans serif, typewriter, and so on). Observe that the word "family" is being used here in a different sense from the technical meaning that it has when it is used to discuss the typesetting of formulas in TeX.

What is new here is not this classification, but that commands have been set up so that changes can be made—in principle—to any of these characteristics while leaving the others untouched. Thus, if one is using the standard 10-point roman font (family: roman; series: medium; shape: normal; size: normal), one can switch from it to a heavier weight (say, series: bold) and a different shape (say, shape: italic) by issuing two commands: one for the series and one for the shape. The other two characteristics are left unaffected by this switch. In some cases, the appropriate fonts may not exist. There are no bold faces in the typewriter fonts, for example, so an attempt to switch simultaneously to the typewriter family and to the bold series will result in a warning. It will also result in an automatic substitution of an available font for the unavailable one.

Switching sizes is also easy, since line-spacing changes are forced to go along automatically with font size changes. One either uses a basic \size command that explicitly insists on a specification of both a new font size and a new \baselineskip, or one uses LaTeX font size commands that have the changes to the line spacing built in.

Changing typefaces in mathematics is, as always, somewhat more complicated. The one type of global change that LaTeX allows is still easy: the switch from the normal style to the bold style (that uses bold letters and symbols). The change is made by typing \mathversion{bold}, and the switch back by typing \mathversion{normal}. Other global options can be added in the form of style files that may be used as input.

Typeface changes may be made to parts of formulas as well. Old commands like \cal and \mit can continue to be used to control typefaces, and things beyond that are also possible. The proposed scheme offers a well-defined procedure for adding new alphabets for use in mathematical formulas; once that procedure is followed, the new typeface can be used.

The 1992 version of the NFSS also includes support for font scaling (thus allowing a specification of arbitrary sizes) and for PostScript fonts.

Another font selection scheme has recently been developed by Michael Spivak to accompany the next version of LAMS-TeX. This scheme is based on two new commands \FontSet (which is used to load new fonts in a uniform way) and \FontMem which allows references to be made to fonts loaded by \FontSet. For example,

```
\FontSet{rm}
  cmr10 \\ cmr7 \\ cmr5 \endFontSet
```

will load the fonts cmr10, cmr7 and cmr5 and it will allow the entire set to be referred to by the name rm. For instance, \FontMem{rm}1 may be used to refer to the first member of this set of fonts, cmr10, \FontMem{2} to the second member, and so on. A name like \rm can be defined (for pure text) through

```
\def\rm{\FontMem{rm}1}
```

If a different roman font is wanted, not the Computer Modern one, all that has to be done (for text) is to replace the fonts `cmr10`, etc., in the `\FontSet` declaration by the new ones. As a concrete example, here are parts of the new L$\mathcal{A}$$_{\mathcal{M}}$$\mathcal{S}$-T$_{\mkern-1.5mu E}$X definition of `\loadmsbm`:

```
\def\loadmsbm{\FontSet{msbm}
  msbm10 \\ msbm7 \\ msbm5 \endFontSet
  ...
  \textfont\msbfam=\FontMem{msbm}1%
  \scriptfont\msbfam=\FontMem{msbm}2%
  \scriptscriptfont\msbfam=\FontMem{msbm}3%
  ... }
```

5.8 PostScript

PostScript is a widely used low-level page description language that instructs printing (and display) devices on the placement of both text and graphics. "Low-level" is used here in the sense of being closer to the actual workings of the device than the level at which most users operate. In fact, few users actually see or employ PostScript directly. What they do instead is to work with some high-level package that draws pictures or places text—the desktop publishing packages are examples. The package internally translates high-level instructions into PostScript instructions, which are then fed to the printer (or, if the system can handle Display PostScript, to the display device).

Samples of PostScript code are shown in Example 10 in Chapter 2, and the end of the Bibliography gives the name of the standard manual.

The PostScript approach to fonts has similarities to T$_{\mkern-1.5mu E}$X's, but it differs from it in some essential ways. Here, too, external files are sought that contain information about the fonts being used. And here, too, there are metrics files with size information (their names have extension `.afm`—for *Adobe Font Metrics*—named after the company that markets PostScript), and files with character shape information (the names have extensions `.pfa` or `.pfb`). But the metrics files do not usually carry the sort of detailed information that T$_{\mkern-1.5mu E}$X expects, especially in fonts for mathematics, and the shape files describe the outline shape of each character, rather than supplying a bitmap (i.e., a precise description of the pattern of dots that makes up the digitized character).

Files that describe characters by their outlines can be used to produce characters at essentially any size by scaling the outlines up or down. If the scaling is done in too simple a way, it doesn't always give satisfactory results. For example, PostScript fonts at small sizes can appear cramped and ill-formed. The current version of PostScript (level II) uses a somewhat more sophisticated scaling procedure than earlier versions do, but work still remains to be done.

Despite the differences in approach, TEX and PostScript can be made to work together. Programs exist to convert `.afm` metrics files to the `.tfm` ones that TEX needs, and further programs exist that convert `.dvi` files to `.ps` files (i.e., PostScript files). The broad procedures used are these:

1. Create `.tfm` files for all the necessary fonts from the `.afm` files that have come with the PostScript package (using an `.afm`-to-`.tfm` conversion program). Of course, this needs to be done only once for every font and not every time a document is processed. Packages that supply PostScript fonts specially for use with TEX usually come automatically with the `.tfm` files, so this step can often be omitted.

2. Next, prepare a TEX document exactly as usual, but load the new fonts and use them to replace the ones that TEX would otherwise automatically use. Again, this typically needs to be done just once. All the new font commands can be stored in a separate file and used again and again. For example, here are some of the definitions from the file `Times.tex`, used as input in Example 8 of Chapter 2:

   ```
   \font\tentmr=ptmr       % 'ptmr' is the roman font file.
   \def\rm{\fam0 \tentmr}
   \font\tentmb=ptmb       % 'ptmb' is the boldface file.
   \def\bf{\fam\bffam\tentmb}
   \font\tentmri=ptmri     % 'ptmri' is the italic file.
   \def\it{\fam\itfam\tentmri}
   \rm
   ```

   These definitions go along with assignments of new fonts to the appropriate mathematics families.

3. Now TEX the document as usual. Then run a `.dvi` to PostScript conversion program on the `.dvi` file to convert it to a `.ps` file. Print the PostScript file according to the instructions that come with the PostScript package being used.

5.9 Virtual (or Composite) Fonts

Since there is now an increasing interest in getting TEX to talk to fonts from other sources (like PostScript), Donald Knuth has developed an idea of David Fuchs into a proposal for a uniform approach to the entire question (see KNUTH [1990a]). His proposal involves the creation of a new font file that acts as a sort of go-between for TEX and the actual font files being used. This font (i.e., font file), is called a *virtual font*, or sometimes a *composite font*. The second name refers to the ability of such a file to call upon more than one real font file, getting some characters from one file and others from other files, possibly even composing individual characters out of pieces from different files.

Such virtual font files (the standard extension is .vf) may be created in many ways. One way is to first make a list of properties that the new font is to have, following some precise rules laid down by Knuth. Such a list, called a *virtual properties list* (and stored in a file with extension .vpl), must, of course, describe properties that real fonts available on the system do indeed have. The .vpl file is then converted into .vf and .tfm files by a new program called VPtoVF, now supplied with standard TeX distributions. The reverse procedure is also possible through a program called VFtoVP (the procedure allows, among other things, the properties of a virtual font to be read). A .tfm file created by VPtoVF is read by TeX just like any other .tfm file. When the document is to be printed, and character shape information is sought, the .vf file will direct the printing program to the appropriate places where it will find the necessary characters; the file may even contain explicit drawing instructions of its own.

For example, the .vpl file for a mathematics symbols font in TeX will list all 22 \fontdimen parameters. These may be changed if global changes are desired to the spacings that they give.

The idea of a virtual font is rather slowly gaining acceptance; the Math-Time fonts of Example 8 in Chapter 2, for example, support this approach. (Some of the characters in the math italic font there are simply selected from text italic fonts.) Since the approach permits extraordinarily flexible access to fonts of all kinds in a systematic and uniform way, it is to be hoped that more font vendors will start supporting it.

5.10 Comments on the Font Examples

The next several sections in this chapter display font samples, font data and font tables. They provide a fairly complete description of what is available by way of standard TeX fonts (i.e., fonts made through METAFONT) at the moment of writing (August 1992). Readers must keep in mind, however, that new fonts regularly become available, so there may well be additional types (or additional sizes of standard types) available in the future. In particular, in the late 1980s the TeX world began to move towards rather different layouts for fonts than the ones in standard use. This work is still not fully complete; § 5.28 summarizes the present situation.

Most TeX archives (addresses are given on the inside back cover) carry a much vaster range of fonts—for text and for mathematics—than the standard ones displayed in this book. Readers are urged to investigate these resources.

5.11 Example: Computer Modern Roman (cmr10)

This is the default typeface of most TEX packages. It is assigned the name \tenrm in Plain TEX, and it may usually be summoned by the command \rm.

- The other available design sizes (as opposed to magnified ones, such as those given by \magstep1, \magstep2, etc.) are

5*, 6, 7*, 8, 9, 12 and 17 points (standard Plain TEX distribution).

The sizes that are marked by an asterisk (*) are automatically available in Plain TEX; access to the others may be gained by using the \font command. Several other Computer Modern faces may be thought of as variations of this one: their character tables contain essentially the same characters—allowing for differences in style—as the table on the facing page. Here is a list:

Italic (cmti): 7, 8, 9, 10* and 12 points.
Bold extended (cmbx): 5*, 6, 7*, 8, 9, 10* and 12 points.
Slanted (cmsl): 8, 9, 10* and 12 points.

There are a few minor differences between the character tables for cmr10 and some of the faces listed above: cmr5 does not contain ligatures (e.g., "affluent flying fish" becomes "affluent flying fish"—the small characters are all distinct), and \$ and & become £ and &, respectively, in the italic faces.

There are still other faces that may be thought of as variations of cmr10, or as associated with it in some way. None of these is automatically loaded by Plain TEX. The most common one is cmcsc10, THE 10-POINT CAPITALS AND SMALL CAPITALS FONT. $\mathcal{AMS}$-TEX also provides 8- and 9-point sizes, and some of its styles automatically load this face and make it available through the command \smc. Other variations include cmb10 (**unextended bold**), cmbxsl10 (***bold extended slanted***) and cmbxti10 (***bold extended italic***), as well as experimental faces like cmu10 (*unslanted italic*), cmdunh10 (dunhill), cmff10 (funny) and cmfi10 (*funny italic*). Tables, like the ones on the right-hand pages here, may be printed out for each (see the last page of this chapter).

- The \fontdimen parameter values for cmr10 are

1	slant per point	0.0pt
2	interword space	3.33333pt
3	interword space stretch	1.66666pt
4	interword space shrink	1.11111pt
5	x-height (value of ex)	4.30554pt
6	quad width (value of em)	10.00002pt
7	extra space (at a sentence-end)	1.11111pt

The numbers to the left of each character on the facing page are its code numbers in decimal, *octal*, and hexadecimal form. (See Sections 2, 4 and 5 for more information.)

cmr10

	0	1	2	3	4	5	6	7
0 (´0x, "0x)	0 Γ	1 Δ	2 Θ	3 Λ	4 Ξ	5 Π	6 Σ	7 Υ
1 (´1x, "x)	8 Φ	9 Ψ	10 Ω	11 ff	12 fi	13 fl	14 ffi	15 ffl
2 (´2x, "1x)	16 ı	17 J	18 `	19 ´	20 ˇ	21 ˘	22 –	23 °
3 (´3x, "1x)	24 ¸	25 ß	26 æ	27 œ	28 ø	29 Æ	30 Œ	31 Ø
4 (´4x, "2x)	32 –	33 !	34 "	35 #	36 $	37 %	38 &	39 '
5 (´5x, "2x)	40 (	41)	42 *	43 +	44 ,	45 -	46 .	47 /
6 (´6x, "3x)	48 0	49 1	50 2	51 3	52 4	53 5	54 6	55 7
7 (´7x, "3x)	56 8	57 9	58 :	59 ;	60 ¡	61 =	62 ¿	63 ?
8 (´10x, "4x)	64 @	65 A	66 B	67 C	68 D	69 E	70 F	71 G
9 (´11x, "4x)	72 H	73 I	74 J	75 K	76 L	77 M	78 N	79 O
10 (´12x, "5x)	80 P	81 Q	82 R	83 S	84 T	85 U	86 V	87 W
11 (´13x, "5x)	88 X	89 Y	90 Z	91 [	92 "	93]	94 ^	95 .
12 (´14x, "6x)	96 `	97 a	98 b	99 c	100 d	101 e	102 f	103 g
13 (´15x, "6x)	104 h	105 i	106 j	107 k	108 l	109 m	110 n	111 o
14 (´16x, "7x)	112 p	113 q	114 r	115 s	116 t	117 u	118 v	119 w
15 (´17x, "7x)	120 x	121 y	122 z	123 –	124 —	125 "	126 ~	127 ˝

5.12 Example: Computer Modern Sans Serif (cmss10)

This is the basic Computer Modern sans serif text typeface.

- The other available design sizes (as opposed to magnified ones, such as those given by \magstep1, \magstep2, etc.) are

    8, 9, 12 and 17 points (standard Plain TEX distribution).

    None of these, 10-point or other, is automatically loaded by Plain TEX; access to any of them has to be gained by using the \font command. Computer Modern faces that may be thought of as variations of this one are

    Italic (cmssi): 8, 9, 10, 12 and 17 points.
    Bold extended (cmssbx): 10 points.
    Demibold condensed (cmssdc): 10 points.
    Quotation (cmssq): 8 points.
    Quotation italic (cmssqi): 8 points.

- The \fontdimen parameter values for cmss10 are

1	slant per point	0.0pt
2	interword space	3.33333pt
3	interword space stretch	1.66666pt
4	interword space shrink	1.11111pt
5	x-height (value of ex)	4.44444pt
6	quad width (value of em)	10.00002pt
7	extra space (at a sentence-end)	1.11111pt

The numbers to the left of each character on the facing page are its code numbers in decimal, *octal*, and hexadecimal form. (See Sections 2, 4 and 5 for more information.)

cmss10

0 ′0 "0	Γ	1 ′1 "1	Δ	2 ′2 "2	Θ	3 ′3 "3	Λ	4 ′4 "4	Ξ	5 ′5 "5	Π	6 ′6 "6	Σ	7 ′7 "7	Υ
8 ′10 "8	Φ	9 ′11 "9	Ψ	10 ′12 "A	Ω	11 ′13 "B	ff	12 ′14 "C	fi	13 ′15 "D	fl	14 ′16 "E	ffi	15 ′17 "F	ffl
16 ′20 "10	ı	17 ′21 "11	ȷ	18 ′22 "12	`	19 ′23 "13	´	20 ′24 "14	ˇ	21 ′25 "15	˘	22 ′26 "16	¯	23 ′27 "17	°
24 ′30 "18	¸	25 ′31 "19	ß	26 ′32 "1A	æ	27 ′33 "1B	œ	28 ′34 "1C	ø	29 ′35 "1D	Æ	30 ′36 "1E	Œ	31 ′37 "1F	Ø
32 ′40 "20	‐	33 ′41 "21	!	34 ′42 "22	”	35 ′43 "23	#	36 ′44 "24	$	37 ′45 "25	%	38 ′46 "26	&	39 ′47 "27	’
40 ′50 "28	(	41 ′51 "29	)	42 ′52 "2A	*	43 ′53 "2B	+	44 ′54 "2C	,	45 ′55 "2D	-	46 ′56 "2E	.	47 ′57 "2F	/
48 ′60 "30	0	49 ′61 "31	1	50 ′62 "32	2	51 ′63 "33	3	52 ′64 "34	4	53 ′65 "35	5	54 ′66 "36	6	55 ′67 "37	7
56 ′70 "38	8	57 ′71 "39	9	58 ′72 "3A	:	59 ′73 "3B	;	60 ′74 "3C	¡	61 ′75 "3D	=	62 ′76 "3E	¿	63 ′77 "3F	?
64 ′100 "40	@	65 ′101 "41	A	66 ′102 "42	B	67 ′103 "43	C	68 ′104 "44	D	69 ′105 "45	E	70 ′106 "46	F	71 ′107 "47	G
72 ′110 "48	H	73 ′111 "49	I	74 ′112 "4A	J	75 ′113 "4B	K	76 ′114 "4C	L	77 ′115 "4D	M	78 ′116 "4E	N	79 ′117 "4F	O
80 ′120 "50	P	81 ′121 "51	Q	82 ′122 "52	R	83 ′123 "53	S	84 ′124 "54	T	85 ′125 "55	U	86 ′126 "56	V	87 ′127 "57	W
88 ′130 "58	X	89 ′131 "59	Y	90 ′132 "5A	Z	91 ′133 "5B	[	92 ′134 "5C	“	93 ′135 "5D	]	94 ′136 "5E	^	95 ′137 "5F	˙
96 ′140 "60	‘	97 ′141 "61	a	98 ′142 "62	b	99 ′143 "63	c	100 ′144 "64	d	101 ′145 "65	e	102 ′146 "66	f	103 ′147 "67	g
104 ′150 "68	h	105 ′151 "69	i	106 ′152 "6A	j	107 ′153 "6B	k	108 ′154 "6C	l	109 ′155 "6D	m	110 ′156 "6E	n	111 ′157 "6F	o
112 ′160 "70	p	113 ′161 "71	q	114 ′162 "72	r	115 ′163 "73	s	116 ′164 "74	t	117 ′165 "75	u	118 ′166 "76	v	119 ′167 "77	w
120 ′170 "78	x	121 ′171 "79	y	122 ′172 "7A	z	123 ′173 "7B	–	124 ′174 "7C	—	125 ′175 "7D	”	126 ′176 "7E	~	127 ′177 "7F	¨

5.13 Example: Computer Modern Fibonacci (`cmfib8`)

This is an experimental text typeface. The Computer Modern meta-font parameters that shape its characters are based on the Fibonacci sequence in the hope—in Donald Knuth's words—"that these numbers will yield an especially pleasing alphabet."

No other sizes of this face have as yet been designed, nor are there any variations. Page numbers in the present book are printed in the Fibonacci face.

- The \fontdimen parameter values for `cmfib8` are

1	slant per point	0.0pt
2	interword space	3.49997pt
3	interword space stretch	1.74998pt
4	interword space shrink	1.16666pt
5	x-height (value of **ex**)	4.0pt
6	quad width (value of **em**)	10.49991pt
7	extra space (at a sentence-end)	1.16666pt

The numbers to the left of each character on the facing page are its code numbers in decimal, *octal*, and `hexadecimal` form. (See Sections 2, 4 and 5 for more information.)

cmfib8

0 '0 "0	Γ	1 '1 "1	Δ	2 '2 "2	Θ	3 '3 "3	Λ	4 '4 "4	Ξ	5 '5 "5	Π	6 '6 "6	Σ	7 '7 "7	Υ
8 '10 "8	Φ	9 '11 "9	Ψ	10 '12 "A	Ω	11 '13 "B	ff	12 '14 "C	fi	13 '15 "D	fl	14 '16 "E	ffi	15 '17 "F	ffl
16 '20 "10	ı	17 '21 "11	J	18 '22 "12	`	19 '23 "13	´	20 '24 "14	ˇ	21 '25 "15	˘	22 '26 "16	¯	23 '27 "17	˚
24 '30 "18	¸	25 '31 "19	ß	26 '32 "1A	æ	27 '33 "1B	œ	28 '34 "1C	ø	29 '35 "1D	Æ	30 '36 "1E	Œ	31 '37 "1F	Ø
32 '40 "20	´	33 '41 "21	!	34 '42 "22	”	35 '43 "23	#	36 '44 "24	$	37 '45 "25	%	38 '46 "26	&	39 '47 "27	’
40 '50 "28	(	41 '51 "29	)	42 '52 "2A	*	43 '53 "2B	+	44 '54 "2C	,	45 '55 "2D	-	46 '56 "2E	.	47 '57 "2F	/
48 '60 "30	0	49 '61 "31	1	50 '62 "32	2	51 '63 "33	3	52 '64 "34	4	53 '65 "35	5	54 '66 "36	6	55 '67 "37	7
56 '70 "38	8	57 '71 "39	9	58 '72 "3A	:	59 '73 "3B	;	60 '74 "3C	¡	61 '75 "3D	=	62 '76 "3E	¿	63 '77 "3F	?
64 '100 "40	@	65 '101 "41	A	66 '102 "42	B	67 '103 "43	C	68 '104 "44	D	69 '105 "45	E	70 '106 "46	F	71 '107 "47	G
72 '110 "48	H	73 '111 "49	I	74 '112 "4A	J	75 '113 "4B	K	76 '114 "4C	L	77 '115 "4D	M	78 '116 "4E	N	79 '117 "4F	O
80 '120 "50	P	81 '121 "51	Q	82 '122 "52	R	83 '123 "53	S	84 '124 "54	T	85 '125 "55	U	86 '126 "56	V	87 '127 "57	W
88 '130 "58	X	89 '131 "59	Y	90 '132 "5A	Z	91 '133 "5B	[	92 '134 "5C	“	93 '135 "5D	]	94 '136 "5E	^	95 '137 "5F	.
96 '140 "60	‘	97 '141 "61	a	98 '142 "62	b	99 '143 "63	c	100 '144 "64	d	101 '145 "65	e	102 '146 "66	f	103 '147 "67	g
104 '150 "68	h	105 '151 "69	i	106 '152 "6A	j	107 '153 "6B	k	108 '154 "6C	l	109 '155 "6D	m	110 '156 "6E	n	111 '157 "6F	o
112 '160 "70	p	113 '161 "71	q	114 '162 "72	r	115 '163 "73	s	116 '164 "74	t	117 '165 "75	u	118 '166 "76	v	119 '167 "77	w
120 '170 "78	x	121 '171 "79	y	122 '172 "7A	z	123 '173 "7B	–	124 '174 "7C	—	125 '175 "7D	”	126 '176 "7E	~	127 '177 "7F	¨

5.14 Example: Computer Modern Typewriter (cmtt10)

This is the basic fixed-width Computer Modern typeface. It is assigned the name \tentt in Plain TEX, and it may usually be summoned by the command \tt.

Each character here has a width of 0.5em. The fixed width makes the face suitable for tasks like reproducing computer files verbatim.

A comparison of the font tables for cmtt10 and cmr10 shows that there are many differences between the two: for example, the ff, ..., ffl ligatures are missing here, but the \ symbol is now available (position 92).

- The other available design sizes (as opposed to magnified ones, such as those given by \magstep1, \magstep2, etc.) are

  8 and 9 points (standard Plain TEX distribution).

Variations here include the *italic* (cmitt10) and *slanted* (cmsltt10) faces. There is also a **variable-width** typewriter face (cmvtt10), but its character table layout is like that of cmr10, not like the one on the facing page. Each of these faces is available only at 10-point size.

- The \fontdimen parameter values for cmtt10 are

1	slant per point	0.0pt
2	interword space	5.24995pt
3	interword space stretch	0.0pt
4	interword space shrink	0.0pt
5	x-height (value of **ex**)	4.30554pt
6	quad width (value of **em**)	10.4999pt
7	extra space (at a sentence-end)	5.24995pt

The numbers to the left of each character on the facing page are its code numbers in decimal, *octal*, and hexadecimal form. (See Sections 2, 4 and 5 for more information.)

cmtt10

0	1	2	3	4	5	6	7
0 Γ '0 "0	1 Δ '1 "1	2 Θ '2 "2	3 Λ '3 "3	4 Ξ '4 "4	5 Π '5 "5	6 Σ '6 "6	7 Υ '7 "7
8 Φ '10 "8	9 Ψ '11 "9	10 Ω '12 "A	11 ↑ '13 "B	12 ↓ '14 "C	13 ' '15 "D	14 ı '16 "E	15 ¿ '17 "F
16 ı '20 "10	17 J '21 "11	18 ` '22 "12	19 ´ '23 "13	20 ˇ '24 "14	21 ˘ '25 "15	22 ¯ '26 "16	23 ˚ '27 "17
24 �¸ '30 "18	25 ß '31 "19	26 æ '32 "1A	27 œ '33 "1B	28 ø '34 "1C	29 Æ '35 "1D	30 Œ '36 "1E	31 Ø '37 "1F
32 ␣ '40 "20	33 ! '41 "21	34 " '42 "22	35 # '43 "23	36 $ '44 "24	37 % '45 "25	38 & '46 "26	39 ' '47 "27
40 ('50 "28	41) '51 "29	42 * '52 "2A	43 + '53 "2B	44 , '54 "2C	45 - '55 "2D	46 . '56 "2E	47 / '57 "2F
48 0 '60 "30	49 1 '61 "31	50 2 '62 "32	51 3 '63 "33	52 4 '64 "34	53 5 '65 "35	54 6 '66 "36	55 7 '67 "37
56 8 '70 "38	57 9 '71 "39	58 : '72 "3A	59 ; '73 "3B	60 < '74 "3C	61 = '75 "3D	62 > '76 "3E	63 ? '77 "3F
64 @ '100 "40	65 A '101 "41	66 B '102 "42	67 C '103 "43	68 D '104 "44	69 E '105 "45	70 F '106 "46	71 G '107 "47
72 H '110 "48	73 I '111 "49	74 J '112 "4A	75 K '113 "4B	76 L '114 "4C	77 M '115 "4D	78 N '116 "4E	79 O '117 "4F
80 P '120 "50	81 Q '121 "51	82 R '122 "52	83 S '123 "53	84 T '124 "54	85 U '125 "55	86 V '126 "56	87 W '127 "57
88 X '130 "58	89 Y '131 "59	90 Z '132 "5A	91 ['133 "5B	92 \ '134 "5C	93] '135 "5D	94 ^ '136 "5E	95 _ '137 "5F
96 ` '140 "60	97 a '141 "61	98 b '142 "62	99 c '143 "63	100 d '144 "64	101 e '145 "65	102 f '146 "66	103 g '147 "67
104 h '150 "68	105 i '151 "69	106 j '152 "6A	107 k '153 "6B	108 l '154 "6C	109 m '155 "6D	110 n '156 "6E	111 o '157 "6F
112 p '160 "70	113 q '161 "71	114 r '162 "72	115 s '163 "73	116 t '164 "74	117 u '165 "75	118 v '166 "76	119 w '167 "77
120 x '170 "78	121 y '171 "79	122 z '172 "7A	123 { '173 "7B	124	'174 "7C	125 } '175 "7D	126 ~ '176 "7E	127 ¨ '177 "7F

5.15 Example: Computer Modern Math Italic (cmmi10)

This is the Computer Modern math italic font. It is assigned the name \teni in
Plain TEX, and it can be explicitly invoked in a formula—if needed—by typing
\mit. In the conventions followed in Plain TEX, ordinary characters—*a*, *b*, *c*,
etc.—are automatically taken from this font when a formula is typeset.

- The other available design sizes (as opposed to magnified ones, such as
those given by \magstep1, \magstep2, etc.) are

    5*, 6, 7*, 8, 9 and 12 points (standard Plain TEX distribution).

The sizes that are marked by an asterisk (*) are automatically available in
Plain TEX. The standard TEX distribution offers only one variation of this face,
cmmib10, which gives **bold** math italic characters, bold Greek letters ($\alpha\beta\gamma$),
etc. The $\mathcal{AMS}$-TEX distribution provides the bold characters also at 5, 6, 7, 8
and 9 points.

- The \fontdimen parameter values for cmmi10 are

1	slant per point	0.25pt
2	interword space	0.0pt
3	interword space stretch	0.0pt
4	interword space shrink	0.0pt
5	x-height (value of ex)	4.30554pt
6	quad width (value of em)	10.00002pt
7	extra space (at a sentence-end)	0.0pt

The numbers to the left of each character on the facing page are its code numbers in decimal,
octal, and hexadecimal form. (See Sections 2, 4 and 5 for more information.)

cmmi10

0 '0 "0 Γ	1 '1 "1 Δ	2 '2 "2 Θ	3 '3 "3 Λ	4 '4 "4 Ξ	5 '5 "5 Π	6 '6 "6 Σ	7 '7 "7 Υ
8 '10 "8 Φ	9 '11 "9 Ψ	10 '12 "A Ω	11 '13 "B α	12 '14 "C β	13 '15 "D γ	14 '16 "E δ	15 '17 "F ϵ
16 '20 "10 ζ	17 '21 "11 η	18 '22 "12 θ	19 '23 "13 ι	20 '24 "14 κ	21 '25 "15 λ	22 '26 "16 μ	23 '27 "17 ν
24 '30 "18 ξ	25 '31 "19 π	26 '32 "1A ρ	27 '33 "1B σ	28 '34 "1C τ	29 '35 "1D υ	30 '36 "1E ϕ	31 '37 "1F χ
32 '40 "20 ψ	33 '41 "21 ω	34 '42 "22 ε	35 '43 "23 ϑ	36 '44 "24 ϖ	37 '45 "25 ϱ	38 '46 "26 ς	39 '47 "27 φ
40 '50 "28 $\leftharpoonup$	41 '51 "29 $\leftarrow$	42 '52 "2A $\rightarrow$	43 '53 "2B $\rightharpoonup$	44 '54 "2C $`$	45 '55 "2D $'$	46 '56 "2E $\triangleright$	47 '57 "2F $\triangleleft$
48 '60 "30 0	49 '61 "31 1	50 '62 "32 2	51 '63 "33 3	52 '64 "34 4	53 '65 "35 5	54 '66 "36 6	55 '67 "37 7
56 '70 "38 8	57 '71 "39 9	58 '72 "3A .	59 '73 "3B ,	60 '74 "3C $<$	61 '75 "3D $/$	62 '76 "3E $>$	63 '77 "3F $\star$
64 '100 "40 ∂	65 '101 "41 A	66 '102 "42 B	67 '103 "43 C	68 '104 "44 D	69 '105 "45 E	70 '106 "46 F	71 '107 "47 G
72 '110 "48 H	73 '111 "49 I	74 '112 "4A J	75 '113 "4B K	76 '114 "4C L	77 '115 "4D M	78 '116 "4E N	79 '117 "4F O
80 '120 "50 P	81 '121 "51 Q	82 '122 "52 R	83 '123 "53 S	84 '124 "54 T	85 '125 "55 U	86 '126 "56 V	87 '127 "57 W
88 '130 "58 X	89 '131 "59 Y	90 '132 "5A Z	91 '133 "5B $\flat$	92 '134 "5C $\natural$	93 '135 "5D $\sharp$	94 '136 "5E $\smile$	95 '137 "5F $\frown$
96 '140 "60 ℓ	97 '141 "61 a	98 '142 "62 b	99 '143 "63 c	100 '144 "64 d	101 '145 "65 e	102 '146 "66 f	103 '147 "67 g
104 '150 "68 h	105 '151 "69 i	106 '152 "6A j	107 '153 "6B k	108 '154 "6C l	109 '155 "6D m	110 '156 "6E n	111 '157 "6F o
112 '160 "70 p	113 '161 "71 q	114 '162 "72 r	115 '163 "73 s	116 '164 "74 t	117 '165 "75 u	118 '166 "76 v	119 '167 "77 w
120 '170 "78 x	121 '171 "79 y	122 '172 "7A z	123 '173 "7B $\imath$	124 '174 "7C $\jmath$	125 '175 "7D $\wp$	126 '176 "7E $\vec{\ }$	127 '177 "7F $\frown$

5.16 Example: Computer Modern Symbols (cmsy10)

This is the basic Computer Modern math symbols font. It is assigned the name \tensy in Plain TEX, and it can be explicitly invoked in a formula—if that is needed—by the command \cal. The .tfm file for it contains 22 \fontdimen parameters, not just the usual 7, and it is the Plain TEX choice for family 2. (The .tfm file of any replacement must also contain 22 parameters.) The extra parameters are used to fix things like the position of indices, etc.

- The other available design sizes (as opposed to magnified ones, such as those given by \magstep1, \magstep2, etc.) are

    5*, 6, 7*, 8 and 9 points (standard Plain TEX distribution).

The sizes that are marked by an asterisk (*) are automatically available in Plain TEX. As with cmmi10, the only variation here is the bold one ($\forall$, $\exists$, etc.). And again, the standard TEX distribution offers only the 10-point size, cmbsy10; the $\mathcal{AMS}$-TEX distribution offers the face at 5, 6, 7, 8 and 9 points.

- The \fontdimen parameter values for cmsy10 are

1	slant per point	0.25pt
2	space	0.0pt
3	interword space stretch	0.0pt
4	interword space shrink	0.0pt
5	x-height (value of ex)	4.30554pt
6	quad width (value of em)	10.00002pt
7	extra space	0.0pt
8	num 1	6.76508pt
9	num 2	3.93732pt
10	num 3	4.4373pt
11	denom 1	6.85951pt
12	denom 2	3.44841pt
13	sup 1	4.12892pt
14	sup 2	3.62892pt
15	sup 3	2.88889pt
16	sub 1	1.49998pt
17	sub 2	2.47217pt
18	sub drop	3.86108pt
19	sub drop	0.5pt
20	delim 1	23.9pt
21	delim 2	10.09999pt
22	axis height	2.5pt

The numbers to the left of each character on the facing page are its code numbers in decimal, *octal*, and **hexadecimal** form. (See Sections 2, 4 and 5 for more information.)

cmsy10

0 '0 "0 $-$	1 '1 "1 $\cdot$	2 '2 "2 $\times$	3 '3 "3 $*$	4 '4 "4 $\div$	5 '5 "5 $\diamond$	6 '6 "6 $\pm$	7 '7 "7 $\mp$
8 '10 "8 $\oplus$	9 '11 "9 $\ominus$	10 '12 "A $\otimes$	11 '13 "B $\oslash$	12 '14 "C $\odot$	13 '15 "D $\bigcirc$	14 '16 "E $\circ$	15 '17 "F $\bullet$
16 '20 "10 $\asymp$	17 '21 "11 $\equiv$	18 '22 "12 $\subseteq$	19 '23 "13 $\supseteq$	20 '24 "14 $\leq$	21 '25 "15 $\geq$	22 '26 "16 $\preceq$	23 '27 "17 $\succeq$
24 '30 "18 $\sim$	25 '31 "19 $\approx$	26 '32 "1A $\subset$	27 '33 "1B $\supset$	28 '34 "1C $\ll$	29 '35 "1D $\gg$	30 '36 "1E $\prec$	31 '37 "1F $\succ$
32 '40 "20 $\leftarrow$	33 '41 "21 $\rightarrow$	34 '42 "22 $\uparrow$	35 '43 "23 $\downarrow$	36 '44 "24 $\leftrightarrow$	37 '45 "25 $\nearrow$	38 '46 "26 $\searrow$	39 '47 "27 $\simeq$
40 '50 "28 $\Leftarrow$	41 '51 "29 $\Rightarrow$	42 '52 "2A $\Uparrow$	43 '53 "2B $\Downarrow$	44 '54 "2C $\Leftrightarrow$	45 '55 "2D $\nwarrow$	46 '56 "2E $\swarrow$	47 '57 "2F $\propto$
48 '60 "30 $\prime$	49 '61 "31 ∞	50 '62 "32 $\in$	51 '63 "33 $\ni$	52 '64 "34 $\triangle$	53 '65 "35 $\triangledown$	54 '66 "36 $/$	55 '67 "37 $\mid$
56 '70 "38 $\forall$	57 '71 "39 $\exists$	58 '72 "3A $\neg$	59 '73 "3B $\emptyset$	60 '74 "3C $\Re$	61 '75 "3D $\Im$	62 '76 "3E $\top$	63 '77 "3F $\bot$
64 '100 "40 $\aleph$	65 '101 "41 $\mathcal{A}$	66 '102 "42 $\mathcal{B}$	67 '103 "43 $\mathcal{C}$	68 '104 "44 $\mathcal{D}$	69 '105 "45 $\mathcal{E}$	70 '106 "46 $\mathcal{F}$	71 '107 "47 $\mathcal{G}$
72 '110 "48 $\mathcal{H}$	73 '111 "49 $\mathcal{I}$	74 '112 "4A $\mathcal{J}$	75 '113 "4B $\mathcal{K}$	76 '114 "4C $\mathcal{L}$	77 '115 "4D $\mathcal{M}$	78 '116 "4E $\mathcal{N}$	79 '117 "4F $\mathcal{O}$
80 '120 "50 $\mathcal{P}$	81 '121 "51 $\mathcal{Q}$	82 '122 "52 $\mathcal{R}$	83 '123 "53 $\mathcal{S}$	84 '124 "54 $\mathcal{T}$	85 '125 "55 $\mathcal{U}$	86 '126 "56 $\mathcal{V}$	87 '127 "57 $\mathcal{W}$
88 '130 "58 $\mathcal{X}$	89 '131 "59 $\mathcal{Y}$	90 '132 "5A $\mathcal{Z}$	91 '133 "5B $\cup$	92 '134 "5C $\cap$	93 '135 "5D $\uplus$	94 '136 "5E $\wedge$	95 '137 "5F $\vee$
96 '140 "60 $\vdash$	97 '141 "61 $\dashv$	98 '142 "62 $\lfloor$	99 '143 "63 $\rfloor$	100 '144 "64 $\lceil$	101 '145 "65 $\rceil$	102 '146 "66 $\{$	103 '147 "67 $\}$
104 '150 "68 $\langle$	105 '151 "69 $\rangle$	106 '152 "6A $\mid$	107 '153 "6B $\|$	108 '154 "6C $\updownarrow$	109 '155 "6D $\Updownarrow$	110 '156 "6E $\backslash$	111 '157 "6F $\wr$
112 '160 "70 $\surd$	113 '161 "71 $\amalg$	114 '162 "72 ∇	115 '163 "73 $\int$	116 '164 "74 $\sqcup$	117 '165 "75 $\sqcap$	118 '166 "76 $\sqsubseteq$	119 '167 "77 $\sqsupseteq$
120 '170 "78 $\S$	121 '171 "79 $\dagger$	122 '172 "7A $\ddagger$	123 '173 "7B $\P$	124 '174 "7C $\clubsuit$	125 '175 "7D $\diamondsuit$	126 '176 "7E $\heartsuit$	127 '177 "7F $\spadesuit$

5.17 Example: Computer Modern Extension (cmex10)

This is the Computer Modern math extension font. It is assigned the name \tenex in Plain TEX. The font contains large operators, as well as the pieces that make up things like large brackets, large radical signs, etc. It is the Plain TEX choice for family 3. Any substitution of this font by another has to be done with care (as with cmsy10): the font has 13 \fontdimen parameters instead of the normal 7. The extra parameters determine the default rule thickness used in fractions, in underlining, etc., and the spacing of large operators.

• The other available design sizes (as opposed to magnified ones, such as those given by \magstep1, \magstep2, etc.) are

7, 8 and 9 points (*AMS*-TEX distribution).

(Note: The 9-point size also comes with some Plain TEX distributions.)

• The \fontdimen parameter values for cmex10 are

1	slant per point	0.0pt
2	interword space	0.0pt
3	interword space stretch	0.0pt
4	interword space shrink	0.0pt
5	x-height (value of ex)	4.30554pt
6	quad width (value of em)	10.00002pt
7	extra space (at a sentence-end)	0.0pt
8	default rule thickness	0.39998pt
9	big op spacing 1	1.11111pt
10	big op spacing 2	1.66666pt
11	big op spacing 3	1.99998pt
12	big op spacing 4	6.0pt
13	big op spacing 5	1.0pt

The numbers to the left of each character on the facing page are its code numbers in decimal, *octal*, and hexadecimal form. (See Sections 2, 4 and 5 for more information.)

cmex10

0 '0 "0	1 '1 "1	2 '2 "2	3 '3 "3	4 '4 "4	5 '5 "5	6 '6 "6	7 '7 "7
8 '10 "8	9 '11 "9	10 '12 "A	11 '13 "B	12 '14 "C	13 '15 "D	14 '16 "E	15 '17 "F
16 '20 "10	17 '21 "11	18 '22 "12	19 '23 "13	20 '24 "14	21 '25 "15	22 '26 "16	23 '27 "17
24 '30 "18	25 '31 "19	26 '32 "1A	27 '33 "1B	28 '34 "1C	29 '35 "1D	30 '36 "1E	31 '37 "1F
32 '40 "20	33 '41 "21	34 '42 "22	35 '43 "23	36 '44 "24	37 '45 "25	38 '46 "26	39 '47 "27
40 '50 "28	41 '51 "29	42 '52 "2A	43 '53 "2B	44 '54 "2C	45 '55 "2D	46 '56 "2E	47 '57 "2F
48 '60 "30	49 '61 "31	50 '62 "32	51 '63 "33	52 '64 "34	53 '65 "35	54 '66 "36	55 '67 "37
56 '70 "38	57 '71 "39	58 '72 "3A	59 '73 "3B	60 '74 "3C	61 '75 "3D	62 '76 "3E	63 '77 "3F
64 '100 "40	65 '101 "41	66 '102 "42	67 '103 "43	68 '104 "44	69 '105 "45	70 '106 "46	71 '107 "47
72 '110 "48	73 '111 "49	74 '112 "4A	75 '113 "4B	76 '114 "4C	77 '115 "4D	78 '116 "4E	79 '117 "4F
80 '120 "50	81 '121 "51	82 '122 "52	83 '123 "53	84 '124 "54	85 '125 "55	86 '126 "56	87 '127 "57
88 '130 "58	89 '131 "59	90 '132 "5A	91 '133 "5B	92 '134 "5C	93 '135 "5D	94 '136 "5E	95 '137 "5F
96 '140 "60	97 '141 "61	98 '142 "62	99 '143 "03	100 '144 "64	101 '145 "65	102 '146 "66	103 '147 "67
104 '150 "68	105 '151 "69	106 '152 "6A	107 '153 "6B	108 '154 "6C	109 '155 "6D	110 '156 "6E	111 '157 "6F
112 '160 "70	113 '161 "71	114 '162 "72	115 '163 "73	116 '164 "74	117 '165 "75	118 '166 "76	119 '167 "77
120 '170 "78	121 '171 "79	122 '172 "7A	123 '173 "7B	124 '174 "7C	125 '175 "7D	126 '176 "7E	127 '177 "7F

5.18 Example: AMS Math Symbols A, Medium (msam10)

This is the first of two sets of extra symbols that come with $\mathcal{AMS}$-TeX.

- The other available design sizes (as opposed to magnified ones, such as those given by \magstep1, \magstep2, etc.) are

    5, 6, 7, 8 and 9 points ($\mathcal{AMS}$-TeX distribution).

- The \fontdimen parameter values for msam10 are

1	slant per point	0.0pt
2	space	0.0pt
3	interword space stretch	0.0pt
4	interword space shrink	0.0pt
5	x-height (value of ex)	4.30554pt
6	quad width (value of em)	10.00002pt
7	extra space	0.0pt
8	num 1	6.76508pt
9	num 2	3.93732pt
10	num 3	4.4373pt
11	denom 1	6.85951pt
12	denom 2	3.44841pt
13	sup 1	4.12892pt
14	sup 2	3.62892pt
15	sup 3	2.88889pt
16	sub 1	1.49998pt
17	sub 2	2.47217pt
18	sub drop	3.86108pt
19	sub drop	0.5pt
20	delim 1	23.9pt
21	delim 2	10.09999pt
22	axis height	2.5pt

The numbers to the left of each character on the facing page are its code numbers in decimal, *octal*, and hexadecimal form. (See Sections 2, 4 and 5 for more information.)

msam10

0 '0 "0	⊡	1 '1 "1	⊞	2 '2 "2	⊠	3 '3 "3	□	4 '4 "4	■	5 '5 "5	·	6 '6 "6	◇	7 '7 "7	◆
8 '10 "8	↻	9 '11 "9	↺	10 '12 "A	⇌	11 '13 "B	⇋	12 '14 "C	⊟	13 '15 "D	⊩	14 '16 "E	⊪	15 '17 "F	⊨
16 '20 "10	→	17 '21 "11	←	18 '22 "12	⇐	19 '23 "13	⇒	20 '24 "14	⇑	21 '25 "15	⇓	22 '26 "16	↾	23 '27 "17	↓
24 '30 "18	↑	25 '31 "19	↓	26 '32 "1A	↣	27 '33 "1B	↢	28 '34 "1C	⇄	29 '35 "1D	⇌	30 '36 "1E	↰	31 '37 "1F	↱
32 '40 "20	⇝	33 '41 "21	↭	34 '42 "22	↫	35 '43 "23	↬	36 '44 "24	≏	37 '45 "25	⋧	38 '46 "26	⪆	39 '47 "27	⪈
40 '50 "28	⊸	41 '51 "29	∴	42 '52 "2A	∵	43 '53 "2B	÷	44 '54 "2C	≙	45 '55 "2D	⋦	46 '56 "2E	⪅	47 '57 "2F	⪇
48 '60 "30	⩽	49 '61 "31	⩾	50 '62 "32	⪯	51 '63 "33	⪰	52 '64 "34	⪯	53 '65 "35	≦	54 '66 "36	⩽	55 '67 "37	≶
56 '70 "38	∖	57 '71 "39	−	58 '72 "3A	≑	59 '73 "3B	≒	60 '74 "3C	≽	61 '75 "3D	≧	62 '76 "3E	⩾	63 '77 "3F	≷
64 '100 "40	⊏	65 '101 "41	⊐	66 '102 "42	▷	67 '103 "43	◁	68 '104 "44	⊳	69 '105 "45	⊲	70 '106 "46	★	71 '107 "47	◊
72 '110 "48	▼	73 '111 "49	▶	74 '112 "4A	◀	75 '113 "4B	➔	76 '114 "4C	⬸	77 '115 "4D	△	78 '116 "4E	▲	79 '117 "4F	▽
80 '120 "50	⊨	81 '121 "51	⋜	82 '122 "52	⋝	83 '123 "53	⋚	84 '124 "54	⋛	85 '125 "55	¥	86 '126 "56	⇒	87 '127 "57	⇐
88 '130 "58	✓	89 '131 "59	⋁	90 '132 "5A	⊼	91 '133 "5B	⊼	92 '134 "5C	∠	93 '135 "5D	∡	94 '136 "5E	◁	95 '137 "5F	∝
96 '140 "60	⌣	97 '141 "61	⌢	98 '142 "62	∈	99 '143 "63	∋	100 '144 "64	⊌	101 '145 "65	⋒	102 '146 "66	⋏	103 '147 "67	⋎
104 '150 "68	⋋	105 '151 "69	⋌	106 '152 "6A	⊑	107 '153 "6B	⊒	108 '154 "6C	≏	109 '155 "6D	≎	110 '156 "6E	⋘	111 '157 "6F	⋙
112 '160 "70	⌐	113 '161 "71	¬	114 '162 "72	®	115 '163 "73	Ⓢ	116 '164 "74	⋔	117 '165 "75	⊹	118 '166 "76	⌣	119 '167 "77	⋍
120 '170 "78	⌞	121 '171 "79	⌟	122 '172 "7A	✠	123 '173 "7B	∁	124 '174 "7C	⊤	125 '175 "7D	⊙	126 '176 "7E	⊛	127 '177 "7F	⊖

5.19 Example: AMS Math Symbols B, Medium (msbm10)

This is the second of two sets of extra symbols that come with $\mathcal{AMS}$-TeX.

• The other available design sizes (as opposed to magnified ones, such as those given by \magstep1, \magstep2, etc.) are

5, 6, 7, 8 and 9 points ($\mathcal{AMS}$-TeX distribution).

• The \fontdimen parameter values for msbm10 are

1	slant per point	0.25pt
2	space	0.0pt
3	interword space stretch	0.0pt
4	interword space shrink	0.0pt
5	x-height (value of ex)	4.62964pt
6	quad width (value of em)	10.0pt
7	extra space	0.0pt
8	num 1	6.76508pt
9	num 2	3.93732pt
10	num 3	4.4373pt
11	denom 1	6.85951pt
12	denom 2	3.44841pt
13	sup 1	4.12892pt
14	sup 2	3.62892pt
15	sup 3	2.88889pt
16	sub 1	1.49998pt
17	sub 2	2.47217pt
18	sub drop	3.86108pt
19	sub drop	0.5pt
20	delim 1	23.9pt
21	delim 2	10.09999pt
22	axis height	2.5pt

The numbers to the left of each character on the facing page are its code numbers in decimal, *octal*, and `hexadecimal` form. (See Sections 2, 4 and 5 for more information.)

msbm10

0 '0 "0	1 '1 "1	2 '2 "2	3 '3 "3	4 '4 "4	5 '5 "5	6 '6 "6	7 '7 "7
8 '10 "8	9 '11 "9	10 '12 "A	11 '13 "B	12 '14 "C	13 '15 "D	14 '16 "E	15 '17 "F
16 '20 "10	17 '21 "11	18 '22 "12	19 '23 "13	20 '24 "14	21 '25 "15	22 '26 "16	23 '27 "17
24 '30 "18	25 '31 "19	26 '32 "1A	27 '33 "1B	28 '34 "1C	29 '35 "1D	30 '36 "1E	31 '37 "1F
32 '40 "20	33 '41 "21	34 '42 "22	35 '43 "23	36 '44 "24	37 '45 "25	38 '46 "26	39 '47 "27
40 '50 "28	41 '51 "29	42 '52 "2A	43 '53 "2B	44 '54 "2C	45 '55 "2D	46 '56 "2E	47 '57 "2F
48 '60 "30	49 '61 "31	50 '62 "32	51 '63 "33	52 '64 "34	53 '65 "35	54 '66 "36	55 '67 "37
56 '70 "38	57 '71 "39	58 '72 "3A	59 '73 "3B	60 '74 "3C	61 '75 "3D	62 '76 "3E	63 '77 "3F
64 '100 "40	65 '101 "41	66 '102 "42	67 '103 "43	68 '104 "44	69 '105 "45	70 '106 "46	71 '107 "47
72 '110 "48	73 '111 "49	74 '112 "4A	75 '113 "4B	76 '114 "4C	77 '115 "4D	78 '116 "4E	79 '117 "4F
80 '120 "50	81 '121 "51	82 '122 "52	83 '123 "53	84 '124 "54	85 '125 "55	86 '126 "56	87 '127 "57
88 '130 "58	89 '131 "59	90 '132 "5A	91 '133 "5B	92 '134 "5C	93 '135 "5D	94 '136 "5E	95 '137 "5F
96 '140 "60	97 '141 "61	98 '142 "62	99 '143 "63	100 '144 "64	101 '145 "65	102 '146 "66	103 '147 "67
104 '150 "68	105 '151 "69	106 '152 "6A	107 '153 "6B	108 '154 "6C	109 '155 "6D	110 '156 "6E	111 '157 "6F
112 '160 "70	113 '161 "71	114 '162 "72	115 '163 "73	116 '164 "74	117 '165 "75	118 '166 "76	119 '167 "77
120 '170 "78	121 '171 "79	122 '172 "7A	123 '173 "7B	124 '174 "7C	125 '175 "7D	126 '176 "7E	127 '177 "7F

5.20 Example: AMS Euler Roman Medium (eurm10)

This is one of six fonts that Hermann Zapf designed in the 1980s for the American Mathematical Society

Together, the six make up the AMS Euler typeface collection. The face here is the "medium-weight roman" (where the term "roman" is being used a little loosely); a companion **boldface** font is also available. The two are designed to have an appearance slightly like handwriting (and may be thought of as cursive faces). For example, the 0 has a point at the top, since handwritten zeros are seldom smooth there. The face is meant as a substitute for math italics; see Example 7 in Chapter 2.

- The other available design sizes (as opposed to magnified ones, such as those given by \magstep1, \magstep2, etc.) are

5, 6, 7, 8 and 9 points ($\mathcal{AMS}$-TEX distribution).

The bold variant also comes in these sizes.

- The \fontdimen parameter values for eurm10 are

1	slant per point	0.0pt
2	space	3.33333pt
3	interword space stretch	0.0pt
4	interword space shrink	0.0pt
5	x-height (value of ex)	4.59137pt
6	quad width (value of em)	10.0pt
7	extra space	0.0pt
8	num 1	3.78113pt
9	num 2	2.7008pt
10	num 3	2.97089pt
11	denom 1	3.78113pt
12	denom 2	1.62048pt
13	sup 1	4.05121pt
14	sup 2	3.78113pt
15	sup 3	3.24097pt
16	sub 1	1.89056pt
17	sub 2	2.43073pt
18	sub drop	4.05121pt
19	sub drop	0.27008pt
20	delim 1	21.98453pt
21	delim 2	9.99298pt
22	axis height	2.56577pt

The numbers to the left of each character on the facing page are its code numbers in decimal, *octal*, and hexadecimal form. (See Sections 2, 4 and 5 for more information.)

eurm10

0 '0 "0 Γ	1 '1 "1 Δ	2 '2 "2 Θ	3 '3 "3 Λ	4 '4 "4 Ξ	5 '5 "5 Π	6 '6 "6 Σ	7 '7 "7 Υ
8 '10 "8 Φ	9 '11 "9 Ψ	10 '12 "A Ω	11 '13 "B α	12 '14 "C β	13 '15 "D γ	14 '16 "E δ	15 '17 "F ε
16 '20 "10 ζ	17 '21 "11 η	18 '22 "12 θ	19 '23 "13 ι	20 '24 "14 κ	21 '25 "15 λ	22 '26 "16 μ	23 '27 "17 ν
24 '30 "18 ξ	25 '31 "19 π	26 '32 "1A ρ	27 '33 "1B σ	28 '34 "1C τ	29 '35 "1D υ	30 '36 "1E φ	31 '37 "1F χ
32 '40 "20 ψ	33 '41 "21 ω	34 '42 "22 ε	35 '43 "23 ϑ	36 '44 "24 ϖ	37 '45 "25	38 '46 "26	39 '47 "27 φ
40 '50 "28	41 '51 "29	42 '52 "2A	43 '53 "2B	44 '54 "2C	45 '55 "2D	46 '56 "2E	47 '57 "2F
48 '60 "30 0	49 '61 "31 1	50 '62 "32 2	51 '63 "33 3	52 '64 "34 4	53 '65 "35 5	54 '66 "36 6	55 '67 "37 7
56 '70 "38 8	57 '71 "39 9	58 '72 "3A .	59 '73 "3B ,	60 '74 "3C <	61 '75 "3D /	62 '76 "3E >	63 '77 "3F
64 '100 "40 ∂	65 '101 "41 A	66 '102 "42 B	67 '103 "43 C	68 '104 "44 D	69 '105 "45 E	70 '106 "46 F	71 '107 "47 G
72 '110 "48 H	73 '111 "49 I	74 '112 "4A J	75 '113 "4B K	76 '114 "4C L	77 '115 "4D M	78 '116 "4E N	79 '117 "4F O
80 '120 "50 P	81 '121 "51 Q	82 '122 "52 R	83 '123 "53 S	84 '124 "54 T	85 '125 "55 U	86 '126 "56 V	87 '127 "57 W
88 '130 "58 X	89 '131 "59 Y	90 '132 "5A Z	91 '133 "5B	92 '134 "5C	93 '135 "5D	94 '136 "5E	95 '137 "5F
96 '140 "60 ℓ	97 '141 "61 a	98 '142 "62 b	99 '143 "63 c	100 '144 "64 d	101 '145 "65 e	102 '146 "66 f	103 '147 "67 g
104 '150 "68 h	105 '151 "69 i	106 '152 "6A j	107 '153 "6B k	108 '154 "6C l	109 '155 "6D m	110 '156 "6E n	111 '157 "6F o
112 '160 "70 p	113 '161 "71 q	114 '162 "72 r	115 '163 "73 s	116 '164 "74 t	117 '165 "75 u	118 '166 "76 v	119 '167 "77 w
120 '170 "78 x	121 '171 "79 y	122 '172 "7A z	123 '173 "7B ι	124 '174 "7C ȷ	125 '175 "7D ℘	126 '176 "7E	127 '177 "7F

5.21 Example: AMS Euler Fraktur Medium (eufm10)

This is one of six fonts that Hermann Zapf designed in the 1980s for the American Mathematical Society.

Together, the six make up the AMS Euler typeface collection. The face shown here is of medium weight and is based on the traditional typeface called "fraktur" or, sometimes, "gothic." A **bold** version is also available (with a very slightly different layout).

• The other available design sizes (as opposed to magnified ones, such as those given by \magstep1, \magstep2, etc.) are

5, 6, 7, 8 and 9 points ($\mathcal{AMS}$-TeX distribution).

The bold variant also comes in these sizes.

• The \fontdimen parameter values for eufm10 are

1	slant per point	0.0pt
2	space	3.33333pt
3	interword space stretch	0.0pt
4	interword space shrink	0.0pt
5	x-height (value of ex)	4.75342pt
6	quad width (value of em)	10.0pt
7	extra space	0.0pt
8	num 1	3.78113pt
9	num 2	2.7008pt
10	num 3	2.97089pt
11	denom 1	3.78113pt
12	denom 2	1.62048pt
13	sup 1	4.05121pt
14	sup 2	3.78113pt
15	sup 3	3.24097pt
16	sub 1	1.89056pt
17	sub 2	2.43073pt
18	sub drop	4.05121pt
19	sub drop	0.27008pt
20	delim 1	21.98453pt
21	delim 2	9.99298pt
22	axis height	2.56577pt

The numbers to the left of each character on the facing page are its code numbers in decimal, *octal*, and **hexadecimal** form. (See Sections 2, 4 and 5 for more information.)

eufm10

0 '0 "0 ð	1 '1 "1 ð	2 '2 "2 f	3 '3 "3 f	4 '4 "4 g	5 '5 "5 t	6 '6 "6 t	7 '7 "7 u
8 '10 "8	9 '11 "9	10 '12 "A	11 '13 "B	12 '14 "C	13 '15 "D	14 '16 "E	15 '17 "F
16 '20 "10	17 '21 "11	18 '22 "12 '	19 '23 "13 '	20 '24 "14	21 '25 "15	22 '26 "16	23 '27 "17
24 '30 "18	25 '31 "19	26 '32 "1A	27 '33 "1B	28 '34 "1C	29 '35 "1D	30 '36 "1E	31 '37 "1F
32 '40 "20	33 '41 "21 !	34 '42 "22	35 '43 "23	36 '44 "24	37 '45 "25	38 '46 "26 &	39 '47 "27 '
40 '50 "28 (	41 '51 "29)	42 '52 "2A *	43 '53 "2B +	44 '54 "2C ,	45 '55 "2D −	46 '56 "2E .	47 '57 "2F /
48 '60 "30 0	49 '61 "31 1	50 '62 "32 2	51 '63 "33 3	52 '64 "34 4	53 '65 "35 5	54 '66 "36 6	55 '67 "37 7
56 '70 "38 8	57 '71 "39 9	58 '72 "3A :	59 '73 "3B ;	60 '74 "3C	61 '75 "3D =	62 '76 "3E	63 '77 "3F ?
64 '100 "40	65 '101 "41 𝔄	66 '102 "42 𝔅	67 '103 "43 ℭ	68 '104 "44 𝔇	69 '105 "45 𝔈	70 '106 "46 𝔉	71 '107 "47 𝔊
72 '110 "48 ℌ	73 '111 "49 ℑ	74 '112 "4A 𝔍	75 '113 "4B 𝔎	76 '114 "4C 𝔏	77 '115 "4D 𝔐	78 '116 "4E 𝔑	79 '117 "4F 𝔒
80 '120 "50 𝔓	81 '121 "51 𝔔	82 '122 "52 ℜ	83 '123 "53 𝔖	84 '124 "54 𝔗	85 '125 "55 𝔘	86 '126 "56 𝔙	87 '127 "57 𝔚
88 '130 "58 𝔛	89 '131 "59 𝔜	90 '132 "5A ℨ	91 '133 "5B [	92 '134 "5C	93 '135 "5D]	94 '136 "5E ^	95 '137 "5F
96 '140 "60	97 '141 "61 a	98 '142 "62 b	99 '143 "63 c	100 '144 "64 ð	101 '145 "65 e	102 '146 "66 f	103 '147 "67 g
104 '150 "68 h	105 '151 "69 i	106 '152 "6A j	107 '153 "6B k	108 '154 "6C l	109 '155 "6D m	110 '156 "6E n	111 '157 "6F o
112 '160 "70 p	113 '161 "71 q	114 '162 "72 r	115 '163 "73 s	116 '164 "74 t	117 '165 "75 u	118 '166 "76 v	119 '167 "77 w
120 '170 "78 x	121 '171 "79 y	122 '172 "7A z	123 '173 "7B	124 '174 "7C	125 '175 "7D "	126 '176 "7E	127 '177 "7F 1

5.22 Example: AMS Euler Script Medium (eusm10)

This is one of six fonts that Hermann Zapf designed in the 1980s for the American Mathematical Society. Together, the six make up the AMS Euler typeface collection. This face mainly provides $\mathcal{UPPERCASE}$ medium-weight script letters.

- The other available design sizes (as opposed to magnified ones, such as those given by \magstep1, \magstep2, etc.) are

   5, 6, 7, 8 and 9 points ($\mathcal{AMS}$-TEX distribution).

Here, too, a $\mathcal{BOLD}$ variant is available at the same sizes.

- The \fontdimen parameter values for eusm10 are

1	slant per point	0.0pt
2	space	3.33333pt
3	interword space stretch	0.0pt
4	interword space shrink	0.0pt
5	x-height (value of ex)	4.59137pt
6	quad width (value of em)	10.0pt
7	extra space	0.0pt
8	num 1	3.78113pt
9	num 2	2.7008pt
10	num 3	2.97089pt
11	denom 1	3.78113pt
12	denom 2	1.62048pt
13	sup 1	4.05121pt
14	sup 2	3.78113pt
15	sup 3	3.24097pt
16	sub 1	1.89056pt
17	sub 2	2.43073pt
18	sub drop	4.05121pt
19	sub drop	0.27008pt
20	delim 1	21.98453pt
21	delim 2	9.99298pt
22	axis height	2.56577pt

The numbers to the left of each character on the facing page are its code numbers in decimal, *octal*, and **hexadecimal** form. (See Sections 2, 4 and 5 for more information.)

eusm10

	0	1	2	3	4	5	6	7
0 '0x "0x	—							
8 '1x "8–F								
16 '2x "10–17								
24 '3x "18–1F	~							
32 '4x "20–27								
40 '5x "28–2F								
48 '6x "30–37								
56 '7x "38–3F			$\neg$		$\Re$	$\Im$		
64 '10x "40–47	$\aleph$	$\mathcal{A}$	$\mathcal{B}$	$\mathcal{C}$	$\mathcal{D}$	$\mathcal{E}$	$\mathcal{F}$	$\mathcal{G}$
72 '11x "48–4F	$\mathcal{H}$	$\mathcal{I}$	$\mathcal{J}$	$\mathcal{K}$	$\mathcal{L}$	$\mathcal{M}$	$\mathcal{N}$	$\mathcal{O}$
80 '12x "50–57	$\mathcal{P}$	$\mathcal{Q}$	$\mathcal{R}$	$\mathcal{S}$	$\mathcal{T}$	$\mathcal{U}$	$\mathcal{V}$	$\mathcal{W}$
88 '13x "58–5F	$\mathcal{X}$	$\mathcal{Y}$	$\mathcal{Z}$				$\wedge$	$\vee$
96 '14x "60–67							{	}
104 '15x "68–6F			|				\	
112 '16x "70–77								
120 '17x "78–7F	§							

Cell code reference (decimal / octal / hexadecimal):

dec	0	1	2	3	4	5	6	7
octal	'0	'1	'2	'3	'4	'5	'6	'7
hex	"0	"1	"2	"3	"4	"5	"6	"7

5.23 Example: AMS Euler Extension (euex10)

This is a partial set of math extension symbols designed by Donald Knuth and Hermann Zapf to accompany the Euler faces eurm and eurb. The symbols in this set are more suited to the generally upright appearance of the Euler faces than the corresponding ones from the standard extension font cmex10. They are also in tune with the Euler fonts in other ways. Compare, for example, the summation symbol here with the capital sigma, Σ, from eurm10. Example 7 in Chapter 2 shows more instances of how the two faces look together.

- The other available design sizes (as opposed to magnified ones, such as those given by \magstep1, \magstep2, etc.) are

  7, 8 and 9 points ($\mathcal{AMS}$-TEX distribution).

- The \fontdimen parameter values for euex10 are

1	slant per point	0.0pt
2	interword space	0.0pt
3	interword space stretch	0.0pt
4	interword space shrink	0.0pt
5	x-height (value of ex)	4.30554pt
6	quad width (value of em)	10.00002pt
7	extra space (at a sentence-end)	0.0pt
8	default rule thickness	0.39998pt
9	big op spacing 1	1.11111pt
10	big op spacing 2	1.66666pt
11	big op spacing 3	1.99998pt
12	big op spacing 4	6.0pt
13	big op spacing 5	1.0pt

The numbers to the left of each character on the facing page are its code numbers in decimal, *octal*, and hexadecimal form. (See Sections 2, 4 and 5 for more information.)

euex10

0	1	2	3	4	5	6	7
0 '0 "0	1 '1 "1	2 '2 "2	3 '3 "3	4 '4 "4	5 '5 "5	6 '6 "6	7 '7 "7
8 { '10 "8	9 } '11 "9	10 { '12 "A	11 } '13 "B	12 { '14 "C	13 } '15 "D	14 { '16 "E	15 } '17 "F
16 '20 "10	17 '21 "11	18 '22 "12	19 '23 "13	20 '24 "14	21 '25 "15	22 '26 "16	23 '27 "17
24 '30 "18	25 ↤ '31 "19	26 ↦ '32 "1A	27 ↦ '33 "1B	28 '34 "1C	29 '35 "1D	30 '36 "1E	31 '37 "1F
32 ← '40 "20	33 → '41 "21	34 ↑ '42 "22	35 ↓ '43 "23	36 ↔ '44 "24	37 ↗ '45 "25	38 ↘ '46 "26	39 '47 "27
40 ⇐ '50 "28	41 ⇒ '51 "29	42 ⇑ '52 "2A	43 ⇓ '53 "2B	44 ⇔ '54 "2C	45 ↖ '55 "2D	46 ↙ '56 "2E	47 '57 "2F
48 '60 "30	49 ∞ '61 "31	50 '62 "32	51 '63 "33	52 '64 "34	53 '65 "35	54 '66 "36	55 '67 "37
56 ⌈ '70 "38	57 ⌉ '71 "39	58 ⌊ '72 "3A	59 ⌋ '73 "3B	60 { '74 "3C	61 } '75 "3D	62 ' '76 "3E	63 '77 "3F
64 '100 "40	65 '101 "41	66 '102 "42	67 '103 "43	68 '104 "44	69 '105 "45	70 '106 "46	71 '107 "47
72 ∮ '110 "48	73 ∮ '111 "49	74 '112 "4A	75 '113 "4B	76 '114 "4C	77 '115 "4D	78 '116 "4E	79 '117 "4F
80 ∑ '120 "50	81 ∏ '121 "51	82 ∫ '122 "52	83 '123 "53	84 '124 "54	85 '125 "55	86 '126 "56	87 '127 "57
88 ∑ '130 "58	89 ∏ '131 "59	90 ∫ '132 "5A	91 '133 "5B	92 '134 "5C	93 '135 "5D	94 '136 "5E	95 '137 "5F
96 ⊔ '140 "60	97 ⊔ '141 "61	98 '142 "62	99 '143 "63	100 '144 "64	101 '145 "65	102 '146 "66	103 '147 "67
104 '150 "68	105 '151 "69	106 '152 "6A	107 '153 "6B	108 ↕ '154 "6C	109 ⇕ '155 "6D	110 '156 "6E	111 '157 "6F
112 '160 "70	113 '161 "71	114 '162 "72	115 '163 "73	116 '164 "74	117 '165 "75	118 '166 "76	119 '167 "77
120 '170 "78	121 '171 "79	122 ⌢ '172 "7A	123 ⌢ '173 "7B	124 ⌣ '174 "7C	125 ⌣ '175 "7D	126 '176 "7E	127 '177 "7F

5.24 Example: Computer Concrete Roman (ccr10)

"Computer Concrete" is a new typeface class, designed in the late 1980s by Donald Knuth. It was used in the book *Concrete Mathematics* (see KNUTH [1989a]), and the name *Concrete* derives both from this as well as from the generally solid appearance of the face.

An initial motivation behind Computer Concrete was the desire to find a text typeface that was a suitable match for Hermann Zapf's rather strong AMS Euler mathematics fonts (see Example 7 in Chapter 2). Like other modern (in the technical sense) faces, Computer Modern has a delicate appearance, and text set in it contrasts a little sharply with mathematical formulas set in AMS Euler. The new face may find another use as well: the fine lines in Computer Modern faces don't always emerge well from the relatively low resolutions of standard laser printers (300 to 600 dots per inch, compared to over 1000 in phototypesetting machines), or from repeated photocopying. It is possible that Computer Concrete will provide a sturdier face for day-to-day use.

- The other available design sizes (as opposed to magnified ones, such as those given by \magstep1, \magstep2, etc.) are

  5, 6, 7, 8 and 9 points.

The available variations of Computer Concrete roman are

  Slanted (ccsl): 10 points.
  Italic (ccti): 10 points.
  Capitals and small capitals (cccsc): 10 points.

There is also a whimsical *slanted condensed 9-point* font (ccsl9); it was used to typeset student graffiti in the margins of the book *Concrete Mathematics*. Missing from the list is a bold face specially designed for Computer Concrete. The normal Computer Modern **bold extended** typeface (the one that usually pops up in response to \bf) can be used with reasonably satisfactory results. A companion math italic font is displayed in the next example.

- The \fontdimen parameter values for ccr10 are

1	slant per point	0.0pt
2	interword space	3.33333pt
3	interword space stretch	1.66666pt
4	interword space shrink	1.11111pt
5	x-height (value of ex)	4.58333pt
6	quad width (value of em)	10.00002pt
7	extra space (at a sentence-end)	1.11111pt

The numbers to the left of each character on the facing page are its code numbers in decimal, *octal*, and hexadecimal form. (See Sections 2, 4 and 5 for more information.)

ccr10

0 '0 "0	1 '1 "1	2 '2 "2	3 '3 "3	4 '4 "4	5 '5 "5	6 '6 "6	7 '7 "7
Γ	Δ	Θ	Λ	Ξ	Π	Σ	Υ
8 '10 "8	**9** '11 "9	**10** '12 "A	**11** '13 "B	**12** '14 "C	**13** '15 "D	**14** '16 "E	**15** '17 "F
Φ	Ψ	Ω	ff	fi	fl	ffi	ffl
16 '20 "10	**17** '21 "11	**18** '22 "12	**19** '23 "13	**20** '24 "14	**21** '25 "15	**22** '26 "16	**23** '27 "17
ı	J	`	´	ˇ	˘	¯	˚
24 '30 "18	**25** '31 "19	**26** '32 "1A	**27** '33 "1B	**28** '34 "1C	**29** '35 "1D	**30** '36 "1E	**31** '37 "1F
¸	ß	æ	œ	ø	Æ	Œ	Ø
32 '40 "20	**33** '41 "21	**34** '42 "22	**35** '43 "23	**36** '44 "24	**37** '45 "25	**38** '46 "26	**39** '47 "27
–	!	"	#	$	%	&	'
40 '50 "28	**41** '51 "29	**42** '52 "2A	**43** '53 "2B	**44** '54 "2C	**45** '55 "2D	**46** '56 "2E	**47** '57 "2F
(	)	*	+	,	-	.	/
48 '60 "30	**49** '61 "31	**50** '62 "32	**51** '63 "33	**52** '64 "34	**53** '65 "35	**54** '66 "36	**55** '67 "37
0	1	2	3	4	5	6	7
56 '70 "38	**57** '71 "39	**58** '72 "3A	**59** '73 "3B	**60** '74 "3C	**61** '75 "3D	**62** '76 "3E	**63** '77 "3F
8	9	:	;	¡	=	¿	?
64 '100 "40	**65** '101 "41	**66** '102 "42	**67** '103 "43	**68** '104 "44	**69** '105 "45	**70** '106 "46	**71** '107 "47
@	A	B	C	D	E	F	G
72 '110 "48	**73** '111 "49	**74** '112 "4A	**75** '113 "4B	**76** '114 "4C	**77** '115 "4D	**78** '116 "4E	**79** '117 "4F
H	I	J	K	L	M	N	O
80 '120 "50	**81** '121 "51	**82** '122 "52	**83** '123 "53	**84** '124 "54	**85** '125 "55	**86** '126 "56	**87** '127 "57
P	Q	R	S	T	U	V	W
88 '130 "58	**89** '131 "59	**90** '132 "5A	**91** '133 "5B	**92** '134 "5C	**93** '135 "5D	**94** '136 "5E	**95** '137 "5F
X	Y	Z	[	"	]	ˆ	˙
96 '140 "60	**97** '141 "61	**98** '142 "62	**99** '143 "63	**100** '144 "64	**101** '145 "65	**102** '146 "66	**103** '147 "67
'	a	b	c	d	e	f	g
104 '150 "68	**105** '151 "69	**106** '152 "6A	**107** '153 "6B	**108** '154 "6C	**109** '155 "6D	**110** '156 "6E	**111** '157 "6F
h	i	j	k	l	m	n	o
112 '160 "70	**113** '161 "71	**114** '162 "72	**115** '163 "73	**116** '164 "74	**117** '165 "75	**118** '166 "76	**119** '167 "77
p	q	r	s	t	u	v	w
120 '170 "78	**121** '171 "79	**122** '172 "7A	**123** '173 "7B	**124** '174 "7C	**125** '175 "7D	**126** '176 "7E	**127** '177 "7F
x	y	z	–	—	"	~	¨

5.25 Example: Computer Concrete Math Italic (ccmi10)

This is the only mathematics font specifically designed to accompany text in Computer Concrete (see the previous example). Since Computer Concrete was itself designed to match the AMS Euler mathematics fonts, it wasn't considered necessary to construct further sets of Concrete mathematics fonts (e.g., symbol or extension faces). In fact, even the one face shown here comes only in 10-point size. It therefore cannot be used as a source for Greek letters and so forth (since it is not possible to make them shrink in indices), but only to provide a few miscellaneous symbols that are to be used in straight text. For example, the book *Concrete Mathematics* draws its equation numbers from the "old style" digits (character numbers 48–57) in this font; see Example 7 in Chapter 2 for an illustration.

- The \fontdimen parameter values for ccmi10 are

1	slant per point	0.25pt
2	interword space	0.0pt
3	interword space stretch	0.0pt
4	interword space shrink	0.0pt
5	x-height (value of **ex**)	4.58333pt
6	quad width (value of **em**)	10.00002pt
7	extra space (at a sentence-end)	0.0pt

The numbers to the left of each character on the facing page are its code numbers in decimal, *octal*, and **hexadecimal** form. (See Sections 2, 4 and 5 for more information.)

ccmi10

0 '0 "0	1 '1 "1	2 '2 "2	3 '3 "3	4 '4 "4	5 '5 "5	6 '6 "6	7 '7 "7
Γ	Δ	Θ	Λ	Ξ	Π	Σ	Υ
8 '10 "8	**9** '11 "9	**10** '12 "A	**11** '13 "B	**12** '14 "C	**13** '15 "D	**14** '16 "E	**15** '17 "F
Φ	Ψ	Ω	α	β	γ	δ	ϵ
16 '20 "10	**17** '21 "11	**18** '22 "12	**19** '23 "13	**20** '24 "14	**21** '25 "15	**22** '26 "16	**23** '27 "17
ζ	η	θ	ι	κ	λ	μ	ν
24 '30 "18	**25** '31 "19	**26** '32 "1A	**27** '33 "1B	**28** '34 "1C	**29** '35 "1D	**30** '36 "1E	**31** '37 "1F
ξ	π	ρ	σ	τ	υ	ϕ	χ
32 '40 "20	**33** '41 "21	**34** '42 "22	**35** '43 "23	**36** '44 "24	**37** '45 "25	**38** '46 "26	**39** '47 "27
ψ	ω	ε	ϑ	ϖ	ϱ	ς	φ
40 '50 "28	**41** '51 "29	**42** '52 "2A	**43** '53 "2B	**44** '54 "2C	**45** '55 "2D	**46** '56 "2E	**47** '57 "2F
$\leftharpoonup$	$\leftharpoondown$	$\rightharpoonup$	$\rightharpoondown$	$\`$	$\'$	$\triangleright$	$\triangleleft$
48 '60 "30	**49** '61 "31	**50** '62 "32	**51** '63 "33	**52** '64 "34	**53** '65 "35	**54** '66 "36	**55** '67 "37
0	1	2	3	4	5	6	7
56 '70 "38	**57** '71 "39	**58** '72 "3A	**59** '73 "3B	**60** '74 "3C	**61** '75 "3D	**62** '76 "3E	**63** '77 "3F
8	9	.	,	$<$	/	$>$	$\star$
64 '100 "40	**65** '101 "41	**66** '102 "42	**67** '103 "43	**68** '104 "44	**69** '105 "45	**70** '106 "46	**71** '107 "47
∂	A	B	C	D	E	F	G
72 '110 "48	**73** '111 "49	**74** '112 "4A	**75** '113 "4B	**76** '114 "4C	**77** '115 "4D	**78** '116 "4E	**79** '117 "4F
H	I	J	K	L	M	N	O
80 '120 "50	**81** '121 "51	**82** '122 "52	**83** '123 "53	**84** '124 "54	**85** '125 "55	**86** '126 "56	**87** '127 "57
P	Q	R	S	T	U	V	W
88 '130 "58	**89** '131 "59	**90** '132 "5A	**91** '133 "5B	**92** '134 "5C	**93** '135 "5D	**94** '136 "5E	**95** '137 "5F
X	Y	Z	$\flat$	$\natural$	$\sharp$	$\smile$	$\frown$
96 '140 "60	**97** '141 "61	**98** '142 "62	**99** '143 "63	**100** '144 "64	**101** '145 "65	**102** '146 "66	**103** '147 "67
ℓ	a	b	c	d	e	f	g
104 '150 "68	**105** '151 "69	**106** '152 "6A	**107** '153 "6B	**108** '154 "6C	**109** '155 "6D	**110** '156 "6E	**111** '157 "6F
h	i	j	k	l	m	n	o
112 '160 "70	**113** '161 "71	**114** '162 "72	**115** '163 "73	**116** '164 "74	**117** '165 "75	**118** '166 "76	**119** '167 "77
p	q	r	s	t	u	v	w
120 '170 "78	**121** '171 "79	**122** '172 "7A	**123** '173 "7B	**124** '174 "7C	**125** '175 "7D	**126** '176 "7E	**127** '177 "7F
x	y	z	$\imath$	$\jmath$	$\wp$	$\vec{}$	$\frown$

5.26 Example: LaTeX Symbols (`lasy10`)

LaTeX comes with a few extra fonts of its own. The font `lasy10` provides a small number of miscellaneous mathematical symbols. The font also comes in 5, 6, 7, 8 and 9 point sizes, as well as in one bold version, `lasyb10`.

The LaTeX font distribution also includes these: sans serif fonts, suitable for slides and (when magnified) for transparencies—`lcmss8`, `lcmssi8` and `lcmssb8`; and fonts whose characters are arcs of circles and lines of different slopes, out of which some basic pictures may be constructed—`line10`, `linew10`, `lcircle10` and `lcirclew10`, where the `w` stands for *wide* (the names may differ slightly from system to system: `lcircle10` and `lcirclew10` were formerly called `circle10` and `circlew10`). See Example 13 in Chapter 2. These fonts may be loaded and used like any other, even if one isn't using LaTeX.

- The \fontdimen parameter values for `lasy10` are

1	slant per point	0.25pt
2	space	0.0pt
3	interword space stretch	0.0pt
4	interword space shrink	0.0pt
5	x-height (value of ex)	4.30554pt
6	quad width (value of em)	10.00002pt
7	extra space	0.0pt
8	num 1	6.76508pt
9	num 2	3.93732pt
10	num 3	4.4373pt
11	denom 1	6.85951pt
12	denom 2	3.44841pt
13	sup 1	4.12892pt
14	sup 2	3.62892pt
15	sup 3	2.88889pt
16	sub 1	1.49998pt
17	sub 2	2.47217pt
18	sub drop	3.86108pt
19	sub drop	0.5pt
20	delim 1	23.9pt
21	delim 2	10.09999pt
22	axis height	2.5pt

The numbers to the left of each character on the facing page are its code numbers in decimal, *octal*, and `hexadecimal` form. (See Sections 2, 4 and 5 for more information.)

lasy10

0 ’0 "0		1 ’1 "1	◁	2 ’2 "2	◀	3 ’3 "3	▷	4 ’4 "4	▶	5 ’5 "5		6 ’6 "6		7 ’7 "7	
8 ’10 "8		9 ’11 "9		10 ’12 "A		11 ’13 "B		12 ’14 "C		13 ’15 "D		14 ’16 "E		15 ’17 "F	
16 ’20 "10		17 ’21 "11		18 ’22 "12		19 ’23 "13		20 ’24 "14		21 ’25 "15		22 ’26 "16		23 ’27 "17	
24 ’30 "18		25 ’31 "19		26 ’32 "1A		27 ’33 "1B		28 ’34 "1C		29 ’35 "1D		30 ’36 "1E		31 ’37 "1F	
32 ’40 "20		33 ’41 "21		34 ’42 "22		35 ’43 "23		36 ’44 "24		37 ’45 "25		38 ’46 "26		39 ’47 "27	
40 ’50 "28	⟨	41 ’51 "29	⟩	42 ’52 "2A	ˆ	43 ’53 "2B	ˇ	44 ’54 "2C		45 ’55 "2D		46 ’56 "2E		47 ’57 "2F	
48 ’60 "30	℧	49 ’61 "31	⋈	50 ’62 "32	□	51 ’63 "33	◇	52 ’64 "34		53 ’65 "35		54 ’66 "36		55 ’67 "37	
56 ’70 "38		57 ’71 "39		58 ’72 "3A	~	59 ’73 "3B	↝	60 ’74 "3C	⊏	61 ’75 "3D	⊐	62 ’76 "3E		63 ’77 "3F	
64 ’100 "40		65 ’101 "41		66 ’102 "42		67 ’103 "43		68 ’104 "44		69 ’105 "45		70 ’106 "46		71 ’107 "47	
72 ’110 "48		73 ’111 "49		74 ’112 "4A		75 ’113 "4B		76 ’114 "4C		77 ’115 "4D		78 ’116 "4E		79 ’117 "4F	
80 ’120 "50		81 ’121 "51		82 ’122 "52		83 ’123 "53		84 ’124 "54		85 ’125 "55		86 ’126 "56		87 ’127 "57	
88 ’130 "58		89 ’131 "59		90 ’132 "5A		91 ’133 "5B		92 ’134 "5C		93 ’135 "5D		94 ’136 "5E		95 ’137 "5F	
96 ’140 "60		97 ’141 "61		98 ’142 "62		99 ’143 "63		100 ’144 "64		101 ’145 "65		102 ’146 "66		103 ’147 "67	
104 ’150 "68		105 ’151 "69		106 ’152 "6A		107 ’153 "6B		108 ’154 "6C		109 ’155 "6D		110 ’156 "6E		111 ’157 "6F	
112 ’160 "70		113 ’161 "71		114 ’162 "72		115 ’163 "73		116 ’164 "74		117 ’165 "75		118 ’166 "76		119 ’167 "77	
120 ’170 "78		121 ’171 "79		122 ’172 "7A		123 ’173 "7B		124 ’174 "7C		125 ’175 "7D		126 ’176 "7E		127 ’177 "7F	

5.27 Example: Washington Cyrillic (wncyr10)

This is a Russian font distributed by the American Mathematical Society. The font and its companions listed below were developed at the University of Washington. In general, access to the correct fonts is just one of the ingredients necessary to handle a language. One also needs access to the correct hyphenation patterns (see page 85 for a discussion of what this means) and to a scheme that conveniently matches keyboard input to output characters. The issues have been discussed by VULIS [1989] and by MALYSHEV *et al.* [1991]. It is possible to get a file of definitions from the AMS that simplifies input for its Russian fonts; further, several characters in these fonts can be summoned as ligatures (i.e., particular combinations of input characters will produce some of the otherwise inaccessible output characters).

Here is a crude way of getting small amounts of Russian output:

```
\font\TenCyr=wncyr10
```

Then,

'{\TenCyr matem\accent38atika}' gives 'матема́тика', and
'{\TenCyr v\char121\char113isl\accent38\char31t\char126}'
gives 'вычисля́ть'.

This is very awkward and cumbersome, even if only a few words are needed in Russian, and it is clearly not the way to go.

- The other available design sizes (as opposed to magnified ones, such as those given by \magstep1, \magstep2, etc.) are

5, 6, 7, 8 and 9 points ($\mathcal{AMS}$-TEX distribution).

The available variations are

Bold (wncyb): 5, 6, 7, 8, 9 and 10 points.
Italic (wncyi): 5, 6, 7, 8, 9 and 10 points.
Small capitals (wncysc): 10 points.
Sans serif (wncyss): 8, 9 and 10 points.

- The \fontdimen parameter values for wncyr10 are

1	slant per point	0.0pt
2	interword space	3.77774pt
3	interword space stretch	1.74997pt
4	interword space shrink	1.16666pt
5	x-height (value of ex)	4.30554pt
6	quad width (value of em)	11.05545pt
7	extra space (at a sentence-end)	1.16666pt

The numbers to the left of each character on the facing page are its code numbers in decimal, *octal*, and hexadecimal form. (See Sections 2, 4 and 5 for more information.)

0 '0 "0 Њ	1 '1 "1 Љ	2 '2 "2 Ц	3 '3 "3 Э	4 '4 "4 I	5 '5 "5 Є	6 '6 "6 Ђ	7 '7 "7 Ћ
8 '10 "8 њ	9 '11 "9 љ	10 '12 "A ц	11 '13 "B э	12 '14 "C і	13 '15 "D є	14 '16 "E ђ	15 '17 "F ћ
16 '20 "10 Ю	17 '21 "11 Ж	18 '22 "12 Й	19 '23 "13 Ё	20 '24 "14 V	21 '25 "15 Ѳ	22 '26 "16 Ѕ	23 '27 "17 Я
24 '30 "18 ю	25 '31 "19 ж	26 '32 "1A й	27 '33 "1B ё	28 '34 "1C ѵ	29 '35 "1D ѳ	30 '36 "1E ѕ	31 '37 "1F я
32 '40 "20 ¨	33 '41 "21 !	34 '42 "22 ″	35 '43 "23 Ђ	36 '44 "24 ˘	37 '45 "25 %	38 '46 "26 ´	39 '47 "27 '
40 '50 "28 (	41 '51 "29)	42 '52 "2A *	43 '53 "2B ѣ	44 '54 "2C '	45 '55 "2D -	46 '56 "2E ·	47 '57 "2F /
48 '60 "30 0	49 '61 "31 1	50 '62 "32 2	51 '63 "33 3	52 '64 "34 4	53 '65 "35 5	54 '66 "36 6	55 '67 "37 7
56 '70 "38 8	57 '71 "39 9	58 '72 "3A :	59 '73 "3B ;	60 '74 "3C «	61 '75 "3D ı	62 '76 "3E »	63 '77 "3F ?
64 '100 "40 ˘	65 '101 "41 А	66 '102 "42 Б	67 '103 "43 Ц	68 '104 "44 Д	69 '105 "45 Е	70 '106 "46 Ф	71 '107 "47 Г
72 '110 "48 Х	73 '111 "49 И	74 '112 "4A J	75 '113 "4B К	76 '114 "4C Л	77 '115 "4D М	78 '116 "4E Н	79 '117 "4F О
80 '120 "50 П	81 '121 "51 Ч	82 '122 "52 Р	83 '123 "53 С	84 '124 "54 Т	85 '125 "55 У	86 '126 "56 В	87 '127 "57 Ш
88 '130 "58 Щ	89 '131 "59 Ы	90 '132 "5A З	91 '133 "5B [	92 '134 "5C "	93 '135 "5D]	94 '136 "5E Ь	95 '137 "5F Ъ
96 '140 "60 '	97 '141 "61 а	98 '142 "62 б	99 '143 "63 ц	100 '144 "64 д	101 '145 "65 е	102 '146 "66 ф	103 '147 "67 г
104 '150 "68 х	105 '151 "69 и	106 '152 "6A j	107 '153 "6B к	108 '154 "6C л	109 '155 "6D м	110 '156 "6E н	111 '157 "6F о
112 '160 "70 п	113 '161 "71 ч	114 '162 "72 р	115 '163 "73 с	116 '164 "74 т	117 '165 "75 у	118 '166 "76 в	119 '167 "77 щ
120 '170 "78 ш	121 '171 "79 ы	122 '172 "7A з	123 '173 "7B –	124 '174 "7C —	125 '175 "7D №	126 '176 "7E ь	127 '177 "7F ъ

5.28 Extended (256-Character) Fonts

After several years of discussion, the T_EX Users Group agreed in 1990 on a standard layout for T_EX's Latin text fonts (see FERGUSON [1990]). This layout was adopted at a T_EX conference in Cork, Ireland, and it is sometimes referred to as the "Cork standard."

The new layout accommodates all of the 256 characters permitted by T_EX in a font, and it differs from previous layouts even in the first 128 places. The principal architects of the new layout are Michael Ferguson, Jan Michael Rynning and Norbert Schwarz. Work on layouts for mathematics fonts is now in progress (1992). Once this work is complete, it is expected that the new layouts will define the standard to which all fonts used with T_EX will adhere. (Though it must be kept in mind that the successful implementation by Donald Knuth of the idea of a *virtual font*—see §5.9—will continue to allow the relatively easy use of fonts with any layout at all.) Here are some features of the new layouts:

- Text and mathematics are kept separate. The text font layout does not play host any longer to stray characters that are used primarily in mathematics. (The Computer Modern text fonts, on the other hand, contain the capital Greek letters that are used in mathematical formulas).

- Several accented characters occur as individual characters in their own right in the new layout. Thus, the hyphenation-suppressing side effect of using \accent is minimized.

- All Latin text fonts are to have essentially identical layouts. (On the other hand, the layouts of different Computer Modern text fonts sometimes differ in several ways: compare the layouts of cmr10 in §5.11 and cmtt10 in §5.14.)

- The new layout includes a respectable number of the characters needed to typeset various European languages.

- The math fonts are expected to contain standard ways of laying out Greek letters, math italic characters and special symbols. The symbols are to include Computer Modern symbols as well as the most useful of the AMS symbols.

- *Preliminary Extended Computer Fonts*

Of course, it is one thing to propose a standard for font layouts and quite another thing to actually design a set of fonts. At least one commercial font vendor now offers fonts that follow the new layout and a preliminary version of a set of public domain fonts is available on computer archives. The design of the public domain fonts is based on that of the standard Computer Modern fonts; the new fonts were largely developed by Norbert Schwarz. The eventual name for these new fonts is to be *Extended Computer* or EC fonts. The preliminary versions are generally referred to as DC fonts (they are currently available from computer archives under that name).

The next two pages display the 10-point roman DC font arranged according to the new font layout. Readers may find it instructive to compare this layout with that of, say, `cmr10` in § 5.11. DC fonts come in all the standard variations (bold, italic, etc.) and in all the standard sizes. Here are a few comments on those parts of the layout that differ from the present arrangement of Computer Modern text fonts:

- *Characters '000–'014*: Accents. They may be used along with TEX's standard accent mechanisms to construct further accented characters beyond the ones already present in the table.

- *Characters '015–'024, '042*: Quotation marks. The straight quotation mark at position *'042* will be greeted with relief by users who tend to mistakenly type ' " ' instead of ' " ' and ' " '. (When using `cmr10`, " gives "; in `dcr10`, on the other hand, it gives ".)

- *Character '027*: An invisible character of zero width and depth, from which its designers expect great things. The character—called a `Compound Word Mark`—can be used to block a ligature, as an invisible hyphen character, etc.

- *Character '030*: A small zero that can be used along with the per cent sign to make per thousand (‰) and per million (‰₀) signs.

- *Character '040*: A "single space" sign.

- *Characters '074 and '076*: "Less than" and "greater than" signs. These are missing from several of the Computer Modern text fonts; see, for example, the table for `cmr10` in § 5.11.

- *Character '200–'377*: Accented and other special characters.

0 `'0` `"0` `	1 `'1` `"1` ´	2 `'2` `"2` ^	3 `'3` `"3` ~	4 `'4` `"4` ¨	5 `'5` `"5` ″	6 `'6` `"6` °	7 `'7` `"7` ˇ
8 `'10` `"8` ˘	9 `'11` `"9` ‾	10 `'12` `"A` ·	11 `'13` `"B` ¸	12 `'14` `"C` ‹	13 `'15` `"D` ›	14 `'16` `"E` ‹	15 `'17` `"F` ›
16 `'20` `"10` "	17 `'21` `"11` "	18 `'22` `"12` „	19 `'23` `"13` «	20 `'24` `"14` »	21 `'25` `"15` –	22 `'26` `"16` —	23 `'27` `"17`
24 `'30` `"18` ◦	25 `'31` `"19` ı	26 `'32` `"1A` J	27 `'33` `"1B` ff	28 `'34` `"1C` fi	29 `'35` `"1D` fl	30 `'36` `"1E` ffi	31 `'37` `"1F` ffl
32 `'40` `"20` ␣	33 `'41` `"21` !	34 `'42` `"22` "	35 `'43` `"23` #	36 `'44` `"24` $	37 `'45` `"25` %	38 `'46` `"26` &	39 `'47` `"27` '
40 `'50` `"28` (	41 `'51` `"29`)	42 `'52` `"2A` *	43 `'53` `"2B` +	44 `'54` `"2C` ,	45 `'55` `"2D` -	46 `'56` `"2E` .	47 `'57` `"2F` /
48 `'60` `"30` 0	49 `'61` `"31` 1	50 `'62` `"32` 2	51 `'63` `"33` 3	52 `'64` `"34` 4	53 `'65` `"35` 5	54 `'66` `"36` 6	55 `'67` `"37` 7
56 `'70` `"38` 8	57 `'71` `"39` 9	58 `'72` `"3A` :	59 `'73` `"3B` ;	60 `'74` `"3C` <	61 `'75` `"3D` =	62 `'76` `"3E` >	63 `'77` `"3F` ?
64 `'100` `"40` @	65 `'101` `"41` A	66 `'102` `"42` B	67 `'103` `"43` C	68 `'104` `"44` D	69 `'105` `"45` E	70 `'106` `"46` F	71 `'107` `"47` G
72 `'110` `"48` H	73 `'111` `"49` I	74 `'112` `"4A` J	75 `'113` `"4B` K	76 `'114` `"4C` L	77 `'115` `"4D` M	78 `'116` `"4E` N	79 `'117` `"4F` O
80 `'120` `"50` P	81 `'121` `"51` Q	82 `'122` `"52` R	83 `'123` `"53` S	84 `'124` `"54` T	85 `'125` `"55` U	86 `'126` `"56` V	87 `'127` `"57` W
88 `'130` `"58` X	89 `'131` `"59` Y	90 `'132` `"5A` Z	91 `'133` `"5B` [	92 `'134` `"5C` \	93 `'135` `"5D`]	94 `'136` `"5E` ^	95 `'137` `"5F` —
96 `'140` `"60` '	97 `'141` `"61` a	98 `'142` `"62` b	99 `'143` `"63` c	100 `'144` `"64` d	101 `'145` `"65` e	102 `'146` `"66` f	103 `'147` `"67` g
104 `'150` `"68` h	105 `'151` `"69` i	106 `'152` `"6A` j	107 `'153` `"6B` k	108 `'154` `"6C` l	109 `'155` `"6D` m	110 `'156` `"6E` n	111 `'157` `"6F` o
112 `'160` `"70` p	113 `'161` `"71` q	114 `'162` `"72` r	115 `'163` `"73` s	116 `'164` `"74` t	117 `'165` `"75` u	118 `'166` `"76` v	119 `'167` `"77` w
120 `'170` `"78` x	121 `'171` `"79` y	122 `'172` `"7A` z	123 `'173` `"7B` {	124 `'174` `"7C` |	125 `'175` `"7D` }	126 `'176` `"7E` ~	127 `'177` `"7F` -

(dcr10)

128 '200 "80 Ă	129 '201 "81 Ą	130 '202 "82 Ć	131 '203 "83 Č	132 '204 "84 Ď	133 '205 "85 Ě	134 '206 "86 Ę	135 '207 "87 Ğ
136 '210 "88 Ĺ	137 '211 "89 Ľ	138 '212 "8A Ł	139 '213 "8B Ń	140 '214 "8C Ň	141 '215 "8D Ŋ	142 '216 "8E Ő	143 '217 "8F Ŕ
144 '220 "90 Ř	145 '221 "91 Ś	146 '222 "92 Š	147 '223 "93 Ş	148 '224 "94 Ť	149 '225 "95 Ţ	150 '226 "96 Ű	151 '227 "97 Ů
152 '230 "98 Ÿ	153 '231 "99 Ź	154 '232 "9A Ž	155 '233 "9B Ż	156 '234 "9C IJ	157 '235 "9D İ	158 '236 "9E đ	159 '237 "9F §
160 '240 "A0 ă	161 '241 "A1 ą	162 '242 "A2 ć	163 '243 "A3 č	164 '244 "A4 ď	165 '245 "A5 ě	166 '246 "A6 ę	167 '247 "A7 ğ
168 '250 "A8 ĺ	169 '251 "A9 ľ	170 '252 "AA ł	171 '253 "AB ń	172 '254 "AC ň	173 '255 "AD ŋ	174 '256 "AE ő	175 '257 "AF ŕ
176 '260 "B0 ř	177 '261 "B1 ś	178 '262 "B2 š	179 '263 "B3 ş	180 '264 "B4 ť	181 '265 "B5 ţ	182 '266 "B6 ű	183 '267 "B7 ů
184 '270 "B8 ÿ	185 '271 "B9 ź	186 '272 "BA ž	187 '273 "BB ż	188 '274 "BC ij	189 '275 "BD ı	190 '276 "BE ¿	191 '277 "BF £
192 '300 "C0 À	193 '301 "C1 Á	194 '302 "C2 Â	195 '303 "C3 Ã	196 '304 "C4 Ä	197 '305 "C5 Å	198 '306 "C6 Æ	199 '307 "C7 Ç
200 '310 "C8 È	201 '311 "C9 É	202 '312 "CA Ê	203 '313 "CB Ë	204 '314 "CC Ì	205 '315 "CD Í	206 '316 "CE Î	207 '317 "CF Ï
208 '320 "D0 Đ	209 '321 "D1 Ñ	210 '322 "D2 Ò	211 '323 "D3 Ó	212 '324 "D4 Ô	213 '325 "D5 Õ	214 '326 "D6 Ö	215 '327 "D7 Œ
216 '330 "D8 Ø	217 '331 "D9 Ù	218 '332 "DA Ú	219 '333 "DB Û	220 '334 "DC Ü	221 '335 "DD Ý	222 '336 "DE Þ	223 '337 "DF SS
224 '340 "E0 à	225 '341 "E1 á	226 '342 "E2 â	227 '343 "E3 ã	228 '344 "E4 ä	229 '345 "E5 å	230 '346 "E6 æ	231 '347 "E7 ç
232 '350 "E8 è	233 '351 "E9 é	234 '352 "EA ê	235 '353 "EB ë	236 '354 "EC ì	237 '355 "ED í	238 '356 "EE î	239 '357 "EF ï
240 '360 "F0 ð	241 '361 "F1 ñ	242 '362 "F2 ò	243 '363 "F3 ó	244 '364 "F4 ô	245 '365 "F5 õ	246 '366 "F6 ö	247 '367 "F7 œ
248 '370 "F8 ø	249 '371 "F9 ù	250 '372 "FA ú	251 '373 "FB û	252 '374 "FC ü	253 '375 "FD ý	254 '376 "FE þ	255 '377 "FF ß

(dcr10)

5.29 Character Table Commands

Here are some commands, defined for this chapter but likely to be of general interest. One set converts non-negative decimal numbers to other radices (not exceeding 36). \Conv{13}{2} reexpresses the decimal number 13 in binary form: 1101. \Hex and \Oct produce hexadecimal and octal numbers, respectively. For instance, \Hex{64} produces "40, and \Oct{64} produces '100.

```
\def\Conv #1#2{{\count0=#1 \count1=#1 \count2=#2 \divide\count1 by\count2
     \ifnum\count1>0 \Conv{\count1}{\count2}\fi \count3=\count1
     \multiply\count3 by-\count2 \advance\count0 by\count3 \Digit}}%
\def\Digit{\ifnum\count0<10 \number\count0
     \else\advance\count0 by-10 \advance\count0 by`A \char\count0 \fi}%
\def\Hex #1{{\tt\char'42\Conv{#1}{16}}}    \def\Oct #1{{\it'\Conv{#1}{8}}}
NOTE: It is important to confine uses of '\count0' within braces.
```

A set of additional commands produces character tables. The main command is \ShowTable{xxx}, which shows a table of 128 characters from font xxx. The table is built up box by box. The character displayed in each box is labelled by the current value of the variable called \CharNo; the starting value is 0, but it can be reset. Numerical labels are put on the left of each character; the position is reversed if the indicator \CharToCode is given the value 1.

```
%NOTE: If 8-pt fonts are not yet loaded, remove the % from the next line:
%\font\Eightrm=cmr8 \font\Eightit=cmti8  \font\Eighttt=cmtt8
\newcount\CharNo
\newcount\CharToCode \CharToCode=0 % Fixes the order of boxes.
\def\OnLine#1{\ifnum\CharToCode=1 \leftline{#1}\else\rightline{#1}\fi}
\def\CodeBox{\vbox to 30pt{\hsize15pt \parindent0pt \baselineskip7pt
               \lineskip=.5pt \let\it\Eightit \let\tt\Eighttt
               \vfil \OnLine{\Eightrm \thinspace \the\CharNo}
               \vfil \OnLine{\kern-.1pt\Oct{\the\CharNo}}
               \vfil \OnLine{\Hex{\the\CharNo}} \vfil}}
\def\CharBox{\vbox to 30pt{\hsize18pt \vfil
               \centerline{\DispFont\char\CharNo}\vfil}}
\def\FrameBox #1#2{\vbox{\hrule height#1pt
        \hbox{\vrule width#1pt#2\vrule width#1pt}\hrule height#1pt}}
\def\FullBox{\ifnum\CharToCode=1 \let\TmpBox=\CharBox
  \let\CharBox=\CodeBox \let\CodeBox=\TmpBox \fi % Boxes are switched.
  \FrameBox{.2}{\kern1pt\CodeBox\kern2pt\vrule width.15pt\kern2pt\CharBox
     \kern2pt}\global\advance\CharNo by 1 }%
\def\ShowTable #1{{\font\DispFont=#1\setbox0=\vbox{}\count0=1 \MakeLns
     \vbox{\hbox{\FrameBox{.2}{\kern1pt\vbox{\box0\kern1pt}}}
            \hbox{\Eighttt#1}}}\CharNo=0 } %  THIS WAS THE MAIN COMMAND.
\def\MakeLns{\setbox0=\vbox{\unvbox0 \kern1pt\OneLine}\advance\count0 by1
     \ifnum\count0<17 \MakeLns\fi}
\def\OneLine{\setbox1=\hbox{}\count1=0 \MakeBox \box1 }
\def\MakeBox{\setbox1=\hbox{\unhbox1 \FullBox\kern1pt}\advance\count1 by1
               \ifnum\count1<8 \MakeBox\fi}
% FOR EXAMPLE:
%    \CharNo=100 \ShowTable{cmsy10}
% will display characters 100 to 227 of the font 'cmsy10'.
```

6

Code

ΓΔΘΛΞΠΣΥ
ΦΩ
δ
θ β
ε ζ
ψ λ
ω φ
1 2 3 4 5 6 7 8
A LMOST ALL the significant formatting commands for this book are displayed here. Those that are omitted fall into five categories: (1) commands that serve simply as abbreviations in a few places; (2) commands that are shown elsewhere in this book, e.g., the character table commands for Chapter 5 that are shown on the previous page; (3) commands that are based on external packages, like the typeface-changing files used in Examples 7–9, information about which is given on the inside back cover; (4) commands that are used to present the Glossary entries in Chapter 8; and (5) commands that are used to form the input pages in the examples in Chapter 2. The last two sets of commands are interesting, but they are also very similar to the corresponding sets of commands in *TEX by Example* (BORDE [1992]); readers with a strong interest in, say, drawing boxes around verbatim input and having the lines automatically numbered may look up the commands there.

Since this book involves several different types of pages, the formatting commands are grouped into separate files according to where they are used. The files are largely self-contained in that they can achieve their stated purpose either entirely independently of the others or with clear indications of what other files they rely on. Occasionally this leads to repetitions and other minor inefficiencies, but the organizational benefits that arise seem to outweigh these considerations. Further benefits that arise from this grouping are that it makes it easier for readers to track down what they want, and it more clearly identifies what they need to copy in order to get a desired effect.

For example, readers with an interest only in small typefaces (possibly for footnotes) need only copy the fairly short file called `EightPt.tex`.

The start of each file provides a little information on what the commands in it do and what part they play in the shaping of this book.

6.1 Utilities (Utility.tex)

```
%%% File: Utility.tex
%%% Contents: Utility commands to allow verbatim reproduction of input, to
%%% allow '"' to be used as an in-line verbatim quoting mechanism, to produce
%%% various logos, etc.

%----------------------------------------------------------------------------
%%% REPRODUCING INPUT:

% The present book is filled with pieces of input files, or even entire input
% files, that are reproduced verbatim. The commands shown here provide
% the basis for this reproduction.

\def\cc{\catcode}    % Abbreviation used when changing category codes.

% The next command reproduces input verbatim by switching off special
% category codes and assigning code 12 to all normal command characters:

\def\Literal {\begingroup \cc'\\=12 \cc'\{=12 \cc'\}=12 \cc'\$=12
    \cc'\&=12 \cc'\#=12 \cc'\%=12 \cc'\~=12 \cc'\¯12 \cc'\^=12
    \cc'\@=0 \cc'\'=\active \obeyspaces \VFont} % A 'space' is now active.

{\obeyspaces\gdef {\hglue.5em\relax}} % Interword spacing.

{\cc'\'=\active \gdef'{\relax\lq}} % To block certain ligatures.

% The next few lines define the role that '"' plays if it is later made
% active (it will be used to reproduce in-line input):

{\cc'\"=\active
\gdef"{\Literal \VQuotingFont \com}
{\cc'\@=0
@gdef@com#1"{@leavevmode@hbox@bgroup#1@egroup@endgroup}}}
\def\VQuotingOn{\cc'\"=\active }
\def\VQuotingOff{\cc'\"=12 }

\let\VFont=\tt        % Default; may be changed as desired.
\let\VQuotingFont=\tt  % Ditto.

%----------------------------------------------------------------------------
```

```
%%% LOGOS:

\def\cal{\fam2 }   % Restore 'Plain' definition, in case it has been altered.
\def\AmS{$\cal A\kern-.1667em\lower.5ex\hbox{$\cal M$}\kern-.125em S$}
\def\AmSTeX{\AmS-\TeX}
\font\Eightsy=cmsy8 \skewchar\Eightsy='60
\def\LamSTeX{L\kern-.4em\raise.3ex\hbox{$\scriptstyle\cal
        A$}\kern-.25em\lower.4ex\hbox{\Eightsy M}\kern-.1em$\cal S$-\TeX}
\font\LogoTen=logo10
\def\MF{{\LogoTen META}\-{\LogoTen FONT}}
\font\LogoTenSl=logosl10
\def\MFsl{{\LogoTenSl META}\-{\LogoTenSl FONT}}
\font\Smallcaps=cmcsc10
\def\LaTeX{L\kern-.36em\raise.35ex\hbox{\Smallcaps a}\kern-.15em\TeX}
\font\Eightit=cmti8 % The logo below is for use in italics.
\def\ItLaTeX{L\kern-.32em\raise.32ex\hbox{\Eightit A}\kern-.22em\TeX}
\def\PiC{P\kern-.12em\lower.5ex\hbox{I}\kern-.075emC}
\def\PiCTeX{\PiC\kern-.11em\TeX}
\def\ssfTeX{T\kern-.15em\lower.5ex\hbox{E}\kern0em X} % When using sans-serif.
% Just to be complete:
% \def\TeX{T\kern-.1667em\lower.5ex\hbox{E}\kern-.125em X}

%----------------------------------------------------------------------------
%%% FOOTNOTES:

\def\OnFootnoterule #1{%
 \def\footnoterule{\kern-3pt \hrule width#1 height.4pt \kern2.6pt }}
\def\OffFootnoterule{\def\footnoterule{}}

%----------------------------------------------------------------------------
%%% MISCELLANEOUS:

\def\raggedleft{\leftskip=0pt plus 2em \parfillskip0pt
               \spaceskip.3333em \xspaceskip=.5em\relax}
\def\Raggedleft{\leftskip=0pt plus 4em minus.6em \parfillskip0pt
               \spaceskip.3333em \xspaceskip=.5em\relax}
\def\Raggedright{\rightskip=0pt plus 4em minus.6em \parfillskip0pt
               \spaceskip.3333em \xspaceskip=.5em\relax}
\def\EndPage{\vfil\eject}
\def\EndLine{\ifhmode \hfil\break \fi}
\def\:{\thinspace }
\def\T {\vrule width0pt\kern-.1em} % To occasionally adjust spacing.
\def\Lcase #1{\lowercase\expandafter{#1}}
\def\Sn #1.{\S\:#1.}

%----------------------------------------------------------------------------
\endinput
```

6.2 Eight-Point Type (`EightPt.tex`)

```
%%% File: EightPt.tex
%%% Contents: Provides a command for loading 8-point fonts, provides a
%%% comprehensive font-switching command, and a command to use 8-point type
%%% in footnotes.

%-----------------------------------------------------------------------------

\def\LoadEight {\font\Eightrm=cmr8  \font\Eightsl=cmsl8
                \font\Eightit=cmti8 \font\Eightbf=cmbx8
                \font\Eighttt=cmtt8 \font\Eighti=cmmi8
                \font\Eightsy=cmsy8
                \font\Eightcsc=cmcsc8
                \font\Sixrm=cmr6    \font\Sixbf=cmbx6
                \font\Sixi=cmmi6    \font\Sixsy=cmsy6
                \def\LoadEight{}}%
% Note: The last line above protects against inefficiencies.

\def\Eightpoint {\def\rm{\fam0\Eightrm}%
  \textfont0=\Eightrm \scriptfont0=\Sixrm \scriptscriptfont0=\fiverm
  \textfont1=\Eighti  \scriptfont1=\Sixi  \scriptscriptfont1=\fivei
  \textfont2=\Eightsy \scriptfont2=\Sixsy \scriptscriptfont2=\fivesy
  \textfont3=\tenex   \scriptfont3=\tenex \scriptscriptfont3=\tenex
  \textfont\itfam=\Eightit  \def\it{\fam\itfam\Eightit}%
  \textfont\slfam=\Eightsl  \def\sl{\fam\slfam\Eightsl}%
  \textfont\ttfam=\Eighttt  \def\tt{\fam\ttfam\Eighttt}%
  \textfont\bffam=\Eightbf  \scriptfont\bffam=\Sixbf
     \scriptfont\bffam=\fivebf  \def\bf{\fam\bffam\Eightbf}%
  \let\Smallcaps=\Eightcsc
  \let\Smallrm=\Eightrm % This is needed just for the present book.
  \normalbaselineskip=9pt
  \setbox\strutbox=\hbox{\vrule height7pt depth2pt width0pt}%
  % Also reset the spacing around displays:
  \abovedisplayskip 8pt plus2pt  minus7pt
  \abovedisplayshortskip 0pt plus2pt
  \belowdisplayskip 8pt plus2pt  minus7pt
  \belowdisplayshortskip 5pt plus2pt  minus3pt
  \normalbaselines \rm }
% For Footnotes:
\let\FFFFFF=\footstrut  % Store the standard 'value'.
\def\SmallFootnotes {\LoadEight
  \def\footnoterule{}\def\footstrut{\Eightpoint\FFFFFF}}%
\def\NormalFootnotes{\def\footstrut{\FFFFFF}%
  \def\footnoterule{\kern-3pt \hrule width2truein height.4pt \kern2.6pt}}%
%-----------------------------------------------------------------------------
\endinput
```

6.3 Standard Pages (StdPage.tex)

```
%%% File: StdPage.tex
%%% Contents: The page layout, standard fonts, etc., used across the board in
%%% all the files of 'Mathematical TeX by Example'.

\input Utility

%---------------------------------------------------------------------------
%%% TYPEFACE NAMES:

\font\TitleFont=cmb10 scaled 2986       % For chapter titles.
\def\TitleFonts{\TitleFont}             % Needed for redefinition purposes later.
\font\NumFont=cmb10 scaled 5160         % For the large chapter numbers.
\font\SubTitleFont=cmbx12               % For subsection titles.
\font\SubTitlett=cmtt10 scaled 1200     % Also used in section titles.
\def\SubTitleFonts{\SubTitleFont}       % Needed for redefinition purposes later.
\font\SubSectionTitleFont=cmti10 scaled\magstephalf % For subsection titles.
\font\Bigsy=cmsy10 scaled 1200          % For 'bullets' in subsection titles.
\font\SmallPnoFont=cmfib8               % For page numbers at the bottom.
\font\PnoFont=cmfib8 scaled\magstep1    % For page numbers in the margin.
\font\Smallcaps=cmcsc10                 % Small capitals, used in many places.
\let\sc=\Smallcaps \let\smc=\Smallcaps  % For compatibility w/other packages.
\font\Sansserif=cmss10                  % Sans serif, used in many places.
\let\sf=\Sansserif                      % For LaTeX compatibility.
\font\Ninerm=cmr9
\let\Smallrm=\Ninerm        % Used in places to give unobtrusive numerals, etc.

% For section titles in the margins:
\font\Eightbf=cmbx8
\font\Eighttt=cmtt8
\let\MarginFont\Eightbf

% For chapter and section titles that use several fonts:
\def\UseBigCal{\font\TitleCal=cmbsy10 scaled2986
               \def\TitleFonts{\textfont2=\TitleCal
                    \exhyphenpenalty10000 \pretolerance10000
                    \hbadness10000 \TitleFont}}
\def\Trick#1{{\Ninebf\uppercase{#1}}} % Used for the LaTeX logo in titles.
\def\UseMedCal{\font\SubTitleCalFont=cmbsy10 scaled\magstep1
               \font\SmSubTitleCalFont=cmbsy9
               \font\Tenbsy=cmbsy10
               \font\Ninebf=cmbx9
                \def\SubTitleFonts{\textfont2=\SubTitleCalFont
                    \scriptfont2=\SmSubTitleCalFont
                   \let\Eightsy\Tenbsy \let\Smallcaps\Trick
                   \SubTitleFont}}

%---------------------------------------------------------------------------
```

```
%%% REPRODUCING INPUT:

\let\XXX=\Literal    % Temporary assignment.
\def\Literal {\XXX \cc'\*=12 \losenolines} % Redefine '\Literal' for use here.
{\cc'\^^M=\active %
\gdef\losenolines{\cc'\^^M=\active \def^^M{\leavevmode\endgraf}}}

% Use these to display input:
\def\beginliteral{\penalty1000\vskip3pt plus2pt minus1pt \Literal \cc'\"=12 %
  \parskip0pt \parindent\indsize \baselineskip2.77ex \thatisit}
{\cc'\@=0 \cc'\\=12 @cc'@^^M=@active %
 @gdef@thatisit^^M#1\endliteral{#1@endgroup%
   @vskip.8@smallskipamount@noindent@ignorespaces}}

%------------------------------------------------------------------------
%%% MISCELLANEOUS:

% First, a utility command that gives lowercase small capitals:
\def\LCSmCaps#1{{\let\ZZ=\&\def\&{{\sevenrm\:\ZZ\:}}\Smallcaps
                  \xspaceskip=.1667em\relax
                  \lowercase\expandafter{#1}}}
% Next, utilities to save typing; the first of these typesets references
% in the correct format.
\def\Ref#1[#2]{\LCSmCaps{#1}{\Smallrm [#2]}}
\def\[#1]{{\Sansserif #1}}

%------------------------------------------------------------------------
%%% CHAPTER AND SECTION FORMATTING:

\newskip\BelowTitleSkip \BelowTitleSkip=4.5pc plus .5pc % Note: 1pc=12pt
\newcount\StartAnew
\newtoks\NameofChapter
\newtoks\SectionLabel
\newcount\SectionNumber
\newcount\ChapterNumber
\newcount\NumberSectionLbl   \NumberSectionLbl=1   % An indicator; see below.

\def\NoSections{\NumberSectionLbl=0} % For chapters without sections.

% The command that creates the format for chapter titles:
\outer\def\Title #1/#2\par{\ChapterNumber=#1
              {\parindent0pt
               \ifnum\ChapterNumber>0 {\NumFont\the\ChapterNumber}\medskip\fi
               \pretolerance1000 \TitleFonts \righthyphenmin=50
               \raggedright \baselineskip2.75ex  #2\par}
              \NameofChapter={\tensl#2}\StartAnew=1 \SectionNumber=0
              \SectionLabel={}\mark{}\vskip\BelowTitleSkip\relax
              \noindent\ignorespaces}%
```

```
% The command that follows does several things. It tests the current
% position on the page by comparing '\pagetotal' and a reduced value of
% '\pagegoal'. If there is enough space available to start a new section on
% the current page, the section begins there (after a vertical skip);
% otherwise, the section begins on the next page. The command also defines
% what is to go in the 'token lists' called '\LMargStuff' and '\RMargStuff'
% (see the output routine at the end of this file) and it 'marks' the material
% that is to go in the margin. Further, the command typesets the section title
% and section number.
\def\NewSection #1\par{\par \dimen0=\pagegoal \advance\dimen0 by-120pt
                \ifdim\pagetotal>\dimen0
                    \ifdim\pagetotal<\pagegoal \EndPage
                    \else \vskip 24pt plus 10pt minus 6pt \fi
                \else \vskip 24pt plus 10pt minus 6pt \fi
                \LMargStuff={\MarginFont
                    \iftrue\firstmark\fi}
                \RMargStuff={\MarginFont
                    \iftrue\botmark\fi}
                \global\advance\SectionNumber by1
                \leftline{\SubTitleFonts \let\tt\SubTitlett
                    \the\ChapterNumber.\the\SectionNumber\ \ #1}
                \ifnum\NumberSectionLbl=1
                    \SectionLabel={\tenrm \quad
                    \S\:\the\ChapterNumber.\the\SectionNumber}
                \else
                    \SectionLabel={}
                \fi
                \mark{#1\noexpand\else \the\SectionLabel}
                \vskip 12pt plus2pt minus4pt\noindent\ignorespaces}%
% The next command is used in starting subsections:
\def\NewSub #1\par{\par \dimen0=\pagegoal \advance\dimen0 by-40pt
                \ifdim\pagetotal>\dimen0
                    \ifdim\pagetotal<\pagegoal \EndPage
                    \else \vskip 12pt plus6pt minus4pt\fi
                \else \vskip 12pt plus6pt minus4pt\fi
                \leftline{\Bullet\SubSectionTitleFont\ \:#1}
                \nobreak\vskip6pt plus1pt minus2pt\noindent\ignorespaces}%
\def\Bullet{\leavevmode\raise.5pt\hbox{\Bigsy\char15 }}%
```

```
% The next few commands go towards creating the large, 'illuminated'
% opening letter that starts each chapter. The commands are crude,
% 'one-shot' ones, meant just to serve for this book. But, they may
% help suggest how more general ones may be constructed.

% The first command, '\Patbox', fills the edges of a box of fixed size
% with nonletter characters from a given font (the height of each character
% is first examined, and only ones of height above a certain value are used).

\def\Patbox #1#2{\vbox to 44pt{\hsize42pt \lineskip 0pt
    \parindent0pt \baselineskip4.75pt \parfillskip0pt \hbadness10000
    \count255=#2\font\temp=#1 \fontdimen2 \temp=0pt \temp \vfil
    \def\X{\loop
            \ifnum\count255=60 \global\count255=63 \fi   % Skip <, /, >.
            \ifnum\count255=65 \global\count255=91 \fi   % Skip letters.
            \ifnum\count255=97 \global\count255=123 \fi  % Skip letters.
            \ifnum\count255=126 \global\count255=0 \fi   % Start over.
            \setbox0=\hbox{\char\count255}\ifdim\ht0<3pt
                \global\advance\count255 by 1
        \repeat % Use only the tall guys.
        \unhbox0 \hskip0pt minus.5pt\global\advance\count255 by 1 }
    \hfuzz.2pt
    \line{\X\hfil\X\hfil\X\hfil\X\hfil\X\hfil\X\hfil\X\hfil\X}
    \line{\X\hfil\X}\line{\X\hfil\X}\line{\X\hfil\X}
    \line{\X\hfil\X}\line{\X\hfil\X}\line{\X\hfil\X}\line{\X\hfil\X}
    \line{\X\hfil\X\hfil\X\hfil\X\hfil\X\hfil\X\hfil\X\hfil\X}}}

% Next, the font that will be used for the large letter is chosen:

\font\BigFancy=cmr7 scaled 5160

% Finally, the main command is constructed. Its first argument is the
% font that will be used for the boundary pattern (only 5-point fonts really
% work well, and then too not all of them); the second, the starting
% character position in that font; the third, the first letter of the
% paragraph that is to just open; the fourth and fifth, the first two
% words of that paragraph (spaces are used to 'delimit' these arguments;
% the words will appear in small capitals.

\def\Fancy #1#2#3#4 #5 {\noindent \hangindent46pt \hangafter-4
        \llap{\vbox to0pt{\vss
        \hbox{\Patbox{#1}{#2}\kern-42pt
        \vbox to44pt{\hsize42pt\vfil
            \vskip3pt\centerline{\hskip2pt\BigFancy #3}\vfil
        }}\vskip-3\baselineskip}\hskip4pt}{\Smallcaps #4\ #5\ }}

% FOR EXAMPLE:
%      '\Fancy{cmbsy5}{12} Let us begin with a bang.'
% This will cause a large 'L' to appear, surrounded by the taller characters
% of the font cmbsy5, starting with character number 12; the first two
% words will also appear capitalized.

%---------------------------------------------------------------------------
```

```
%%% IN THE MARGINS:

\def\TopofRMarg #1{\vtop{\hsize .75in \parindent0pt \parskip0pt
            \rightskip0pt \leftskip0pt % Just for protection.
            \line{\vrule width1pt \hfil
            \dimen0=\hsize \advance\dimen0 by-1pt \dimen1=10pt
            \vtop to\dimen1{\hsize\dimen0 \rightline{\PnoFont \Folio}\vfil}}
             \Raggedleft \pretolerance10000 \lefthyphenmin=50 \hbadness10000
             \vrule height18pt width0pt #1\baselineskip2.7ex}}%
\def\TopofLMarg #1{\vtop{\hsize .75in \parindent0pt \parskip0pt
            \rightskip0pt \leftskip0pt % Just for protection.
            \line{\dimen0=\hsize \advance\dimen0 by-1pt \dimen1=10pt
            \vtop to\dimen1{\hsize\dimen0
              \leftline{\PnoFont \Folio}\vfil}\hfil \vrule width1pt}
             \Raggedright \pretolerance10000 \lefthyphenmin=50 \hbadness10000
             \vrule height18pt width0pt #1\baselineskip2.7ex}}

%----------------------------------------------------------------------
%%% PAGE NUMBERING:

\countdef\RealPno=1  % To be used as the 'real' page number for this book.
  \RealPno=1         % Default starting value
\def\Folio{\ifnum\RealPno<0 \romannumeral-\RealPno \else \number\RealPno \fi}

%----------------------------------------------------------------------
%%% OUTPUT ROUTINE:

\newtoks\LMargStuff
\newtoks\RMargStuff
\newdimen\Llmargspace  \Llmargspace=.5in
\newdimen\Rrmargspace  \Rrmargspace=.5in
\newdimen\Lhoffset     \Lhoffset=.75in
\newdimen\Rhoffset     \Rhoffset=0in

\def\MTBEoutput {\ifodd\RealPno \hoffset\Rhoffset \else \hoffset\Lhoffset \fi
   \ifnum\StartAnew=1
      \headline={\hfil}
      \footline={\hss\SmallPnoFont\Folio\hss}
   \else
      \footline={\hfil}
      \ifodd\RealPno
         \headline={\hfil \the\NameofChapter\expandafter\iffalse\botmark\fi
            \rlap{\hskip\Rrmargspace\TopofRMarg{\the\RMargStuff}}}
      \else
         \headline={\llap{\TopofLMarg{\the\LMargStuff}\hskip\Llmargspace
           }\the\NameofChapter\hfil}
      \fi
   \fi
   \plainoutput \Advancepageno \global\StartAnew=0 }%
```

```
\def\UseMTBEoutput{%
  \gdef\Advancepageno{\ifnum\RealPno<0 \global\advance\RealPno by -1
    \else\global\advance\RealPno by1 \fi }
  \gdef\makefootline{\baselineskip36pt \line{\the\footline}}
  \gdef\makeheadline{\vbox to 0pt{\vskip-.5in
    \line{\vbox to 8.5pt{}\the\headline}\vss}\nointerlineskip}
  \output={\MTBEoutput}}

%-------------------------------------------------------------------------------
%%% PARAGRAPH AND PAGE LAYOUT:

\newdimen\indsize    \indsize23pt
\parindent\indsize

\widowpenalty300    \clubpenalty300

\newdimen\StdHsize \StdHsize=4.75 in
\newdimen\StdVsize \StdVsize=7in
\hsize\StdHsize   \vsize\StdVsize

%-------------------------------------------------------------------------------
\endinput
```

6.4 Output Format for the Examples (OutForm.tex)

```
%%% File: OutForm.tex
%%% Contents: The format commands for the 'output' pages (i.e., right hand)
%%% of the Examples in Chapter 2 of 'Mathematical TeX by Example'.

\input StdPage   % The file containing the standard MTBE page specifications.

%-------------------------------------------------------------------------------
%%% TYPEFACE NAMES:

% Note: most typeface names used in this file have been defined in 'StdPage'.
\font\sevenit=cmti7                          % For book titles in the margin.
\font\ExampleTitleFont=cmbx12 scaled\magstep1 % For Example Titles.
\font\ExampleTypeFont=cmsl12 scaled \magstep2 % For Example Types.
\font\ExampleTypeCal=cmsy10 scaled \magstep3  % For 'Cal' in Example Types.
\font\SmExampleTypeCal=cmsy7 scaled \magstep3 % For scriptsize Cal.
\font\BigEightsy=cmsy8 scaled \magstep3       % For small Cal.
\def\ExampleTypeFonts{\textfont2=\ExampleTypeCal
                \scriptfont2=\SmExampleTypeCal
                \let\Eightsy\BigEightsy \ExampleTypeFont}

%-------------------------------------------------------------------------------
%%% MISCELLANEOUS:

\def\EnoughAlready{$$\vdots$$ \EndPage}

%-------------------------------------------------------------------------------
```

```
%%% EXAMPLE COMMANDS:

\newtoks\Author             % To store the name of the author of the source.
\newtoks\BookTitle          % To store the name of the book/journal.
\newtoks\PublishInfo        % To store publisher information.
\newtoks\CopyrightDate      % To store information about the copyright date.
\newtoks\Example            % To store information about the example.
\newtoks\ExampleType        % To store information about the example type.
\newcount\ExampleNumber     % For the example number.
\newcount\StartExample      % Start-of-example indicator.
\newdimen\LMarginOdd        % Left margin on odd-numbered pages.
\newdimen\LMarginEven       % Left margin on even-numbered pages.

\outer\def\NewExample{\hsize\StdHsize  \vsize\StdVsize  \parindent\indsize
        \StartExample=1
        \LMarginOdd=.625in  \LMarginEven=.625in  \voffset0in
        \headline={\hfil}\global\advance\ExampleNumber by1
        \leftline{\ExampleTypeFonts\the\ExampleType}
        \vskip2pc\relax
        \message{ \the\ExampleNumber }
        \setbox0=\hbox{\ExampleTitleFont 1}\dimen0=-\ht0
        \divide\dimen0 by2 \advance\dimen0 by.75pt
        \dimen1=-\dimen0 \advance\dimen1 by1pt
        \centerline{\vrule width8pc height\dimen1 depth\dimen0
        {\ExampleTitleFont \ \ Example \the\ExampleNumber\ \ }\vrule
         width8pc height\dimen1 depth\dimen0 }
         \Example={\the\ExampleNumber}%
        \vskip1pc\relax \noindent
        From {\tenit \the\BookTitle}
        \edef\Test{\the\Author}%
        \ifx\Test\empty \else \EndLine
           {\spaceskip.3333em\relax
           \LCSmCaps{\the\Author}} \fi
        \edef\Test{\the\PublishInfo}%
        \ifx\Test\empty \else \EndLine {\tenrm\the\PublishInfo}\fi
        \edef\Test{\the\CopyrightDate}%
        \ifx\Test\empty \else\unskip{\tenrm,\ \the\CopyrightDate}\fi
        \vskip1pc\relax \noindent \ignorespaces}
\def\ShortExample <#1>{\medbreak
      \line{\vrule width.4pt depth2.6pt height.4pt\hrulefill
            \vrule width.4pt depth2.6pt height.4pt}\nointerlineskip
      \leftline{\Eightbf (Page #1)}
      \nobreak}
\def\EndShortExample{\nobreak\vskip-1pt
      \line{\vrule width.4pt depth0pt height3pt\hrulefill
            \vrule width.4pt depth0pt height3pt}\medbreak}
\def\NegVspace{\ifvmode\nointerlineskip\fi
        \nobreak\vskip-3pt\relax} % Reduces space.

%--------------------------------------------------------------------------
```

```
%%% TOWARDS THE OUTPUT ROUTINE:

\newdimen\RMargin          % The right margin size.
\newdimen\BotMargin        % The bottom margin size.
\newcount\ChapIndicator    % For compatibility with BookForm.tex

\def\Issue{}
\def\Volume #1,#2(#3){\def\Issue{{\fivebf #1},#2(#3)}}
\def\lcopyr{\leavevmode\hbox{{\fivesy\char'15}\llap{\fiverm c\kern.3em }}}

\def\StandardPage{\vbox{\makeheadline\pagebody\makefootline
        \advance\BotMargin by-24pt\vglue\BotMargin}}
\def\RightMarginStuff{\vbox{\parskip0pt \hsize.75in \advance\hsize by-.4pt
                    \parindent0pt \hbadness5000 \raggedleft \pretolerance10000
                    \vbox to 0pt{\vskip-.5in
                      \rightline{\TopofRMarg{\MarginFont
                      Example~\:\the\Example}}\vss}
                    \dimen0=\vsize \advance\dimen0 by \BotMargin
                    \advance\dimen0 by\voffset
                    \nointerlineskip
                    \vbox to \dimen0{\null \vskip.25in
                                \baselineskip 6 pt \fiverm Source:
                                \smallskip
                                {\sevenit \baselineskip8pt \the\BookTitle
                                \smallskip }
                                \edef\Test{\the\Author}
                                {\ifx\Test\empty \removelastskip \else
                                  \spaceskip=.1667em\relax
                                  \let\ZZ=\& \def\&{\:\ZZ\:}
                                  \uppercase\expandafter{\the\Author} \fi
                                \smallskip}
                                \edef\Test{\the\CopyrightDate}
                                \ifx\Test\empty \else
                                    \lcopyr\ \the\CopyrightDate\break \fi
                                \edef\Test{\the\PublishInfo}
                                \ifx\Test\empty \Issue \else
                                    \the\PublishInfo \fi
                                \advance\dimen0 by-\StdVsize
                                \vskip\dimen0 \vfill}}}
\def\AugmentedPage{\vbox{\ifnum\StartExample=1
        \headline={\hfil}\footline={\hfil\SmallPnoFont\Folio\hfil}%
        \def\makefootline{\baselineskip36pt \line{\the\footline}}%
        \StandardPage
    \else
        \ifnum\ChapIndicator=1 \else\footline={\hfil}\fi
        \ifodd\pageno \dimen0=-\LMarginOdd \else \dimen0=-\LMarginEven \fi
        \advance\dimen0 by6.625in \hbox to \dimen0
        {\StandardPage \hskip\RMargin \BotCornerVrule \hss
        \RightMarginStuff }\nointerlineskip
        {\dimen0=\hsize \advance\dimen0 by\RMargin \advance\dimen0 by.4pt
            \hbox to \dimen0{\hfil \BotCornerHrule}}
        \fi}\global\StartExample=0 }%
```

```
\def\MarkPageCorner{\def\BotCornerVrule{\vrule height36pt width.4pt depth0pt}
              \def\BotCornerHrule{\vrule width36.4pt height.4pt depth0pt}}
\def\DontMarkPageCorner{\def\BotCornerVrule{\hskip-\RMargin}
              \def\BotCornerHrule{}}

% Here comes a small cheat:
\def\plainoutput{\ifodd\pageno \RMargin=\LMarginEven \else
              \RMargin=\LMarginOdd \fi % Assuming mirror symmetry.
          \shipout\AugmentedPage
          \advancepageno \global\advance\RealPno by2
          \ifnum\outputpenalty>-20000 \else\dosupereject\fi}%

% Defaults:
\DontMarkPageCorner
\BotMargin=0pt

%--------------------------------------------------------------------------
\catcode`\@=14   % Need this in places.
\endinput
```

6.5 Simple Book Formatting (BookForm.tex)

```
%%% File: BookForm.tex
%%% Contents: The formatting commands for the 'inner' right hand pages
%%% of some of the Examples in 'Mathematical TeX by Example'.

% The commands shown here should suggest how a relatively simple
% TeX file may be used to create a variety of different output formats:
% Examples 2--5 were all formatted with the commands given here, yet
% they each have a different 'look'.

\input EightPt  % To use small fonts.
\nopagenumbers  % Switch off normal page numbering.

%--------------------------------------------------------------------------
%%% TYPEFACE NAMES:

\font\Smallcaps=cmcsc10
\font\Sansserif=cmss10

%--------------------------------------------------------------------------
```

```
%%% FOR DISPLAYED FORMULAS:

\newcount\EquationNo      \EquationNo=1

% Some convolutions are needed in order to allow the equation number to
% 'leak out' and to make it correctly to the headline:

\def\EqLbl #1{(\the\EquationNo #1)\xdef\Store{.$\>$\the\EquationNo #1}%
   \global\postdisplaypenalty10000
   \global\skip0=\belowdisplayskip
   \global\skip1=\belowdisplayshortskip
   \global\belowdisplayskip=-.8\baselineskip
      \global\advance\belowdisplayskip by\skip0
   \global\belowdisplayshortskip=-.8\baselineskip
      \global\advance\belowdisplayshortskip by\skip1
   \aftergroup\PlaceMark
   \def\AA{#1}\def\BB{}\ifx\AA\BB
            \global\advance\EquationNo by1 \fi}%
\def\PlaceMark{\mark{\Store}\par \global\postdisplaypenalty0
            \global\belowdisplayskip=\skip0
            \global\belowdisplayshortskip=\skip1
            \noindent\ignorespaces}%
\def\FirstEquationOnPage{\edef\Testfm{\firstmark}%
               \edef\Testtm{\topmark}%
               \edef\Testbm{\botmark}%
               \ifx\Testfm\Testtm\else\firstmark\fi}%
\def\LastEquationOnPage{\edef\Testfm{\firstmark}%
               \edef\Testtm{\topmark}%
               \edef\Testbm{\botmark}%
               \ifx\Testbm\Testtm\else\botmark\fi}%
\def\PrevEq {{\advance\EquationNo by -1 \the\EquationNo}}
\def\PrevEqs #1{{\advance\EquationNo by -#1 \the\EquationNo}}
\def\ShiftEquationNo #1{\advance\EquationNo by #1}


%---------------------------------------------------------------------------
%%% FOR FOOTNOTES:

\newcount\FootnoteNo      \FootnoteNo=1
\let\OldFootnoteCommand=\footnote

\def\NumberFootnotes{\def\footnote{\OldFootnoteCommand
      {$^{\the\FootnoteNo}$\aftergroup\global\aftergroup\advance
         \aftergroup\FootnoteNo\aftergroup b\aftergroup y\aftergroup
         1\aftergroup\relax}}}
\def\DontNumberFootNotes{\def\footnote{\OldFootnoteCommand}}
\def\FootnoteSize #1{\dimen\footins=#1} % To set maximum vertical size.

%---------------------------------------------------------------------------
```

```
%%% FORMATTING CHAPTERS, SECTIONS, SUBSECTIONS:

\newcount\ChapIndicator    % New chapter indicator.
\newcount\ChapNo           % Chapter number.
\newcount\SectNo           % Section number.
\newcount\SubSectNo        % Subsection number.

\def\ChapterFonts #1#2{\font\ChpFont=#1
                       \font\ChpNameFont=#2}
\def\SectionFont #1{\font\SctFont=#1}
\def\SubSectionFont #1{\font\SbSctFont=#1}

\newskip\StartofChapSkip         \StartofChapSkip=.75in
\newskip\InChapTitleSkip         \InChapTitleSkip=3\bigskipamount
\newskip\AfterChapTitleSkip      \AfterChapTitleSkip=2\bigskipamount
\newskip\BetweenSectionSkip      \BetweenSectionSkip=2\bigskipamount
\newskip\AfterSectionTitleSkip   \AfterSectionTitleSkip=\bigskipamount
\newtoks\ChapName            % To store the chapter name.
\newtoks\SectionName         % To store the section name.
\newtoks\EveryChapter        % The end of this file shows the default.
\newtoks\EverySection        % Ditto
\newtoks\EverySubSection     % Ditto

\newif\ifResetSectNo         \ResetSectNotrue
\def\DontResetSectNos{\ResetSectNofalse}

\def\StartBook{\ChapNo=0 \SectNo=0 \SubSectNo=0 \ChapIndicator=0 \pageno=1
               \ChapName{}\SectionName{}}%
\def\Chapter #1{\ChapIndicator=1 \ifResetSectNo \SectNo=0 \fi
                \advance\ChapNo by 1
                \topglue\StartofChapSkip\relax \ChapName={#1}%
                \ChPosition{\ChpFont\the\EveryChapter}%
                \nobreak\vskip\InChapTitleSkip\relax
                \Position{\ChpNameFont #1}%
                \nobreak\vskip\AfterChapTitleSkip\relax}%
\def\Section #1{\goodbreak \vskip\BetweenSectionSkip\relax \SectionName={#1}%
                \advance\SectNo by1  \EquationNo=1
                \Position{\SctFont\the\EverySection\ #1}%
                \nobreak\vskip\AfterSectionTitleSkip\relax}%
\def\SubSection #1{\goodbreak \vskip\BetweenSectionSkip\relax
                   \advance\SubSectNo by1
                   \Position{\SbSctFont\the\EverySubSection\ #1}%
                   \nobreak\vskip\AfterSectionTitleSkip\relax}%
\def\SemiLeftTitles {\let\Position=\Leftlines \let\ChPosition=\rightline }
\def\LeftTitles {\let\Position=\Leftlines \let\ChPosition=\leftline }
\def\CenteredTitles {\let\Position=\Centerlines \let\ChPosition=\Centerlines }
\def\Centerlines #1{{\parindent0pt \pretolerance10000
                     \rightskip 0pt plus 1fil \leftskip 0pt plus 1fil
                     \parfillskip 0pt #1 \par}}
\def\Leftlines #1{{\parindent0pt \pretolerance10000
                   \rightskip0pt plus 1fil \parfillskip 0pt #1 \par}}

%--------------------------------------------------------------------------
```

```
%%% FOR THEOREMS, ETC.:

\def\PropositionFonts #1#2{\font\PropLabelFont=#1
                          \font\PropStateFont=#2}
\def\ProofFont #1{\font\PrfFont=#1}
\def\Proposition #1.#2\par{\removelastskip \vskip 1.5\bigskipamount
    \noindent {\PropLabelFont #1.\ } {\PropStateFont #2} \endgraf}
\def\ProofsEtc #1.{\removelastskip \medskip
    \noindent {\PrfFont #1.\quad}}

%-------------------------------------------------------------------------------
%%% MISCELLANEOUS:

\def\EndLine{\hfil\break}
\def\EndPage{\vfil\eject}
\def\Lcase #1{\lowercase\expandafter{#1}}
\def\RomanNumeral #1{\uppercase\expandafter{\romannumeral #1}}
\def\Hfil {\aftergroup\hfil}

%-------------------------------------------------------------------------------
%%% THE HEADLINE, ETC.:

\newtoks\LeftRunningHead
\newtoks\RightRunningHead
\def\TopLRPno {\headline{\ifnum\ChapIndicator=1 \hfil
  \else \ifodd\pageno \the\RightRunningHead\llap{\folio}\else
                      \rlap{\folio}\the\LeftRunningHead\fi \fi}
    \footline{\ifnum\ChapIndicator=1 \hfil\sevenrm\folio\hfil \else \hfil\fi}}

%-------------------------------------------------------------------------------
%%% THE OUTPUT ROUTINE:

\newdimen\LMarginOdd      \LMarginOdd=1 in
\newdimen\LMarginEven     \LMarginEven=1 in
\output{\ifodd\pageno \hoffset=\LMarginOdd \else \hoffset=\LMarginEven \fi
        \advance\hoffset by -1in % Because 1in is the Plain TeX left margin.
        \advance\hoffset by.375in % To conform to the publisher's rules.
        \plainoutput \global\ChapIndicator=0 }

%-------------------------------------------------------------------------------
%%% DEFAULTS:

\EveryChapter={C$\>$H$\>$A$\>$P$\>$T$\>$E$\>$R\ \ \the\ChapNo}
\EverySection={\the\ChapNo.\the\SectNo}
\EverySubSection={\the\ChapNo.\the\SectNo.\the\SubSectNo.}
\SemiLeftTitles
\TopLRPno
\RightRunningHead={\hfil\the\ChapNo.\the\SectNo.\ \the\SectionName\hfil}
\LeftRunningHead={\hfil\the\ChapNo.\ {\Smallcaps\the\ChapName}\hfil}
\SectNo=0  \SubSectNo=0
\OnFootnoterule{.8125in}

%-------------------------------------------------------------------------------
\endinput
```

Bibliography

EVERAL BOOKS and articles were consulted when the present account was being written. An annotated list follows. The list isn't meant to cover all the literature on the typesetting of mathematics, or even on TₑX, merely those items that were used as references for this book. The development of TₑX is an important step in the tradition of typesetting, a tradition which—lying as it does at the intersection of art, technology and commerce—has been marked throughout by conflicting pulls, by tensions between the old and the new. Some of the references given below are not directly about the subject of the present book, but they help in conveying a sense of the tradition to which TₑX belongs.

Many of the articles listed here were published in *TUGboat*, the official "Communications of the TₑX Users Group." The journal, which began publication in 1980, is a generally good source of information on TₑX. Unfortunately, it isn't always easy to get. The TₑX Users Group can help with current and back issues. Their address is given on the inside back cover. Another very lively source of ideas, this time related to broader issues, is the journal *Visible Language*. Currently available from the Rhode Island School of Design, it is a "quarterly concerned with all that is involved in our being literate."

In addition to published material, electronic bulletin boards and discussion groups were also used as sources of information. They provided, as always, rich mixtures of hard fact and soft opinion (the latter usually strongly worded). The inside back cover says a little more about electronic resources.

ARVIND BORDE [1992]. *TₑX by Example: A Beginner's Guide*, Academic Press, Boston, Massachusetts.

An introductory companion to the present book.

D. J. R. BRUCKNER [1990]. *Frederick Goudy*, Harry N. Abrams, Inc., New York, New York.

A biography of the typographer Frederick Goudy. (See also GOUDY [1918].) In his review of the biography (published in The New York Times Book Review, December 16, 1990), John Updike made, in passing, a pointed observation on computerized typesetting and type design: "computer setting [is] a process of pure imaging whose virtually infinite resources have as yet done little to change the look of the printed page. The letters that are still used are closely based upon

233

prototypes of roman and italic faces developed in the Renaissance, by printers imitating the calligraphy of manuscripts."

VICTOR EIJKHOUT and ANDRIES LENSTRA [1991]. *The Document Style Designer as a Separate Entity*, TUGboat, **12** (1), 31–34.

A proposal for "meta-formats" that document style designers can use with little knowledge of TEX. The article contains an example from such a format, called *Lollipop* by the authors. The format was also used in the production of *TEX by Topic* (EIJKHOUT [1992]).

VICTOR EIJKHOUT [1992]. *TEX by Topic: A TEXnician's Reference*, Addison-Wesley, Reading, Massachusetts.

An advanced book that systematically covers all the basic mechanisms of TEX. The "universe of TEX" (as the author calls it) is presented through a series of short, but very clear, discussions augmented by some telling examples.

MICHAEL J. FERGUSON [1990]. *Report on Multilingual Activities*, TUGboat, **11** (4), 514–516.

A description of the 256-character extended font layout, adopted in 1990.

FREDERICK W. GOUDY [1918, 1942]. *The Alphabet* and *Elements of Lettering*, Dorset Press, New York, New York (1989 reprint).

A short study of the development of the English alphabet and of printing types.

ANDRÉ GÜRTLER and CHRISTIAN MENGELT [1985]. *Fundamental Research Methods and Form Innovations in Type Design Compared to Technological Developments in Type Production*, Visible Language, **XIX** (1), 123–147.

An historical study of type design and its connections with the available technologies of type production.

ALAN HOENIG [1991]. *Labelling Figures in TEX documents*, TUGboat, **12** (1), 125–128.

A discussion of the uses of METAFONT for drawing pictures for inclusion in TEX documents, and for correctly labelling these pictures.

DONALD E. KNUTH [1979a]. *Mathematical Typography*, Bulletin of the American Mathematical Society, **1** 337–372.

The text of the Josiah Willard Gibbs Lecture given to the American Mathematical Society on January 4, 1978. The article is reprinted, with corrections, as part of the book *TEX and METAFONT* (KNUTH [1979b]).

DONALD E. KNUTH [1979b]. *TEX and METAFONT: New Directions in Typesetting*, The American Mathematical Society and Digital Press, Bedford, Massachusetts.

A three-part book, describing early work on TEX. Part 1 reprints a corrected version of *Mathematical Typography* (KNUTH [1979a]). Parts 2 and 3 are, respectively, the manuals for early versions of TEX and METAFONT. There is a lot of fascinating material in this book, historical and otherwise, and it is worth trying to track down a copy.

DONALD E. KNUTH [1980]. *The Letter S*, The Mathematical Intelligencer, **2** (3), 114–122.

A detailed description of the difficulties that were encountered in drawing an

'S' that would fit satisfactorily into the general scheme of the then-developing Computer Modern typefaces.

DONALD E. KNUTH [1982]. *The Concept of a Meta-Font*, Visible Language, **XVI** (1), 3–27.

A discussion of the concept of a meta-font (not to be confused with the program called METAFONT), with a short annotated reading list of earlier work on similar ideas. Responses to this article and Donald Knuth's response to the responses were published in issue No. 4 of the same volume of the journal.

DONALD E. KNUTH [1984, 1986]. *The TEXbook*, Addison-Wesley, Reading, Massachusetts.

The basic source of information on TEX, straight from the composer's keyboard. The book is an expansive, opulently detailed discourse on TEX, more a piece of literature than a conventional manual. As with all good literature, it bears—indeed requires—re-readings: one comes away with something new from it every time. The book is absolutely indispensable for a serious understanding of TEX. It is available either by itself, or as Volume A of a five-volume series called *Computers and Typesetting*. (Note: TEX underwent major internal revision in 1989, so it is important to use a post-1989 printing of *The TEXbook*. The copyright page will say if the book describes the new versions—3.0 or later—of TEX.)

DONALD E. KNUTH [1984]. *TEX Incunabula*, TUGboat, **5** (1), 4–11.

A history of the earliest uses of TEX for book production.

DONALD E. KNUTH [1985]. *Lessons learned from METAFONT*, Visible Language, **XIX** (1), 35–53.

An account of the several years of work that went into producing acceptable typefaces with METAFONT.

DONALD E. KNUTH [1986a]. *TEX: The Program*, Addison-Wesley, Reading, Massachusetts.

Volume B of *Computers & Typesetting*. It contains a complete program listing of TEX, along with detailed discussions. It also provides an example of Knuth's idea of "literate programming."

DONALD E. KNUTH [1986b]. *The METAFONTbook*, Addison-Wesley, Reading, Massachusetts.

This book is to METAFONT what *The TEXbook* is to TEX. It is also available as Volume C of *Computers & Typesetting*.

DONALD E. KNUTH [1986c]. *METAFONT: The Program*, Addison-Wesley, Reading, Massachusetts.

Volume D of *Computers & Typesetting*. It contains the full program listing of METAFONT, and it provides another example of "literate programming."

DONALD E. KNUTH [1986d]. *Computer Modern Typefaces*, Addison-Wesley, Reading, Massachusetts.

Volume E of *Computers & Typesetting*. It contains discussions and listings of the programs that draw all the characters of the Computer Modern class of typefaces.

DONALD E. KNUTH [1986e]. *Remarks to Celebrate the Publication of Computers & Typesetting*, *TUGboat*, **7** (2), 95–97.

The remarks were made at a coming out party for the just-published series, held on May 21, 1986, at the Boston Computer Museum. They contain an account of the factors that motivated the author's work (which began, according to a diary entry quoted in the talk, on Thursday, May 5, 1977).

DONALD E. KNUTH [1989a]. *Typesetting Concrete Mathematics*, *TUGboat*, **10** (1), 31–36.

An account of the typesetting of the book *Concrete Mathematics*. The project was apparently the first large-scale use of the *AMS Euler* typefaces, and it eventually led to the creation of a complementary new text face (called *Computer Concrete*). Both sets of faces are shown in Example 7 and in Chapter 5 of the present book. (A correction to some erroneous parameter values in the original article appears in issue No. 3, page 342, of the same volume.)

DONALD E. KNUTH [1989b]. *The New Versions of TEX and METAFONT*, *TUGboat*, **10** (3), 325–328.

A discussion of the changes made to TEX for version 3.0 of the program. (A few corrections to the original article appear on page 12 of Volume 11, No. 1, of the same journal.)

DONALD E. KNUTH [1989c]. *The Errors of TEX*, *Software—Practice and Experience*, **19** 607–685.

An overview of the author's work on TEX, describing the forces that motivated him, the evolution of his approach to designing the program, the difficulties that he encountered along the way and all the mistakes that he made. The article ends with a remarkable "Error Log of TEX" that lists every mistake and explains how it was fixed. The total count of errors until September 20, 1991—this count is from an extended version of the article that appears in the book *Literate Programming* (KNUTH [1992])—was 1,259. The article is highly recommended, even to those with no special interest in the development of TEX itself, as a fascinatingly honest account of how scientific work is actually done.

DONALD E. KNUTH [1990a]. *Virtual Fonts: More Fun for Grand Wizards*, *TUGboat*, **11** (1), 13–23.

A description of a method of combining TEX with fonts from sources other than METAFONT.

DONALD E. KNUTH [1990b]. *The Future of TEX and METAFONT*, *TUGboat*, **11** (4), 489.

Donald Knuth's farewell to TEX, METAFONT and Computer Modern.

DONALD E. KNUTH [1992]. *Literate Programming*, Center for the Study of Language and Information.

A reprint volume, mostly on the concept of literate programming. There are several articles here of direct interest to TEX.

DONALD E. KNUTH, TRACY LARRABEE, and PAUL M. ROBERTS [1989]. *Mathematical Writing*, The Mathematical Association of America.

Lecture notes from a course of the same name, given by Donald Knuth at Stanford University in the autumn quarter of 1987. This is a compendium of pieces of advice—how to write clearly about technical matters, how to present algorithms and proofs, how to get rich by writing books—supplemented by several entertaining anecdotes. The course included guest lectures by Paul Halmos, Leslie Lamport, Mary-Claire van Leunen and others.

DONALD E. KNUTH and MICHAEL F. PLASS [1981]. *Breaking Paragraphs into Lines, Software—Practice and Experience*, 1119–1184.

A detailed discussion of the line-breaking algorithm in TeX, and related matters.

DONALD E. KNUTH, TOMAS G. ROKICKI, and ARTHUR SAMUEL [1990]. *META-FONTware*, TeXniques Number 13.

TeXniques is series of reports and documents distributed by the TeX Users Group. Number 13 is a technical report on METAFONT-related programs, containing "literate" program listings and discussions.

LESLIE LAMPORT [1986]. *LaTeX: A Document Preparation System*, Addison-Wesley, Reading, Massachusetts.

The primary source of information on using LaTeX, written by the author of the package. It contains, among other things, a short discussion of the notion of logical design (in section 1.4).

LESLIE LAMPORT [1987]. *Document Production: Visual or Logical*, Notices of the American Mathematical Society, **34** 621–624.

A forceful argument in favor of logical design.

BASIL MALYSHEV, ALEXANDER SAMARIN, and DIMITRI VULIS [1991]. *Russian TeX*, TUGboat, **12** (2), 212–214.

A discussion of the preliminary procedures that were undertaken to get TeX to handle Russian. See also VULIS [1989].

FRANK MITTELBACH and RAINER SCHÖPF [1989]. *A New Font Selection Scheme for TeX Macro Packages*, TUGboat, **10** (2), 222–238.

A description of the commands that make up NFSS.

FRANK MITTELBACH and RAINER SCHÖPF [1990]. *The New Font Family Selection: User Interface to Standard LaTeX*, TUGboat, **11** (2), 297–305.

A description of how to use NFSS commands in LaTeX.

HUBERT PARTL [1988]. *German TeX*, TUGboat, **9** (1), 70–72.

A description of the ingredients of a German style file. The file was used in simplifying the input for Example 6 of the present book.

TOMAS ROKICKI [1985]. *Packed (PK) Font File Format*, TUGboat, **6** (3), 115–120.

A technical article introducing the now widely used ".pk format" for files where character shape information is stored.

MICHAEL SPIVAK [1982, 1990]. *The Joy of TeX* (2nd ed.), The American Mathematical Society, Providence, Rhode Island.

The official guide to AMS-TeX by the principal author of the package. Editions of this book came out in 1982, 1983 and 1986, after a preliminary version was circulated in 1981, making this the first widely used book on TeX.

MICHAEL SPIVAK [1989]. *A$_{\mathcal{M}}$S-T$_E$X: The Synthesis*, The T$_E$Xplorators Corporation, Houston, Texas.

The official guide to A$_{\mathcal{M}}$S-T$_E$X, by the author of the package.

ELLEN SWANSON [1979]. *Mathematics into Type*, The American Mathematical Society, Providence, Rhode Island.

Contains a lot of useful information on conventions for typesetting mathematics.

ELLEN SWANSON [1980]. *Publishing and T$_E$X*, TUGboat, **1** (1), 7–9.

A discussion of the times and costs involved in composing and printing pages of technical material. The article includes estimates of the savings that might result from both electronic submission and computer typesetting. It ends with these words: "It is unrealistic to assume that all articles in a publication will be submitted in T$_E$X. Not all authors have the temperament or desire to learn the T$_E$X codes and not all university departments will have a T$_E$X expert. However, there would still be a considerable saving in the cost of publication if only part of the papers for a book or journal were submitted by the authors on magnetic tape produced by the T$_E$X system."

JAN TSCHICHOLD [1991]. *The Form of the Book: Essays on the Morality of Good Design*, Hartley & Marks, Point Roberts, Washington.

The original book was published in German in 1975; the 1991 English translation is by Hajo Hadeler. This is an opinionated, often provocative book, but with much factual information also strewn about its pages.

PIET TUTELAERS [1992]. *A Font and a Style for Typesetting Chess Using L^{A}T$_E$X or T$_E$X*, TUGboat, **13** (1), 85–90.

A discussion of a chess font developed by the author, and of a chess style file that makes it very easy to typeset profusely illustrated chess games. The chess figures on page 115 of the present book are from one of the author's fonts.

MARY-CLAIRE VAN LEUNEN [1978, 1992]. *A Handbook for Scholars*, Oxford University Press, Oxford, U.K.

An astonishingly lively book, for one that runs the gamut from "Citation" to "Appendix 2: Federal Documents of the United States." It provides information on conventions for footnotes, references, etc., conventions that need to be more widely appreciated as more and more authors do their own typesetting. The author says at the start that the book is "arranged seductively, not chronologically," and she's right.

DIMITRI VULIS [1989]. *Notes on Russian T$_E$X*, TUGboat, **10** (3), 332–336.

A description of the author's work on a Russian-language version of T$_E$X. See also MALYSHEV *et al.* [1991].

MICHAEL J. WICHURA [1987]. *The P$_I$CT$_E$X manual*, T$_E$Xniques Number 6.

The official manual for P$_I$CT$_E$X, by the author of the package. It is available from the T$_E$X Users Group.

R. E. YOUNGEN, W. B. WOOLF, and D. C. LATTERNER [1989]. *Migration from Computer Modern Fonts to Times Fonts*, TUGboat, **10** (4), 513–519.

A discussion of the reasons behind the shift back to Times Roman fonts at the American Mathematical Society, and of the obstacles encountered on the way.

HERMANN ZAPF [1985]. *Future Tendencies in Type Design: The Scientific Approach to Letterforms*, Visible Language, **XIX** (1), 23–34.

A discussion of new type-designing technologies, particularly computerized design, from one of the foremost type designers of the day. Hermann Zapf has had a long and influential association with the TEX project, and he has designed new typefaces (the *AMS Euler* family) for the American Mathematical Society.

HERMANN ZAPF [1987]. *Hermann Zapf & His Design Philosophy*, The Society of Typographic Arts, Chicago, Illinois.

A reprint volume containing several of Hermann Zapf's articles as well as many examples of his designs. The articles include ones on the effects of technological development on type design.

———— [1969, 1982]. *The Chicago Manual of Style*, The University of Chicago Press, Chicago, Illinois.

A weighty, authoritative guide to style; it includes a chapter on "Mathematics in Type."

———— [1985–1990]. *PostScript Language Reference Manual* (2nd ed.), Adobe Systems Incorporated and Addison-Wesley, Reading, Massachusetts.

The official PostScript reference manual.

Glossary/Index

EVERY COMMAND from Plain TeX and $\mathcal{AMS}$-TeX that is likely to be useful in typesetting mathematics is listed here. Many individual topics dealing with the typesetting of mathematics are also listed and discussed. Page numbers are given against commands and topics if they have been used or discussed elsewhere in the book.

An attempt has been made here to provide some information on how TeX does the behind-the-scenes calculations on which it bases its typesetting decisions. A few definitions of standard commands are also shown. It is hoped that these details will provide an introduction to such matters for readers who are interested in designing their own formats—without unduly distracting those who primarily want to use TeX, not shape it.

Primitive commands of TeX are marked by an asterisk (*). $\mathcal{AMS}$-TeX commands are marked thus: $\mathcal{AMS}$. Unless otherwise indicated, special $\mathcal{AMS}$-TeX symbols come from one of the two $\mathcal{AMS}$-TeX symbols fonts. The $\mathcal{AMS}$-TeX fonts have to be specially loaded when needed; this is discussed in Chapter 3.

Unmarked commands are all part of the Plain TeX package and unmarked symbols all come from the fonts in the basic TeX distribution (in a couple of cases different TeX and $\mathcal{AMS}$-TeX commands have to be used to invoke them). In cases where Plain TeX commands might clash with ones from $\mathcal{AMS}$-TeX, it is so noted.

Some commands are sandwiched between $ signs when they head an entry: this indicates that they work only in math **mode**. Things that work outside math **mode** (e.g., punctuation), but which follow special spacing rules within it, are also sandwiched between $ signs. (See **spacing**.) Unless otherwise noted, commands listed without $ signs will work both outside and inside math mode.

▷ **\ ⁎** (i.e., \ followed by a space.) Gives a single space.

▷ **$** Starts and ends math mode.

▷ **\\$** ⌈$⌉

▷ **$$** Starts and ends the display math mode.

▷ **&** The standard choice for a "tabbing character" in alignments. It signals a jump to the next column.

▷ **\\&** ⌈&⌉ Also see \and.

▷ **$($** ⌈(⌉ (Class 4: opening.) Also see delimiters.

▷ **$)$** ⌈)⌉ (Class 5: closing.) Also see delimiters.

▷ **$[$** ⌈[⌉ (Class 4: opening.) \lbrack yields the same result. Also see delimiters.

▷ **$]$** ⌈]⌉ (Class 5: closing.) \rbrack yields the same result. Also see delimiters.

▷ **$\\{$** ⌈{⌉ (Class 4: opening.) \lbrace yields the same result. Also see delimiters. In $\mathcal{AMS}$-TeX the command works outside math mode as well.

▷ **$\\}$** ⌈}⌉ (Class 5: closing.) \rbrace yields the same result. Also see delimiters. In $\mathcal{AMS}$-TeX the command works outside math mode as well.

▷ **$+$** ⌈+⌉ (Class 2: binary operation.) Also see spacing.

▷ **\\+** See \settabs and \cleartabs.

▷ **$-$** ⌈−⌉ (Class 2: binary operation.) Also see spacing. Note: outside math mode it gives ⌈-⌉, i.e., a hyphen.

▷ **\\⁎** Provides a "discretionary ×": see line breaks for an example of its use and \discretionary for its definition.

▷ **$/$** ⌈/⌉ (Class 0: ordinary.) Also see spacing. May be used in text, with the same effect: for example, 'yes/no' gives 'yes/no'. If a line break is to be permitted at the /, then \slash must be used in its place.

▷ **\\/ ⁎** Allows users to make an "italic correction": see \it.

▷ **$|$** ⌈|⌉ (Class 0: ordinary.) Alternate name: \vert. If used outside math mode, | gives ⌈—⌉. Also see absolute value and delimiters.

▷ **$\\|$** ⌈∥⌉ (Class 0: ordinary.) Alternate name: \Vert. Also see absolute value and delimiters.

▷ **** Plays the role of Plain TeX's \cr in forming alignments of formulas in $\mathcal{AMS}$-TeX. See, for example, \align.

▷ **$<$** ⌈<⌉ (Class 3: relation.) Warning: outside math mode it usually gives ⌈¡⌉ (but see §5.28).

▷ **$=$** ⌈=⌉ (Class 3: relation.) Also see spacing.

▷ **\\=** In Plain TeX this gives a text accent: \=a gives ā. In $\mathcal{AMS}$-TeX it is undefined. See \B.

▷ **$>$** ⌈>⌉ (Class 3: relation.) Warning: outside math mode it usually gives ⌈¿⌉ (but see §5.28).

▷ $\>$ Gives a medium space in Plain TeX, but is undefined in $\mathcal{AMS}$-TeX. See medspace and keywords.

▷ $,$ ⌈,⌋ (Class 6: punctuation.) Also see spacing.

▷ $\,$ Gives a thin space. See keywords. In $\mathcal{AMS}$-TeX the command works outside math mode as well.

▷ $.$ ⌈.⌋ (Class 6: punctuation.) It may also be used to represent an invisible delimiter. See the discussion under \left.

▷ \. In Plain TeX, \.o gives ȯ. In $\mathcal{AMS}$-TeX the command is replaced by \D (in both packages, \d gives a "dot under" accent in text), and \. is redefined to give a period with a normal space after it. Thus 'Proc\. Roy\. Soc\. of' will give 'Proc. Roy. Soc. of' instead of the 'Proc. Roy. Soc. of' that would be obtained from using '.' in place of '\.'. See \spacefactor for a discussion of interword spacing.

▷ $;$ ⌈;⌋ (Class 6: punctuation.) Also see spacing.

▷ $\;$ Gives a thick space. See keywords. In $\mathcal{AMS}$-TeX the command works outside math mode as well.

▷ $:$ ⌈:⌋ (Class 3: relation.) Also see \colon, punctuation and spacing.

▷ $\!$ Gives a negative thin space. See keywords. In $\mathcal{AMS}$-TeX the command works outside math mode as well.

▷ $_$ Produces subscripts: X_2 gives X_2, and Y_{ab} gives Y_{ab}.

▷ _ ⌈_⌋

▷ $^$ Produces superscripts: X^2 gives X^2, and Y^{ab} gives Y^{ab}.

▷ \^ ⌈^⌋

▷ ^^ A device used to represent unprintable characters: for example, ^^M represents a carriage return. A full discussion is given in §5.2.

▷ ~ Used as a "tie" in preventing awkward or inelegant line breaks. See penalties.

▷ \~ ⌈~⌋

▷ @| $\mathcal{AMS}$ See \CD.

▷ @<<< $\mathcal{AMS}$ See arrows and \CD.

▷ @= $\mathcal{AMS}$ See \CD.

▷ @>>> $\mathcal{AMS}$ See arrows and \CD.

▷ @,, @., @;, @:, @?, @! $\mathcal{AMS}$ See periods.

▷ @AAA $\mathcal{AMS}$ See \CD.

▷ @vert $\mathcal{AMS}$ Works like @|; see \CD.

▷ @VVV $\mathcal{AMS}$ See \CD.

▷ \aa ⌈å⌋

▷ \AA ⌈Å⌋

▷ $\above$ * Gives a fraction the thickness of whose fraction bar can be controlled: ${a\above1pt b}$ gives $\frac{a}{b}$, ${a\above.5pt b}$ gives $\frac{a}{b}$, ${a\above0pt b}$ gives $\frac{a}{b}$, etc.

▷ $\abovewithdelims$ * Allows the construction of general fraction-like structures; it permits control over the thickness of the fraction bar, and specification of the left and right delimiters. For example:

$\{0 \abovewithdelims \lbrace \rbrace .4pt 1\}$ gives $\{\frac{0}{1}\}$,

$\{\alpha \abovewithdelims ( ) 0pt \beta\}$ gives $\binom{\alpha}{\beta}$, and

$\{a \abovewithdelims \Vert . 1pt b\}$ gives $\left\|\frac{a}{b}\right.$.

If the constructed symbol is to be used often, it is usually convenient to define an appropriate abbreviation. For example, the definition

```
\def\Ket #1#2{{#1 \abovewithdelims\vert\rangle0pt #2}}
```

allows one to say $\Ket ab$ to get $\left|\frac{a}{b}\right\rangle$. Also see \atopwithdelims.

▷ **absolute value** 63, 76, 79, 81. $|x|$ is the quick way to get the absolute value of x, $|x|$. A more cumbersome way—but necessary if the | key is disabled or if | is being used as a reserved symbol for other purposes—is to use \vert x\vert. There's a twist, however: neither gives quite the right **spacing** (see the discussion under delimiters). The preferred way is to use \left and \right; for example, $\left| -y\right|$ gives $|-y|$ (whereas $|-y|$ gives $|-y|$, spacing more appropriate at the start for the difference of | and y).

▷ **Acc** See atom.

▷ \accent * 156–157, 210, 212. Used in defining names for text accents.

▷ \accentedsymbol $\mathcal{AMS}$ The correct positioning of double accents is usually tricky. See the discussion of \skew under **spacing**. $\mathcal{AMS}$-TEX comes with some built-in commands that may be used to compose double accents automatically: \Hat, \Check, \Tilde, \Acute, \Grave, \Dot, \Ddot, \Breve, \Bar and \Vec. These commands carry out positioning calculations behind the scenes, and they therefore eat computer time. If a doubly accented character is going to be used frequently, the command \accentedsymbol can be used to store away the newly created character for later use (so that it does not have to be specially constructed each time). For example,

```
\accentedsymbol\AAcute{\Acute{\Acute X}}
```

allows the use of the made-up name \AAcute to call upon the new character.

▷ **accents** 76, 84, 156–157, 165–167, 169, 170, 210–215. The following mathematics accents are available in Plain and $\mathcal{AMS}$-TEX:

$\hat{o}$ $\hat o$	$\check{o}$ $\check o$	$\tilde{o}$ $\tilde o$	$\acute{o}$ $\acute o$
$\grave{o}$ $\grave o$	$\dot{o}$ $\dot o$	$\ddot{o}$ $\ddot o$	$\breve{o}$ $\breve o$
$\bar{o}$ $\bar o$	$\vec{o}$ $\vec o$		

If i or j are to be accented, they must first be "undotted." $\imath$ and $\jmath$ give $\imath$ and $\jmath$, respectively. In addition, $\mathcal{AMS}$-TEX provides \dddot and \ddddot:

$\dddot x$ gives $\dddot{x}$; $\ddddot x$ gives $\ddddot{x}$.

Wide accents are also available; they can change size to match the expression underneath:

$\bar{o}$	$\overline o$	$\overline{abcde}$	$\overline{abcde}$
$\underline{o}$	$\underline o$	$\underline{abcde}$	$\underline{abcde}$
$\widehat{ab}$	$\widehat{ab}$	$\widehat{abcde}$	$\widehat{abcde}$
$\widetilde{ab}$	$\widetilde{ab}$	$\widetilde{abcde}$	$\widetilde{abcde}$

\overline and \underline provide lines that can grow to accommodate arbitrarily large expressions. The other two wide accents have a maximum size that can only comfortably accommodate three characters. If, however, the msbm font has been loaded when using $\mathcal{AMS}$-TeX, these wide accents can go a step further and embrace four characters. It is also possible to accent accented expressions, though one has to watch out for the positions of the accents. See \accentedsymbol and spacing.

The various other \over... and \under... commands, listed separately, can also be used to give accented expressions.

The vertical position of an accent normally adjusts itself to the height of the character underneath (see the discussion under \fontdimen5 in § 5.6). If a uniform height is desired, that can be achieved as shown in Example 1c in Chapter 2.

Finally, $\mathcal{AMS}$-TeX has a facility that allows the use of accents as superscripts: i.e., it allows one to get expressions like $X^\vee$. The commands that achieve this are \sphat ... \spvec; one types $X\spcheck$, for example, to get the symbol just shown.

Plain TeX's accent commands are mostly defined in terms of a primitive command called \mathaccent.

▷ $\acute$ 165. $\acute a$ produces $\acute{a}$.

▷ $\Acute$ $\mathcal{AMS}$ See \accentedsymbol.

▷ \acuteaccent $\mathcal{AMS}$ May be used in place of \' to get an acute accent in text.

▷ \adjustfootnotemark $\mathcal{AMS}$ See \footnotemark.

▷ \advance * 66, 70, 74, 98, 216. Advances values stored in TeX's internal registers. For example, \advance\count13 by7 advances by 7 the current value stored in the register named \count13 (i.e., if the original value was -5, the new one will be 2).

▷ \ae [æ]

▷ \AE [Æ] 156, 157.

▷ \after $\mathcal{AMS}$ Part of a combination of commands that gives dots in matrices after some specified entity. See matrices.

▷ \aftergroup * 230, 232. Places the next single item (more precisely, token) after the current group. For example, '{X \aftergroup\AE Y Z }' gives 'X Y Z Æ'. If several \aftergroups are encountered, the items are placed in the order in which they occurred.

▷ $\aleph$ ⌐ℵ⌐ 90–91, 157. (Class 0: ordinary.)

▷ $\align$ ᴀₘₛ 118–123. An equation alignment command, similar in some ways to Plain TEX's \eqalignno. The input

```
$$\align
A&=B+C+D\\
e&=1+D.
\endalign$$
```

will produce

$$A = B + C + D$$
$$e = 1 + D.$$

Using \tag at the end of a line (before \\) will place an equation number for that line. See \tag for a discussion of the placement of equation numbers. Also, see \gather for information on including an alignment inside a larger structure.

   Page breaks are not normally permitted within such an alignment, but that can be changed. Typing \allowdisplaybreaks within the formula (i.e., after the opening $$ signs), but before \align, will allow a page break to occur, if needed, after any line of the display. Typing \allowdisplaybreak after an occurrence of \\ will allow a page break after the just-completed line, and typing \displaybreak will force a page break there. Textual matter may be introduced within an alignment by typing \intertext{textual matter}. Vertical spacing can be adjusted in an alignment by using \vspace or \spreadlines.

▷ $\alignat$ ᴀₘₛ 124–125. Allows alignments that are aligned in more than one place. For example,

$$A + B = C, \qquad X = x - y, \qquad Q < P - M^2;$$
$$D = 2E, \qquad YZ > z, \qquad N = 0.$$

is obtained from

```
$$\alignat{3}
A+B & =C, & \qquad X & =x-y, & \qquad Q&<P-M^2; \\
D & =2E, & \qquad YZ & >z, & \qquad N&=0.
\endalignat $$
```

The number specified after \alignat is the number of *pairs* of formulas to be aligned. An & is used both to separate pairs and to indicate a tab within each pair. Additional commands like \tag, \vspace, \allowdisplaybreak and \intertext perform the same functions here as they do in \align.

Alignments can be stretched by using \xalignat and \xxalignat in place of \alignat. The first of these stretches the alignment out across the page, but leaves space for the equation numbers that a \tag command might leave; the second stretches the alignment all the way from margin to margin. The effect of saying \xxalignat in place of \alignat in the example shown above would be:

$$A + B = C, \qquad X = x - y, \qquad Q < P - M^2;$$
$$D = 2E, \qquad YZ > z, \qquad N = 0.$$

▷ $\aligned$ $\mathcal{A_MS}$ 118–119. Allows several different alignments in the same display, much like Plain TEX's \eqalign. For example,

```
$$\aligned A&=B \\ C+D&=E \\ F&=G-H \endaligned
\qquad
\aligned \alpha&=0 \\ \beta+\gamma&=1. \endaligned$$
```

produces

$$\begin{aligned} A &= B \\ C + D &= E \\ F &= G - H \end{aligned} \qquad \begin{aligned} \alpha &= 0 \\ \beta + \gamma &= 1. \end{aligned}$$

An equation number may be assigned to the whole display by using \tag after the last \endaligned. The different sets of formulas in such a display may be made to line up along their bottoms (instead of their centers) by using \botaligned in place of \aligned, and along their tops by using \topaligned.

▷ $\alignedat$ $\mathcal{A_MS}$ This is to \aligned what \alignat is to \align; it is used like \alignat.

▷ **alignments**   20, 85, 117.   Familiarity with TEX's basic alignment mechanism, \halign, and with \settabs is assumed in this book. Alignments of mathematical material can be achieved in a variety of ways: see \eqalign, \eqalignno, \leqalignno for some Plain TEX mechanisms, and \align, \alignat, \aligned and \alignedat for $\mathcal{A_MS}$-TEX ones. Also see matrices.

▷ \allowbreak   54–55.   Suggests a possible break at a place that TEX would normally not have chosen on its own. Consider, for example, this: $(a - b)(c - d)$ $(e - f)(g - h)$; the input for it is

```
    ... this: $(a-b)(c-d)\allowbreak(e-f)\allowbreak(g-h)$; the ...
```

Without the \allowbreak, TEX would not have broken the formula at the point that it did. The definition of the command is simply '\penalty0 '; see penalties.

▷ $\allowdisplaybreak$ $\mathcal{A_MS}$ See \align.

▷ $\allowdisplaybreaks$ $\mathcal{A_MS}$ See \align.

▷ \allowlinebreak $\mathcal{A_MS}$ Suggests a possible line break in the middle of text. It is defined in terms of Plain TEX's \allowbreak and used similarly, but—unlike that command—it will not work in math mode. (It has refinements built in that check the mode and deliver error messages when needed.) Also see \linebreak.

▷ **$\allowmathbreak$** $\mathcal{AMS}$ Suggests a possible line break in the middle of formula within text (i.e., not displayed). Like \allowlinebreak, it is based on \allowbreak and used similarly, but it cannot be used outside a formula within text. Also see \mathbreak.

▷ **α** $\lceil\alpha\rceil$ 18–19, 24–25, 28–35, 80–83, 128–131, 158, 162, 166–167.

▷ **$\amalg$** $\lceil\text{II}\rceil$ (Class 2: binary operation.)

▷ **AMS Euler typefaces** 92–95, 135, 196–207, 236, 239. A class of typefaces designed for the American Mathematical Society by Hermann Zapf. The class consists of medium and bold weights of three different typeface styles: "roman" (more accurately a handwriting-like, or cursive, style), fraktur and script. The fraktur alphabet is shown under \frak, but here are the medium weights of the other two:

$$A\,B\,C\,D\,E\,F\,G\,H\,I\,J\,K\,L\,M\,N\,O\,P\,Q\,R\,S\,T\,U\,V\,W\,X\,Y\,Z$$
$$a\,b\,c\,d\,e\,f\,g\,h\,i\,j\,k\,l\,m\,n\,o\,p\,q\,r\,s\,t\,u\,v\,w\,x\,y\,z$$
$$\mathcal{A\,B\,C\,D\,E\,F\,G\,H\,I\,J\,K\,L\,M\,N\,O\,P\,Q\,R\,S\,T\,U\,V\,W\,X\,Y\,Z}$$

Only uppercase letters are available in the script face. The bold characters in both faces, along with full font tables, are displayed in Chapter 5.

▷ **AMS preprint style** 134, 135. A document style encouraged by the American Mathematical Society for the formatting of preprints. The commands that define the style are supplied in a style file called amsppt.sty. The style is discussed in Chapter 3.

▷ **AMS symbols** 135, 167–169, 192–195, 212. $\mathcal{AMS}$-TEX comes with two special fonts of symbols. Access to these symbols can be arranged either through a comprehensive command like \UseAMSsymbols or symbol by symbol through the \newsymbol command. In the latter case, one needs to know the code numbers assigned to each symbol. Here is a complete list of the symbols, their code numbers and their standard names. All but the first four must be used in mathematics mode. A "(U)" next to a name indicates that it is already in use in Plain TEX; thus \undefine must first be used if the AMS symbol is preferred over the Plain version.

- *Text symbols*

✓	\checkmark	®	\circledR
✠	\maltese	¥	\yen

Note: These may be used in formulas as well.

- *Delimiters*

⌜	\ulcorner	⌝	\urcorner
⌞	\llcorner	⌟	\lrcorner

Note: Larger versions of these are not obtainable with \left and \right.

- *Constructed symbols*

    --→ \dashrightarrow, \dasharrow   ←-- \dashleftarrow

- *Lowercase Greek letters*

  Ϝ   207A \digamma   ϰ   207B \varkappa

- *Hebrew letters*

  ℶ   2069 \beth   ℷ   206A \gimel
  ℸ   206B \daleth

- *Miscellaneous symbols*

  ℏ   207E \hbar (U)   \   2038 \backprime
  ℏ   207D \hslash   ∅   203F \varnothing
  △   234D \vartriangle   ▲   204E \blacktriangle
  ▽   204F \triangledown   ▼   2048 \blacktriangledown
  □   2003 \square   ■   2004 \blacksquare
  ◊   2006 \lozenge   ♦   2007 \blacklozenge
  Ⓢ   2073 \circledS   ★   2046 \bigstar
  ∠   205C \angle (U)   ⊲   205E \sphericalangle
  ∡   205D \measuredangle   𝕜   207C \Bbbk
  ∄   2040 \nexists   ∁   207B \complement
  ℧   2066 \mho   ð   2067 \eth
  ⅃   2060 \Finv   ╱   231E \diagup
  ℽ   2061 \Game   ╲   231F \diagdown

- *Binary operators*

  ∔   2275 \dotplus   ⋉   226E \ltimes
  ╲   2272 \smallsetminus   ⋊   226F \rtimes
  ⋒   2265 \Cap, \doublecap   ⋋   2268 \leftthreetimes
  ⋓   2264 \Cup, \doublecup   ⋌   2269 \rightthreetimes
  ⊼   225A \barwedge   ⋏   2266 \curlywedge
  ⊻   2259 \veebar   ⋎   2267 \curlyvee
  ⩞   225B \doublebarwedge   ⊺   227C \intercal
  ⊟   220C \boxminus   ⊝   227F \circleddash
  ⊠   2202 \boxtimes   ⊛   227E \circledast
  ⊡   2200 \boxdot   ⊚   227D \circledcirc
  ⊞   2201 \boxplus   .   2205 \centerdot
  ⋇   223F \divideontimes

- *Binary relations*

  ≦   2335 \leqq   ≧   233D \geqq
  ⩽   2336 \leqslant   ⩾   233E \geqslant
  ⪕   2330 \eqslantless   ⪖   2331 \eqslantgtr
  ≲   232E \lesssim   ≳   2326 \gtrsim
  ⪅   232F \lessapprox   ⪆   2327 \gtrapprox
  ≊   2375 \approxeq
  ⋖   236C \lessdot   ⋗   236D \gtrdot
  ⋘   236E \lll, \llless   ⋙   236F \ggg, \gggtr

AMS symbols

⋚	2337	\lessgtr	⋛	233F	\gtrless
⋚	2351	\lesseqgtr	⋛	2352	\gtreqless
⪋	2353	\lesseqqgtr	⪌	2354	\gtreqqless
≐	232B	\doteqdot, \Doteq	≖	2350	\eqcirc
≑	233A	\risingdotseq	≗	2324	\circeq
≒	233B	\fallingdotseq	≜	232C	\triangleq
∽	2376	\backsim	∼	2373	\thicksim
≏	2377	\backsimeq	≈	2374	\thickapprox
⊆	236A	\subseteqq	⊇	236B	\supseteqq
⋐	2362	\Subset	⋑	2363	\Supset
⊏	2340	\sqsubset	⊐	2341	\sqsupset
≼	2334	\preccurlyeq	≽	233C	\succcurlyeq
⋞	2332	\curlyeqprec	⋟	2333	\curlyeqsucc
≾	232D	\precsim	≿	2325	\succsim
⪷	2377	\precapprox	⪸	2376	\succapprox
⊲	2343	\vartriangleleft	⊳	2342	\vartriangleright
⊴	2345	\trianglelefteq	⊵	2344	\trianglerighteq
⊨	230F	\vDash	⊩	230D	\Vdash
⊪	230E	\Vvdash			
⌣	2360	\smallsmile	∣	2370	\shortmid
⌢	2361	\smallfrown	∥	2371	\shortparallel
≏	236C	\bumpeq	≬	2347	\between
≎	236D	\Bumpeq	⋔	2374	\pitchfork
∝	235F	\varpropto	϶	237F	\backepsilon
◀	234A	\blacktriangleleft	▶	2349	\blacktriangleright
∴	2329	\therefore	∵	232A	\because

- *Negated relations*

≮	2304	\nless	≯	2305	\ngtr
≰	2302	\nleq	≱	2303	\ngeq
⪇	230A	\nleqslant	⪈	230B	\ngeqslant
≦̸	2314	\nleqq	≧̸	2315	\ngeqq
⪇	230C	\lneq	⪈	230D	\gneq
≨	2308	\lneqq	≩	2309	\gneqq
⪇	2300	\lvertneqq	⪈	2301	\gvertneqq
⋦	2312	\lnsim	⋧	2313	\gnsim
⪉	231A	\lnapprox	⪊	231B	\gnapprox
⋠	2306	\nprec	⋡	2307	\nsucc
⋠	230E	\npreceq	⋡	230F	\nsucceq
⪵	2316	\precneqq	⪶	2317	\succneqq
⋨	2310	\precnsim	⋩	2311	\succnsim
⪹	2318	\precnapprox	⪺	2319	\succnapprox

≁	231C	\nsim		≇	231D	\ncong
∤	232E	\nshortmid		∦	232F	\nshortparallel
†	232D	\nmid		∦	232C	\nparallel
⊬	2330	\nvdash		⊭	2332	\nvDash
⊮	2331	\nVdash		⊯	2333	\nVDash
⋪	2336	\ntriangleleft		⋫	2337	\ntriangleright
⋬	2335	\ntrianglelefteq		⋭	2334	\ntrianglerighteq
⊄	232A	\nsubseteq		⊅	232B	\nsupseteq
⊈	2322	\nsubseteqq		⊉	2323	\nsupseteqq
⊊	2328	\subsetneq		⊋	2329	\supsetneq
⊊	2320	\varsubsetneq		⊋	2321	\varsupsetneq
⊊	2324	\subsetneqq		⊋	2325	\supsetneqq
⊊	2326	\varsubsetneqq		⊋	2327	\varsupsetneqq

- *Arrows*

⇇	2312	\leftleftarrows		⇉	2313	\rightrightarrows
⇆	231C	\leftrightarrows		⇄	231D	\rightleftarrows
⇚	2357	\Lleftarrow		⇛	2356	\Rrightarrow
↞	2311	\twoheadleftarrow		↠	2310	\twoheadrightarrow
↢	231B	\leftarrowtail		↣	231A	\rightarrowtail
↫	2322	\looparrowleft		↬	2323	\looparrowright
⇋	230B	\leftrightharpoons		⇌	230A	\rightleftharpoons (U)
↶	2378	\curvearrowleft		↷	2379	\curvearrowright
↺	2309	\circlearrowleft		↻	2308	\circlearrowright
↰	231E	\Lsh		↱	231F	\Rsh
⇈	2314	\upuparrows		⇊	2315	\downdownarrows
↿	2318	\upharpoonleft		↾	2316	\upharpoonright,
⇃	2319	\downharpoonleft				\restriction
⊸	2328	\multimap		⇁	2317	\downharpoonright
↭	2321	\leftrightsquigarrow		↝	2320	\rightsquigarrow

- *Negated arrows*

↚	2338	\nleftarrow		↛	2339	\nrightarrow
⇍	233A	\nLeftarrow		⇏	233B	\nRightarrow
↮	233D	\nleftrightarrow		⇎	233C	\nLeftrightarrow

▷ $\mathcal{AMS}$-T_EX 1, 2, 4, 6, 8, 9, 11, 12, 59, 75, 116–129, 133–138, 148, 159, 166, 167–169, 178, 186, 188, 190, 192–203, 219, 237. The special package of T_EX commands adopted by the American Mathematical Society. See the chapter summarizing the package, and references scattered throughout this book and this Glossary. The input that produces the $\mathcal{AMS}$-T_EX logo is \AmSTeX defined by:

```
%%% From the amstex.tex file:
\def\AmSTeX{{\textfontii A\kern-.1667em%
  \lower.5ex\hbox{M}\kern-.125emS}-\TeX}
%%% The symbol can also be specially defined in Plain TeX:
\def\AmSTeX{$\cal A\kern-.1667em\lower.5ex\hbox{$
     \cal M$}\kern-.125em S$-\TeX}
```

▷ `amsppt.sty` 134–136. The file that defines the AMS preprint style.

▷ `amssym.def` 74, 84, 135. A file that takes the minimum steps necessary in order to make available the **AMS** symbols. It may be used in place of the full `amstex` file (see § 3.3).

▷ `amssym.tex` 74, 84, 135, 169. The file that assigns names to new symbols that come from the special AMS mathematics **fonts**. The file may be used either with $\mathcal{A}_{\mathcal{M}}\mathcal{S}$-TeX or without it; see Chapter 3.

▷ `\AmSTeX` $_{\mathcal{A_{MS}}}$ 219. See $\mathcal{A}_{\mathcal{M}}\mathcal{S}$-TeX.

▷ `amstex.tex` 116, 133, 135. The file that contains the definitions of $\mathcal{A}_{\mathcal{M}}\mathcal{S}$-TeX commands. See Chapter 3.

▷ `$\and$` $_{\mathcal{A_{MS}}}$ $\lceil \& \rceil$

▷ `$\angle$` $\lceil \angle \rceil$ (Class 0: ordinary.) In Plain TeX the symbol is a composite, defined in a very complicated way as a table in which the angled line given by `\not` is carefully placed on top of an `\hrule`:

```
\def\angle{{\vbox{\ialign{$\m@th\scriptstyle##$\crcr
      \not\mathrel{\mkern14mu}\crcr \noalign{\nointerlineskip}
      \mkern2.5mu\leaders\hrule height.34pt\hfill\mkern2.5mu\crcr}}}}
```

The command `\ialign` is essentially the same as `\halign`, but appropriately initialized; `\m@th` cancels extra spacing around formulas. Both are discussed in their own entries. The rest of the commands are, for the most part, spacing adjustments to make everything look correct. The command `\mathrel` forces its argument (just some space in this case) to act as if it were in the relation class.

In $\mathcal{A}_{\mathcal{M}}\mathcal{S}$-TeX, if `\undefine` is used first to sweep away the old definition (see that entry for details), then the same `\angle` command can be made to produce ∠, a single character from the `msam` font. See `\newsymbol` for information on the procedure. The symbol now shrinks correctly in subscripts and superscripts (if, for any reason, it is used there); the composite one doesn't.

▷ `$\approx$` $\lceil \approx \rceil$ 40–41, 54–55. (Class 3: relation.)

▷ `$\approxeq$` $_{\mathcal{A_{MS}}}$ $\lceil \approxeq \rceil$ (Class 3: relation.)

▷ `$\arccos$` $\lceil \arccos \rceil$

▷ `$\arcsin$` $\lceil \arcsin \rceil$

▷ `$\arctan$` $\lceil \arctan \rceil$

▷ `$\arg$` $\lceil \arg \rceil$

▷ **arrays** See alignments.

▷ **arrows** 43, 57, 71, 98–101, 112–113, 128–131, 137. See accents, AMS symbols, commutative diagrams, delimiters, \overleftarrow, \overrightarrow and relations. $\mathcal{AMS}$-TeX also makes available a special arrow mechanism:

`$@>a>>$`	gives	$\xrightarrow{a}$,
`$@<A+b<<$`	gives	$\xleftarrow{A+b}$,
`$@>>\alpha>$`	gives	$\xrightarrow[\alpha]{}$,
`$@<{<}<b<$`	gives	$\xleftarrow[b]{<}$,

and so on.

▷ `$\arrowvert$` 100–101. See delimiters.

▷ `$\Arrowvert$` See delimiters.

▷ `$\ast$` $\lceil * \rfloor$ 16–19, 24–25, 30–31, 36–37, 78–79, 82–83, 98–99, 110, 126–127. (Class 2: binary operation.)

▷ `$\asymp$` $\lceil \asymp \rfloor$ (Class 3: relation.)

▷ **at** 142. A keyword that is used to specify the size of a font. For example, '\font\Test=cmmi10 at 25pt' asks for the name \Test to be assigned to cmmi10, the 10-point Computer Modern math italic font, at a size of 25 points. For this to work, the font must be available at the specified size. See §5.1.

▷ **atom** In TeX, a formula is made up of entities that are either "atoms" or "others." The "others" are any of these: spacing commands (\hskip, \mskip, \nonscript, \kern, \mkern); commands that introduce horizontal or vertical material (penalties, rules, \mark, \insert, \vadjust, etc.); style change commands (\displaystyle, etc.); commands that produce fraction-like structures (\over, \above, etc.); boundary commands (arising from using \left or \right); commands constructed through \mathchoice; or higher-level commands defined in terms of the preceding entities. Entities not on this list are atoms.

Atoms are classified into 13 types:

Ord (ordinary),	e.g., a, x, etc.
Op (large operator),	e.g., $\int$, $\sum$, etc.
Bin (binary operation),	e.g., $+$, $\div$, etc.
Rel (relation),	e.g., $=$, $\sim$, etc.
Open (opening),	e.g., $($, $\lfloor$, etc.
Close (closing),	e.g., $)$, $]$, etc.
Punct (punctuation),	e.g., $;$, $:$, etc.
Inner (inner),	e.g., $\frac{1}{2}$, $\frac{a}{b}$, etc.
Over (overline),	e.g., $\bar{b}$, $\bar{x}$, etc.
Under (underline),	e.g., $\underline{a}$, $\underline{y}$, etc.
Acc (accented),	e.g., $\vec{b}$, $\tilde{x}$, etc.
Rad (radical),	e.g., $\sqrt{z}$, etc.
Vcent (centered vbox),	e.g., a box produced by \vcenter.

The space between adjacent atoms in a formula is determined by their types. See spacing for more details. An atom is considered to have three constituents, called its *fields*: a *nucleus*, a *superscript* and a *subscript*. A field may be empty, may consist of a single symbol, may be a box, or may itself be a formula. For example, the input `[a^2+y_i^3]^{1+\sqrt 2}` when typed as part of a formula (i.e., between $ signs) produces $[a^2 + y_i^3]^{1+\sqrt 2}$. Here is the atomic analysis of this formula (empty fields are left blank):

Atom	Type	Nucleus	Superscript	Subscript
[	Open	[		
a^2	Ord	a	2	
+	Bin	+		
y_i^3	Ord	y	3	i
$]^{1+\sqrt 2}$	Close	]	$1 + \sqrt 2$	

The final superscript field is itself a formula that can be broken down into further atomic constituents 1, $+$ and $\sqrt 2$.

▷ $\atop$ ∗ 66, 94–95. `${x\to 0 \atop y\to 1}$` gives $\genfrac{}{}{0pt}{}{x\to 0}{y\to 1}$. Structures like this are useful for things like double limits on integrals and sums.

▷ $\atopwithdelims$ ∗ Similar to \abovewithdelims, discussed above, but less general than it: there is no fraction bar and one can only specify the delimiters that are to be used. The Plain TEX structures given by \brace, \brack and \choose are defined in terms of this primitive command:

```
\def\brace{\atopwithdelims\{\}}
\def\brack{\atopwithdelims[]}
\def\choose{\atopwithdelims()}
```

Separate entries under each of the three commands show examples of their use.

▷ axis 172, 188, 192, 194, 196, 198, 200, 208. A horizontal line that is imagined to pass through every formula in TEX. Components of the formula are vertically centered about the axis, as described below. The vertical location of this line with respect to the baseline is given by the value of the font parameter \fontdimen22 from family 2 in the current style (§5.5 explains the notion of font families, and §5.6 that of font parameters).

The axis is used to place these constituents of a formula: boxes created by using \vcenter (centered about the axis); entities in the class called large operator, like $\sum$ and $\prod$ (centered); fractions created with \over (the center of each fraction bar is placed along the axis); and delimiters (centered; see the discussion under \delimitershortfall). For example, in the formula $A + \sum_{n\in N} \frac{1}{n} = \Sigma$, the axis is 2.5 pt above the baseline. (The Σ at the end is the Greek letter obtained by saying \Sigma; observe that it is placed on the baseline, unlike the summation symbol.)

▷ \B $\mathcal{AMS}$ Plays the role of the Plain TEX command '\=': \B a gives $\bar{a}$. Note: \= is undefined in $\mathcal{AMS}$-TEX; and in both packages, \b a gives $\underline{a}$.

▷ $\backepsilon$ $\mathcal{A_MS}$ ⌈϶⌉ (Class 3: relation.)

▷ $\backprime$ $\mathcal{A_MS}$ ⌈\⌉ (Class 0: ordinary.)

▷ $\backsim$ $\mathcal{A_MS}$ ⌈∽⌉ (Class 3: relation.)

▷ $\backsimeq$ $\mathcal{A_MS}$ ⌈⋍⌉ (Class 3: relation.)

▷ $\backslash$ ⌈\⌉ (Class 0: ordinary.)

▷ $\bar$ 20–21, 46–49, 52–53, 78–81, 120–123, 126–127, 130–131. $\bar a$ produces $\bar{a}$.

▷ $\Bar$ $\mathcal{A_MS}$ See \accentedsymbol.

▷ $\barwedge$ $\mathcal{A_MS}$ ⌈$\overline{\wedge}$⌉ (Class 2: binary operation.)

▷ \baselineskip 40–41, 46–47, 74, 100–101, 112–113, 126–127, 143, 153–154, 170, 174, 216, 222, 224, 225, 226, 228, 230. Sets the gap between baselines of successive lines of text; the value assigned in Plain TeX is 12pt. Standard conventions for this distance vary; common choices are "20% more than the stated size of the typeface being used" (thus a 10-point font would be set on baselines 12 points apart, a 6-point font on baselines 7.2 points apart) and "2 points more than the size of the font" (here a 6-point font would be set on baselines 8 points apart). The final choice depends also on factors like the weight of the type: a heavy-looking typeface sometimes requires a little more space than a light one of the same stated size.

▷ $\Bbb$ $\mathcal{A_MS}$ 118–121. Produces "blackboard bold" characters, if the msbm font has been loaded. See \load..., and Chapter 3. For example, \Bbb{AB} gives $\mathbb{AB}$ (the braces aren't necessary if only a single such character is needed). With one exception (see \Bbbk), only uppercase letters are available:

$$\mathbb{A\,B\,C\,D\,E\,F\,G\,H\,I\,J\,K\,L\,M\,N\,O\,P\,Q\,R\,S\,T\,U\,V\,W\,X\,Y\,Z}$$

▷ $\Bbbk$ $\mathcal{A_MS}$ ⌈k⌉ This is the only lowercase blackboard bold letter available. (Class 0: ordinary.)

▷ $\because$ $\mathcal{A_MS}$ ⌈∵⌉ (Class 3: relation.)

▷ \begingroup * 36, 74, 218. See grouping.

▷ $\belowdisplayshortskip$ * 84, 220, 230. See displays.

▷ $\belowdisplayskip$ * 84, 220, 230. See displays.

▷ β ⌈β⌉ 32–35, 82–83, 94–95, 130–131, 156.

▷ $\beth$ $\mathcal{A_MS}$ ⌈ℶ⌉ (Class 0: ordinary.)

▷ $\between$ $\mathcal{A_MS}$ ⌈≬⌉ (Class 3: relation.)

▷ \bf 8, 12, 20–25, 28–29, 32–37, 48–49, 52–53, 58, 62–63, 65–67, 70–71, 78–81, 93–99, 102–103, 110–111, 126–127, 136, 139, 142, 149–153, 159, 162, 163, 167, 176, 204, 220. ($\mathcal{A_MS}$-TeX alert: don't \redefine.)
{\bf bold type} gives **bold type**. In Plain TeX this continues to work in math mode (except that some characters may not have bold versions), but in $\mathcal{A_MS}$-TeX the command \bold must be used instead in mathematical expressions. The \bold command isn't used as \bf is: see the discussion

under it. §5.3 and §5.5 together give a full discussion of how commands like \bf work.

▷ \bgroup An implicit left brace. The Plain TEX definition of the command is '\let\bgroup={'; see grouping.

▷ $\big$ See delimiters.

▷ $\Big$ ($\mathcal{AMS}$-TEX alert: don't \redefine.) See delimiters.

▷ \bigbreak 34–35. Suggests a good page break point to TEX. If the program does not take the suggestion, it will leave a \bigskip worth of space there (unless it immediately follows a vertical space of a size already bigger than a \bigskip).

▷ $\bigcap$ $\lceil\bigcap\rfloor$ in a line, $\lceil\bigcap\rfloor$ in a display. 78–79. (Class 1: large operator.)

▷ $\bigcirc$ $\lceil\bigcirc\rfloor$ (Class 2: binary operation.)

▷ $\bigcup$ $\lceil\bigcup\rfloor$ in a line, $\lceil\bigcup\rfloor$ in a display. 78–79. (Class 1: large operator.)

▷ $\bigg$ 100–101. See delimiters.

▷ $\Bigg$ See delimiters.

▷ $\biggl$ 92–93. See delimiters.

▷ $\Biggl$ 118–119. See delimiters.

▷ $\biggm$ See delimiters.

▷ $\Biggm$ See delimiters.

▷ $\biggr$ 92–93. See delimiters.

▷ $\Biggr$ 118–119. See delimiters.

▷ $\bigl$ 26–27, 34–35, 72–73, 100–101, 120–121. See delimiters.

▷ $\Bigl$ 28–29, 72–73. See delimiters.

▷ $\bigm$ 26–27. See delimiters.

▷ $\Bigm$ 122–123. See delimiters.

▷ $\bigodot$ $\lceil\bigodot\rfloor$ in a line, $\lceil\bigodot\rfloor$ in a display. (Class 1: large operator.)

▷ $\bigoplus$ $\lceil\bigoplus\rfloor$ in a line, $\lceil\bigoplus\rfloor$ in a display. (Class 1: large operator.)

▷ $\bigotimes$ $\lceil\bigotimes\rfloor$ in a line, $\lceil\bigotimes\rfloor$ in a display. (Class 1: large operator.)

▷ \bigpagebreak $\mathcal{AMS}$ Functions like Plain TEX's \bigbreak, but it also checks to see if the command is being issued in an appropriate place (i.e., in vertical mode).

▷ $\bigr$ 26–27, 34–35, 72–73, 100–101, 120–121. See delimiters.

▷ $\Bigr$ 28–29, 72–73. See delimiters.

▷ \bigskip 11, 12, 34–35, 70–71, 76–77. Gives a vertical space of a certain standard size. The relevant definitions are

```
\newskip\bigskipamount % A new 'skip' register
\bigskipamount=12pt plus 4pt minus 4pt
\def\bigskip{\vskip\bigskipamount}
```

▷ $\bigsqcup$ ⌐⌐⌐ in a line, ⌐⌐⌐ in a display. (Class 1: large operator.)

▷ $\bigstar$ 𝒜ℳ𝒮 ⌐★⌐ 126–127. (Class 0: ordinary.)

▷ $\bigtriangledown$ ⌐▽⌐ 124–125. (Class 2: binary operation.)

▷ $\bigtriangleup$ ⌐△⌐ (Class 2: binary operation.)

▷ $\biguplus$ ⌐⊎⌐ in a line, ⌐⊎⌐ in a display. (Class 1: large operator.)

▷ $\bigvee$ ⌐∨⌐ in a line, ⌐∨⌐ in a display. (Class 1: large operator.)

▷ $\bigwedge$ ⌐∧⌐ in a line, ⌐∧⌐ in a display. (Class 1: large operator.)

▷ **Bin** See atom and \mathbin.

▷ **binary operations**

$+$	$+$	$\ddagger$	$\ddagger$	$\dagger$	$\dagger$	$\odot$	$\odot$
$-$	$-$	$\backslash$	$\setminus$	$\cdot$	$\cdot$	$\oplus$	$\oplus$
$/$	$/$	$\times$	$\times$	$\star$	$\star$	$\ominus$	$\ominus$
$.$	$.$	$\cap$	$\cap$	$\cup$	$\cup$	$\otimes$	$\otimes$
$\pm$	$\pm$	$\sqcap$	$\sqcap$	$\sqcup$	$\sqcup$	$\oslash$	$\oslash$
$\mp$	$\mp$	$\triangleleft$	$\triangleleft$	$\bullet$	$\bullet$	$\bigcirc$	$\bigcirc$
$\wr$	$\wr$	$\triangleright$	$\triangleright$	$\uplus$	$\uplus$	$\circ$	$\circ$
$\div$	$\div$	$\bigtriangleup$	$\bigtriangleup$	$\vee$	$\vee$	$\diamond$	$\diamond$
$\ast$	$\ast$	$\bigtriangledown$	$\bigtriangledown$	$\wedge$	$\wedge$	$\amalg$	$\amalg$

The command $\setminus$ produces a '\', and $\backslash$ (listed under symbols below) also produces a '\'. But the spacings are different: e.g., $A\setminus B$ gives $A \setminus B$, whereas $A\backslash B$ gives $A\backslash B$. There are alternate names for $\wedge$ and $\vee$: $\land$ and $\lor$. Similarly, $\lnot$ is an alternate for $\neg$ (listed under symbols).

See AMS symbols for several further binary operator symbols.

▷ **binary relations** See relations.

▷ $\binom$ 𝒜ℳ𝒮 $\binom{n}{k}$ is produced by $\binom nk$. Also see \dbinom and \tbinom.

▷ **binomial coefficients** Produced by using \choose or, in 𝒜ℳ𝒮-TEX, \binom.

▷ \binoppenalty * The penalty assigned to breaking a line at a binary operation, like a +, in a math formula. The Plain TEX assignment is \binoppenalty=700; this can be changed if it is desired to make line breaks at operations more, or less, attractive. See line breaks and penalties.

▷ **blackboard bold** See \Bbb.

▷ \BlackBoxes 𝒜ℳ𝒮 First see \NoBlackBoxes; \BlackBoxes switches the black "overfull box" indicators back on.

▷ $\blacklozenge$ 𝒜ℳ𝒮 ⌐◆⌐ (Class 0: ordinary.)

▷ $\blacksquare$ $\mathcal{AMS}$ ⌈■⌉ 124–125. (Class 0: ordinary.)

▷ $\blacktriangle$ $\mathcal{AMS}$ ⌈▲⌉ (Class 0: ordinary.)

▷ $\blacktriangledown$ $\mathcal{AMS}$ ⌈▼⌉ (Class 0: ordinary.)

▷ $\blacktriangleleft$ $\mathcal{AMS}$ ⌈◀⌉ (Class 3: relation.)

▷ $\blacktriangleright$ $\mathcal{AMS}$ ⌈▶⌉ (Class 3: relation.)

▷ $\bmatrix$ $\mathcal{AMS}$ 124–125. Used in conjunction with \endbmatrix to get a matrix with brackets around it. See matrices.

▷ $\bmod$ 102–103. '$x\bmod y$' gives '$x \bmod y$'; see \mathbin for the definition. Also see \pmod, \mod and \pod.

▷ $\bold$ $\mathcal{AMS}$ 118–127, 130–131, 167. The $\mathcal{AMS}$-TeX way to get some boldface characters in a mathematical expression: instead of ${\bf x}$ (the Plain TeX method), one says $\bold x$. Also see \boldkey and \boldsymbol. If more than a single character at a time is to be put in boldface, say stuff, it is possible to say \bold{stuff} within a formula. This is not, however, a good idea: for example, the combination ff will come out as **ff**, which is not what is usually wanted if each f is a distinct entity. It is better, therefore, to apply \bold separately to each individual letter.

▷ **bold type** 8, 12, 43, 135, 140–142, 149–153, 159, 163, 166–167, 173–174, 176, 178, 180, 186, 188, 196, 198, 200, 204, 208, 210, 213. See \bf, \bold, \boldkey and \boldsymbol.

▷ **boldface** See the directions under **bold type**.

▷ $\boldkey$ $\mathcal{AMS}$ This command works like \bold, but it also gives bold versions of nonletter keyboard characters. For example, $\boldkey+$ gives $+$. The command works with the following input characters:

- $+$ $-$ $=$ $<$ $>$ $($ $)$ $[$ $]$ $|$ $/$ $*$ $.$ $,$ $:$ $;$ $!$ $?$
- All 52 letters, A, . . . , z; with these it gives bold italics: A, . . . , z.
- All 10 numbers, 0, . . . , 9; with these it gives **0**, . . . , **9**, just as \bold does.

The command is effective in certain instances only if bold symbol fonts have been loaded by using \loadbold after the \input amstex statement. For example, $\boldkey x$ will not work without the extra fonts.

▷ $\boldsymbol$ $\mathcal{AMS}$ This command works like \bold and \boldkey, but it is to be used in order to get bold versions of characters like α that are invoked by special commands (like α). For example, $\boldsymbol\alpha$ will give $\boldsymbol\alpha$. Bold versions of the following symbols can be obtained with \boldsymbol:

- All the lowercase and uppercase Greek letters.
- All 52 letters (it has the same effect here as \boldkey).
- All the other standard symbols: e.g., $\boldsymbol\wp$ gives $\wp$.
- Delimiters: e.g., $\boldsymbol\langle$ gives $\langle$. Bold versions of large delimiters—those obtained from using \left, \right, \big, etc.—are, however, not available.

The command is effective in certain instances only if bold symbol fonts have been loaded by using \loadbold after the \input amstex statement. For example, $\boldsymbol\alpha$ will not work without the extra fonts.

▷ $\bordermatrix$ See matrices.

▷ $\bot$ ⌈⊥⌋ (Class 0: ordinary.)

▷ $\botaligned$ 𝒜ℳ𝒮 See \aligned.

▷ \botcaption 𝒜ℳ𝒮 A command that comes with a style file like the one for the AMS preprint style, not a part of 𝒜ℳ𝒮-TEX proper. It is used inside an insert (one given by \midinsert, for example) to specify a caption that will go at the bottom of the insert. Here is some sample input:

```
\midinsert
The actual insert goes here, or space (using '\vspace')
for one to be pasted in.
\captionwidth{3in} % Optional, to override the default.
\botcaption{Diagram A} Optional descriptive stuff goes here.
\endcaption
\endinsert
```

A similar command, \topcaption, is used for captions at the top. See \pagewidth for more information on caption widths.

▷ $\botfoldedtext$ 𝒜ℳ𝒮 See \foldedtext.

▷ \botmark * 223, 225, 230. See \mark.

▷ $\botshave$ 𝒜ℳ𝒮 See \shave.

▷ $\botsmash$ 𝒜ℳ𝒮 (𝒜ℳ𝒮-TEX alert: don't \redefine.) Like \smash, but it only smashes the depth of its argument (i.e., assigns it zero depth).

▷ $\bowtie$ ⌈⋈⌋ (Class 3: relation.)

▷ \box * 66, 70–71, 74, 216. Extracts the contents of box registers. See \copy.

▷ **box operator** 68. A passable symbol may be constructed as follows

```
\font\Bxfnt=cmsy10 scaled\magstep1
\def\BoxOp{\mathord{\hbox{\Bxfnt\char116 }\llap{\Bxfnt\char117 }}}
```

Then $\BoxOp\phi$ will give $\Box\phi$. Also see \square.

▷ $\boxdot$ 𝒜ℳ𝒮 ⌈⊡⌋ (Class 2: binary operation.)

▷ $\boxed$ 𝒜ℳ𝒮 Draws a framed box around a formula. For example, $$\boxed{e^{i\pi}+1=0.}$$ gives

$$\boxed{e^{i\pi} + 1 = 0.}$$

▷ $\boxminus$ 𝒜ℳ𝒮 ⌈⊟⌋ (Class 2: binary operation.)

▷ $\boxplus$ 𝒜ℳ𝒮 ⌈⊞⌋ 124–125. (Class 2: binary operation.)

▷ $\boxtimes$ 𝒜ℳ𝒮 ⌈⊠⌋ (Class 2: binary operation.)

▷ bp 104–105. See keywords and units.

▷ $\brace$ $\left\{{\alpha,\beta \atop x}\right\}$ is produced by ${\alpha,\beta \brace x}$. This is useful in showing things like two-row matrices within a line (and it may also be used in displays). See \atopwithdelims for the definition of the command.

▷ $\bracevert$ ⌜ ┃ ⌝ See delimiters.

▷ $\brack$ $\left[{\alpha \atop \beta}\right]$ is produced by ${\alpha \brack \beta}$. This is useful in showing things like two-row matrices within a line (and it may also be used in displays). See \atopwithdelims for the definition of the command.

▷ \break 14, 52–55, 60–61, 82–83, 219, 228, 232. Forces a line or a page break, depending on whether it is encountered in horizontal or vertical mode. See **line breaks** for more information and **penalties** for the definition.

▷ $\breve$ ă is produced by $\breve a$.

▷ $\Breve$ *AₘS* See \accentedsymbol.

▷ \buffer *AₘS* Represents the size of the extra space that TₑX puts above superscripts and below subscripts. See the brief discussion of \fontdimen13 under *family 3* in §5.6. Its value can be adjusted by using \ChangeBuffer. This command changes the font parameter \fontdimen13: thus, the change is global. The original value can be restored by using \ResetBuffer.

▷ $\buildrel$ 54–57, 98–99. See **special effects**.

▷ $\bullet$ ⌜•⌝ 66, 110–111. (Class 2: binary operation.)

▷ $\bumpeq$ *AₘS* ⌜≏⌝ (Class 3: relation.)

▷ $\Bumpeq$ *AₘS* ⌜≎⌝ (Class 3: relation.)

▷ \bye 82, 85, 90, 94, 100, 102, 108, 114, 126, 130. A better way to end a file than \end.

▷ \c 124–125, 128–129. Normally, in Plain and *AₘS*-TₑX, the command produces a cedilla accent: \c a gives ą. In *AₘS*-TₑX the command also performs the separate function of centering elements of a matrix, when used in conjunction with \format: see **matrices**.

▷ $\cal$ 30–31, 36–39, 76–77, 80–83, 96–97, 136, 157, 163, 174, 188, 219. Produces 𝒞alligraphic 𝒮cript in Plain TₑX (see \Cal for the *AₘS*-TₑX version). Available only in the mathematics mode, and just for uppercase letters. $\cal X$ produces 𝒳. The available letters are

$$\mathcal{A\,B\,C\,D\,E\,F\,G\,H\,I\,J\,K\,L\,M\,N\,O\,P\,Q\,R\,S\,T\,U\,V\,W\,X\,Y\,Z}$$

It is important to remember to confine the effect of this command by using braces around the parts that are to appear in this style.

▷ $\Cal$ 126–128. This is the *AₘS*-TₑX version of \cal. It is used a little differently: $\Cal Z$ will produce 𝒵 and $\Cal{ABC}$ will produce 𝒜ℬ𝒞 (though, as with \bold, it is probably better to apply the command to one letter at a time). Since \Cal acts only on an argument that is fed to it, problems with unrestrained effects of the command do not arise, as they do with \cal.

▷ **calligraphic letters** See \cal and \Cal.

▷ $\cap$ ⌈∩⌉ 38–39, 76–77, 100–101. (Class 2: binary operation.)

▷ $\Cap$ $\mathcal{A}_{\mathcal{M}}S$ ⌈⋒⌉ (Class 2: binary operation.) Alternate name: \doublecap.

▷ **capitals and small capitals** 58, 151, 221. THE FONT THAT INTERPRETS LOW-ERCASE LETTERS AS SMALL CAPITALS. It is also sometimes just called "small capitals," or even "small caps." The font is not automatically loaded with Plain TEX, but it is available in the **AMS preprint style**. See \smc and also \sc. The 10-point **Computer Modern** small caps font file is called cmcsc10.

▷ **captions** Plain TEX provides no special commands for placing captions on figures and tables. The **AMS preprint style** provides a couple: see \botcaption and \topcaption.

▷ \captionwidth $\mathcal{A}_{\mathcal{M}}S$ See \botcaption.

▷ $\cases$ 64–65, 70–71, 94–95, 118–119, 136. Produces displays like:

$$|x| = \begin{cases} +x, & \text{for } x \geq 0; \\ -x, & \text{for } x < 0. \end{cases}$$

The input for this is different in Plain and $\mathcal{A}_{\mathcal{M}}S$-TEX. In Plain TEX, one types

```
$$|x|=\cases{+x,&for $x \geq 0$;\cr
            -x,&for $x < 0$.\cr}$$
```

In $\mathcal{A}_{\mathcal{M}}S$-TEX, one types

```
$$|x|=\cases +x,&\text{for $x \geq 0$};\\
            -x,&\text{for $x < 0$}.\endcases $$
```

In both cases it is possible to use the techniques for adjusting vertical spacing that are discussed under **matrices**.

▷ \catcode * 13, 147, 166, 218, 229. Assigns category codes; see § 5.2. It may also be used to ask what the category code of a character is: for example, \the\catcode`\\\$ gives 3.

▷ **category codes** 13, 147. Code numbers attached to input characters that determine the roles that they (the characters) will play. See § 5.2.

▷ cc See **keywords** and **units**.

▷ $\CD$ $\mathcal{A}_{\mathcal{M}}S$ 128–131. The \CD ... \endCD combination is used in typesetting simple rectangular commutative diagrams in $\mathcal{A}_{\mathcal{M}}S$-TEX. The same combination is also used for more complex diagrams in $\text{L}\mathcal{A}_{\mathcal{M}}S$-TEX (see Examples 17 and 18 in Chapter 2). Here are some $\mathcal{A}_{\mathcal{M}}S$-TEX examples:

$$\begin{array}{ccc} A & \xrightarrow{f} & B \\ \alpha\uparrow & & \downarrow\beta \\ C & \xleftarrow{\quad} & D \\ & {\scriptstyle g} & \end{array}$$

is obtained from

```
$$\CD
A @>f>> B\\
@A\alpha AA @VV\beta V \\
C @<<g< D
\endCD $$
```

It is possible to leave out "vertices" and arrows:

$$
\begin{array}{ccc}
F & \xrightarrow{\ \ g\ \ } & G \\
f \downarrow & & \\
A \times B & &
\end{array}
$$

is obtained from

```
$$\CD
F @>>g> G \\
@VfVV @. \\          % '@.' specifies a missing arrow.
A \times B @.
\endCD $$
```

Long labels may be used if needed. But to get arrows to match in length, some explicit work has to be done:

$$
\begin{array}{ccc}
A \odot B & \xrightarrow{\ \alpha\ } & C \\
h \circ g \downarrow & & \uparrow A \\
D & \xleftarrow[\text{Long label}]{} & E'
\end{array}
$$

is created by

```
$$\define\MatchArr{@>\pretend\alpha\haswidth
        {\text{Long label}}>>}
\CD
A\odot B \MatchArr C \\
@Vh\circ gVV @AA{A}A \\
D @<<\text{Long label}< E'
\endCD $$
```

Finally, it is possible to use double lines instead of arrows, and it is possible to reduce the width of arrows:

$$
\begin{array}{ccc}
X & \to & Y \\
\| & & \downarrow \\
Z & = & U
\end{array}
$$

is produced by

```
$$\minCDarrowwidth{12pt}  % Changes the minimum width.
\CD
X @>>> Y \\
```

```
@| @VVV \\
Z @= U
\endCD $$
```

▷ $\cdot$ ⌈·⌋ 16–17, 30–33. (Class 2: binary operation.)

▷ $\cdots$ ⌈···⌉ 18–19, 24–25, 30–31, 36–41, 60–61, 72–73, 86–91, 96–101. A constructed symbol, assigned the "Inner" atomic type. See atom, spacing and \mathinner.

▷ $\centerdot$ $_{\mathcal{AMS}}$ ⌈·⌋ (Class 2: binary operation.)

▷ \CenteredTagsOnSplits $_{\mathcal{AMS}}$ A global command (i.e., it affects the look of the entire document) that forces the AMS preprint style to center tags (like equation numbers) vertically when they label formulas that are split over many lines. See \tag and also \TopOrBottomTagsOnSplits.

▷ centering Displayed mathematics is automatically centered horizontally in Plain TEX. See displays. Other mechanisms for horizontal centering of stuff include using '\centerline{stuff}' after breaking the preceding line in the paragraph, or by using '\hfil stuff \par' after ending the previous paragraph. And stuff may be vertically centered on a full page by typing '\null \vfil stuff \vfil \eject'; the method also works on the remaining fragment of an already-begun page (\null isn't necessary in that case). Also see axis and \vcenter.

▷ \centerline 12, 24–25, 104–105, 112–113, 216, 224, 227. Centers material horizontally; see centering.

▷ $\cfrac$ $_{\mathcal{AMS}}$ See continued fractions.

▷ \ChangeBuffer $_{\mathcal{AMS}}$ First see \buffer. The size assigned to \buffer can be changed to, say, 3pt by typing \ChangeBuffer{3pt}.

▷ \char * 124–125, 155–156, 161, 210, 216, 223, 224, 228. Calls up characters when fed numbers; e.g., \char10 produces Ω. See § 5.4 for a full discussion.

▷ character codes 143–147. Every character read by TEX from a file is assigned a character code number (essentially its ASCII code). This number defines TEX's internal (i.e., to itself) representation of the character. § 5.2 discusses character codes in detail.

▷ \chardef * 156, 161. Allows names to be defined for characters; see § 5.4.

▷ $\check$ a is produced by $\check a$.

▷ $\Check$ $_{\mathcal{AMS}}$ See \accentedsymbol.

▷ \checkmark $_{\mathcal{AMS}}$ ⌈✓⌋ May be used in math mode as well.

▷ χ ⌈χ⌉ 157.

▷ $\choose$ 34–35, 82–83. ${k \choose m}$ gives $\binom{k}{m}$. Like \brace and \brack, this may be done both within a line and in displays, and it is useful in a variety of contexts. See \atopwithdelims for the definition of the command.

▷ $\circ$ ⌈∘⌋ 26–27, 34–35, 56–57, 78–79, 110–111. (Class 2: binary operation.)

▷ `$\circeq$` AMS ⌈≐⌉ (Class 3: relation.)

▷ `$\circlearrowleft$` AMS ⌈↺⌉ (Class 3: relation.)

▷ `$\circlearrowright$` AMS ⌈↻⌉ (Class 3: relation.)

▷ `$\circledast$` AMS ⌈⊛⌉ (Class 2: binary operation.)

▷ `$\circledcirc$` AMS ⌈⊚⌉ (Class 2: binary operation.)

▷ `$\circleddash$` AMS ⌈⊝⌉ (Class 2: binary operation.)

▷ `\circledR` AMS ⌈®⌉ May be used in math mode as well.

▷ `$\circledS$` AMS ⌈Ⓢ⌉ (Class 0: ordinary.)

▷ **class** See classes.

▷ **classes** 30, 160–161. TₑX regards every character it typesets in math mode as belonging to one of eight classes, numbered from 0 to 7. Classes are listed and discussed in detail in § 5.5, but here is a summary:

Class 0: *Ordinary* (examples: $/$, $\hbar$, $\triangle$).

Class 1: *Large Operator* (examples: $\int$, $\prod$, $\bigcup$).

Class 2: *Binary operation* (examples: $-$, $\times$, $\cup$).

Class 3: *Relation* (examples: $\leq$, $\succeq$, $\doteq$).

Class 4: *Opening* (examples: $($, $[$).

Class 5: *Closing* (examples: $)$, $]$).

Class 6: *Punctuation* (examples: $:$, $;$).

Class 7: *Variable family* (examples: x, y).

For characters in classes 0–6, the class identifies what type of atom it is considered to be, and therefore what the appropriate spacing is around it. Class 7 is special: characters in it are allowed to switch families. (As far as spacing goes, Class 7 characters behave like ones in Class 0.) In addition to the character code and family number assigned to a character, the class code is also assigned to it. The three numbers together comprise the character's `\mathcode`. All these code numbers are discussed in greater detail in § 5.5.

Many entries (in this Glossary) for individual symbols state what class they belong to; if the statement is missing, it can be assumed that the symbol is in class 0.

▷ `\cleartabs` Clears tab settings in an alignment in a way that is useful for typesetting structures like computer programs. For example,

FOR $i = 1$ to n

    IF $i < 5$ THEN

        LET $k = k + i$

    ELSE

        LET $k = k + 2i$

    END IF

NEXT i

IF $k > l$ THEN

    LET $d = k - l$

ELSE
   LET $d = l - k$
END IF

is obtained from this input:

```
\smallskip
\+ FOR \cleartabs&$i=1$ to $n$\cr
\+ &IF &$i<5$ THEN\cr
\+ &&LET $k=k+i$\cr
\+ &ELSE\cr
\+ &&LET $k=k+2i$\cr
\+ &END IF\cr
\+ NEXT $i$\cr
\+ IF \cleartabs&$k>l$ THEN\cr % Clears the old tab positions.
\+ &LET $d=k-l$\cr
\+ ELSE\cr
\+ &LET $d=l-k$\cr
\+ END IF\cr
\smallskip
```

▷ **Close** See atom and \mathclose.

▷ \clubpenalty * 226 The penalty for breaking a page after the first line of a paragraph. See penalties.

▷ $\clubsuit$ ♣ (Class 0: ordinary.)

▷ cm 110, 112. See keywords and units.

▷ cm... 40–41, 42, 58, 74, 84, 124–125, 128–129, 141–142, 149–150, 154, 156, 158, 164, 166, 170, 174–175, 178–191, 202, 212–213, 216, 219–221, 224, 226, 229. The first two letters of the external names of Computer Modern fonts. For example, cmr10 stands for *Computer Modern roman, 10-point* and cmsy10 stands for *Computer Modern symbols, 10-point*. See §5.3 for more information on fonts in general, and on names in the Computer Modern system.

▷ $\colon$: 28–29, 96–97, 114–115. (Class 6: punctuation.) See punctuation.

▷ \columns 46 47. See \settabs.

▷ **comments** A single-line comment—i.e., a note that is to be kept hidden from TEX's processing—may be made by placing % to the left of the comment. Multiline comments may be made in $\mathcal{AMS}$-TEX by sandwiching the lines between \comment and an explicit \endcomment (on a line by itself).

▷ \comment 𝒜ℳ𝒮 See comments.

▷ **commutative diagrams** 56–57, 100–101, 128–131. See \CD and Examples 8, 17 and 18 in Chapter 2.

▷ $\complement$ 𝒜ℳ𝒮 C 76–77. (Class 0: ordinary.)

▷ **compound fractions** This display

$$\frac{\frac{x}{z}}{\frac{u}{v}} \qquad \frac{x/z}{u/v}$$

was obtained from the following input:

$$ \text{\$\$ \{\{x\textbackslash over z\} \textbackslash above.7pt \{u\textbackslash over v\}\} \textbackslash qquad \{x/z \textbackslash over u/v\} \$\$} $$

If fractions of larger size are required in the numerator and denominator of the first sample, each must be prefaced by \displaystyle: for example, one would type \displaystyle{x\over z} to get larger sizes in the numerator.

▷ **compound matrices** 124. This display

$$\begin{bmatrix} \begin{pmatrix} a & b \\ c & d \end{pmatrix} & \mathbf{I} \\ \begin{pmatrix} 0 & -1 \\ 1 & 0 \end{pmatrix} & \begin{pmatrix} e & f \\ g & h \end{pmatrix} \end{bmatrix}$$

is obtained from the following input:

```
$$\left[\matrix{\pmatrix{a&b\cr c&d} & {\bf I}\cr
          \noalign{\smallskip}
          \pmatrix{0&-1\cr 1&0} & \pmatrix{e&f\cr g&h}}\right]$$
```

The \smallskip is employed to adjust the vertical spacing; \noalign is necessary whenever nonaligned material is to be introduced into an alignment. See matrices for more details of matrix-making.

▷ **Computer Modern** 93, 94–95, 142, 148–150, 153, 155, 178–191, 204, 212, 235–236, 238. The name given to the standard family of fonts that go with TeX. See §5.3.

▷ **computer programs** See \cleartabs for one example of how to typeset a simple computer program. More complicated solutions are needed for more complicated programs, especially if a certain degree of automatic formatting (say, keywords in boldface, complex indentation, etc.) is desired. Many TeX archives carry style files that are aimed at automatically formatting computer programs; information about such archives is given on the inside back cover.

▷ **conditionals** TeX provides a rich variety of conditional commands, listed separately in various \if... entries.

▷ **$\cong$** $\lceil \cong \rceil$ (Class 3: relation.)

▷ **continued fractions** The first part of the discussion applies to both Plain and $\mathcal{AMS}$-TeX. The display

$$x_1 + \cfrac{1}{x_2 + \cfrac{1}{x_3 + \cfrac{1}{x_4 + \cdots}}}$$

is obtained from the following input:

```
$$ x_1 + {1\hfill \over \displaystyle x_2 +
        {\strut 1\hfill \over \displaystyle x_3 +
            {\strut 1\hfill \over x_4 + \cdots}}} $$
```

Without the \hfills, the numerators would have appeared centered (a style that is also used). The \struts are necessary: without them the vertical spacing is terrible. The \displaystyle commands block the shrinking in size that would normally occur in such structures. It is instructive to experiment a little with this input (leave out, say, the \struts or the \displaystyles) to see how the output changes.

To save some of this work, $\mathcal{AMS}$-TEX provides the special continued fraction commands \cfrac (centers numerators), \lcfrac (places numerators flush left) and \rcfrac (places numerators flush right). All three automatically adjust the vertical spacing. For example,

```
$$ x_1 + \cfrac 1 \\
   x_2 + \lcfrac 1 \\
   x_3 + \rcfrac 1 \\
   x_4 + \cdots\endcfrac $$
```

will produce

$$ x_1 + \cfrac{1}{x_2 + \cfrac{1}{x_3 + \cfrac{1}{x_4 + \cdots}}} $$

though the three styles are not usually mixed in the same continued fraction in this way.

▷ **control sequence** The expression isn't used much in this book, but it is standard TEX-ese for many of what are called "commands" here.

▷ $\coprod$ $\lceil\coprod\rceil$ in a line, $\lceil\coprod\rceil$ in a display. (Class 1: large operator.)

▷ \copy ∗ Extracts a copy of the contents of a box **register** without erasing them. For example, \setbox0=\hbox{\it Stuff} stores *stuff* in box 0. Then, typing '\copy0, \copy0' will produce '*Stuff, Stuff*'. The command \copy0 can be used as often as the box contents are needed, whereas \box0 can be used just once (the contents are erased after the first use).

▷ \copyright Produces ©. The symbol is a composite, built by superposing the character 'c' and a circle from one of TEX's math **fonts** (the same circle character that is invoked by the math command \bigcirc in text or display styles). As with many superpositions that call on specific characters from specific **fonts**, the definition has to be adjusted if a non-standard size is needed.

▷ $\cos$ $\lceil\cos\rceil$ 62–65, 70–71, 76, 82–83.

▷ $\cosh$ $\lceil\cosh\rceil$ 104–105.

▷ $\cot$ $\lceil\cot\rceil$

▷ $\coth$ ⌈coth⌋

▷ \count * 98, 216, 224. Represents registers where whole numbers may be stored. For example, \count255=-13 stores the number −13 in the register called \count255. The contents of the register may be printed by saying \the\count255.

▷ \cr * 16–17, 20–21, 24–27, 30–31, 38–41, 46–49, 52–57, 60–65, 70–73, 86–89, 94–95, 98–103. Indicates the end of a line in all of Plain TEX's alignments. (It stands for "carriage return.") Also see \\ and matrices.

▷ \crcr * This is used in building high-level commands that involve alignments: it provides protection against a forgotten final \cr when the high level command is used. The \crcr command has the same effect as \cr, except when it follows either \cr or \noalign, in which case it does nothing. The standard \matrix command of Plain TEX is constructed out of more basic ("low-level") alignment mechanisms, and it has a \crcr built in. That is why it is all right to leave out the final \cr when using \matrix. See matrices for examples.

▷ $\csc$ ⌈csc⌋

▷ $\cup$ ⌈∪⌋ 76–77. (Class 2: binary operation.)

▷ $\Cup$ AMS ⌈⋓⌋ (Class 2: binary operation.) Alternate name: \doublecup.

▷ $\curlyeqprec$ AMS ⌈⋞⌋ (Class 3: relation.)

▷ $\curlyeqsucc$ AMS ⌈⋟⌋ (Class 3: relation.)

▷ $\curlyvee$ AMS ⌈⋎⌋ (Class 2: binary operation.)

▷ $\curlywedge$ AMS ⌈⋏⌋ (Class 2: binary operation.)

▷ $\curvearrowleft$ AMS ⌈↶⌋ (Class 3: relation.)

▷ $\curvearrowright$ AMS ⌈↷⌋ (Class 3: relation.)

▷ \D AMS See '\.'.

▷ \dag 18–19, 24–25. Produces †. Part of the cmsy Computer Modern symbols font, but is also available outside math mode. In Plain TEX this symbol (along with ‡, ¶ and §) does not change size in subscripts and superscripts. In AMS-TEX they all do.

▷ $\dagger$ ⌈†⌋ (Class 2: binary operation.)11, 143. The same symbol is produced by \dag but with different spacings. For example $a\dag b$ gives $a†b$, whereas $a\dagger b$ gives $a † b$. Further, the symbol now shrinks in indices.

▷ $\daleth$ AMS ⌈ℸ⌋ (Class 0: ordinary.)

▷ $\dasharrow$ ⌈--→⌋ AMS An alternate name for \dashrightarrow.

▷ $\dashleftarrow$ ⌈←--⌋ AMS

▷ $\dashrightarrow$ ⌈--→⌋ AMS Alternate name: \dasharrow.

▷ $\dashv$ ⌈⊣⌋ (Class 3: relation.)

▷ $\dbinom$ AMS Like \binom but it automatically uses display size.

▷ dd See keywords and units.

▷ `\ddag` Produces ‡. Part of the `cmsy` Computer Modern symbols font, but is available outside math `mode`. Also see `\dag`.

▷ `$\ddagger$` ⌈‡⌉ (Class 2: binary operation.) The difference between `\dag` and `\dagger`, discussed above, holds for `\ddag` and `\ddagger` as well.

▷ `$\ddddot$` *AℳS* `$\ddddot a$` gives $\ddddot{a}$.

▷ `$\dddot$` *AℳS* `$\dddot a$` gives $\dddot{a}$.

▷ `$\ddot$` $\ddot{a}$ is produced by `$\ddot a$`.

▷ `$\Ddot$` *AℳS* See `\accentedsymbol`.

▷ `$\ddots$` ⌈⋱⌉ 102–103, 128–129. A constructed symbol, assigned the "Inner" atomic type. See atom, spacing and `\mathinner`.

▷ `\def` * 2, 12, 14, 20–23, 26–41, 74, 54–55, 66, 68–69, 70–71, 74, 76–77, 98–103, 112–113, 135, 150, 152–154, 156, 163, 165–169, 174–176, 216, 218–232. TEX's basic definition mechanism. The basic procedure involved in defining a new command called, say, `\NewCom` is to type

> `\def\NewCom {`*replacement text*`}`

in one's file, where the *replacement text* is the text, other commands, etc., that the new command now summarizes. Examples of the use of `\def` are scattered throughout this book. Variations of `\def` are also available. An important one is `\edef`, which expands the replacement text of the new command right away. For example,

> `\edef\A {\ifmmode B\else C\fi}`

will make `\A` consistently produce either B or C, depending on whether TEX was in math `mode` or not when the definition was encountered. (Whereas, using `\def` will lead `\A` to give different results in different places, depending on what the conditions are when the command is actually used.) Other variations are `\gdef`, which works like `\def` except that the definition transcends the group in which it is made ("global definition"), and `\xdef`, which is a similar global version of `\edef`. Also see `\outer`.

▷ `\defaultskewchar` * See `\skewchar`.

▷ `\define` *AℳS* 2, 118–119, 124–131. A definition facility of *AℳS*-TEX that does some preliminary checking: it does not, for example, accept a new command name if that name has already been defined. It is used in a way similar to the primitive command `\def`:

> `\define\Einstein {G_{\mu\nu}=KT_{\mu\nu}}`

makes `$\Einstein$` produce $G_{\mu\nu} = KT_{\mu\nu}$. The use of `\define` over `\def` is generally encouraged in *AℳS*-TEX. Also see `\redefine`.

▷ `$\deg$` ⌈deg⌉

▷ `\delcode` * 164, 166. See §5.5.

▷ **delimiter** See delimiters.

▷ `\delimiter` * 164–166, 168. See §5.5.

▷ \delimiterfactor * When TEX sees a balanced pair of \left and \right commands in the input for a display, it selects (or, in some cases, constructs) delimiters of the right size. For example, if it sees \left[and \right], it will sandwich the formula between them by brackets of the correct vertical size. How does it calculate the correct size?

The \delimiterfactor represents an integer that is used internally by TEX in this calculation. It is one of two quantities used—the other is a dimensional quantity called \delimitershortfall. The reason for these names should become a little clearer from this summary of how the calculation is done:

Assume that \delimiterfactor has the value f (an integer) and that \delimitershortfall has the value l units. Every formula has an axis—an invisible horizontal line roughly through its middle. Suppose that the formula is to be flanked by delimiters and suppose that it extends y_1 units above the axis and y_2 units below it. Let $y = \max(y_1, y_2)$. Though $2y$ roughly represents the desired size (i.e., height plus depth) of the delimiter, it is usually considered more pleasing to try and use a slightly smaller size (otherwise the delimiter can visually dominate the formula). A search is conducted through the sets of relevant fonts (see §5.5) for the required delimiter in a size that is at least $\max(\lfloor y/500 \rfloor f, 2y - l)$ (since arbitrary sizes are not available for all delimiters, one has to be content with just this bound). If the delimiter does not come in a large enough size, and cannot be constructed out of smaller pieces, the largest available one is selected. Once a delimiter has been chosen in an appropriate size, it is placed centered with respect to the axis.

Plain TEX makes the following assignments:

```
\delimiterfactor=901
\delimitershortfall=5pt
```

Thus, a formula that has a size $2y$ (as computed above) of 80 pt will be flanked by delimiters of size at least 75 pt (if they are available in that size).

▷ **delimiters** 157, 164–166, 171–172. These are symbols like parentheses, brackets, curly braces, etc., that indicate where expressions begin and end.

((	))	[[	]]
{ $\{$	} $\}$	| $\|$	‖ $\|$
⌊ $\lfloor$	⌋ $\rfloor$	⌈ $\lceil$	⌉ $\rceil$
\ $\backslash$	/ /	⟨ $\langle$	⟩ $\rangle$

{ and } may also be obtained from $\lbrace$ and $\rbrace$, respectively, [and] from $\lbrack$ and $\rbrack$, and | and ‖ from $\vert$ and $\Vert$. The vertical arrow symbols listed under **relations** may also be used as delimiters. All these delimiters will automatically become larger when they enclose large expressions if they are prefixed by \left (for opening delimiters) and \right (for closing delimiters). See separate entries for \right and, especially, \left in this Glossary. Except for the ones on the last line, they can all become arbitrarily large. See `\delimiterfactor` for a sketch of how TEX internally performs delimiter size calculations.

The use of `\left` and `\right` is sometimes also necessary for another reason: some of the entities that are used as delimiters (those obtained from | and \| in the table above and the arrow symbols) are technically not in a delimiter class at all in Plain TEX's internal classification of characters. (The individual entries in this Glossary for the commands in the table above specify what class each belongs to; the notion of a class is itself discussed in its own entry.) Without the use of `\left` and `\right` alongside such pseudo-delimiters, the spacing may not be quite perfect: see absolute value for an example.

If the sizes of the delimiters that are automatically provided are not considered satisfactory, they can be asked for in several specific larger sizes. This is done by prefacing the appropriate command by `\big`, `\Big`, `\bigg`, or `\Bigg`. The effects of these are illustrated by the sequence

where the first line is in "normal size." Apart from giving control over delimiter size, these commands may be used singly, whereas `\left` and `\right` must occur in balanced pairs in an equation. It is also possible, and indeed desirable, to specify whether a big delimiter is at the left of an expression, the right, or in the middle, by typing '`\bigl`', '`\bigr`', '`\bigm`', etc. This is important for fine spacing in formulas because TEX leaves slightly different spaces around opening, closing and middle delimiters.

The precise definitions of `\big` ... `\Bigg` are

```
\def\big#1{{\hbox{$\left#1\vbox to 8.5pt{}\right.\n@space$}}}
\def\Big#1{{\hbox{$\left#1\vbox to 11.5pt{}\right.\n@space$}}}
\def\bigg#1{{\hbox{$\left#1\vbox to 14.5pt{}\right.\n@space$}}}
\def\Bigg#1{{\hbox{$\left#1\vbox to 17.5pt{}\right.\n@space$}}}
\def\n@space{\nulldelimiterspace=0pt \m@th}
```

The command `\m@th` is discussed in its own entry, as is the use of @ in command names. The definitions of the variants `\bigl`, `\Bigm`, etc., are repetitive; here are samples:

```
\def\bigl{\mathopen\big}
\def\bigm{\mathrel\big}
\def\bigr{\mathclose\big}
```

The others are defined in exactly the same way. As an illustration, here are the `\Bigg` sizes of all entities that may be used as delimiters:

`\delimitershortfall`

Finally, there are a few extra delimiters, some of which come only in large sizes; the minimum sizes of these are

$$| \quad | \quad \| \quad \| \quad | \quad | \quad \Big(\quad \Big) \quad \int \quad \bigg\rangle$$

The commands for these are `\arrowvert`, `\Arrowvert` and `\bracevert` (the first three pairs, respectively); `\lgroup` and `\rgroup`; `\lmoustache` and `\rmoustache`. They are made up of parts of other symbols, and are to be prefixed by `\left` or `\right` or by `\big`, etc. For example, the input for the last part of the preceding display was

```
...
\left\lgroup{\quad}\right\rgroup\qquad
\left\lmoustache{\quad}\right\rmoustache$$
```

Also see AMS symbols for a few additional delimiters.

▷ `\delimitershortfall` 124–125. See `\delimiterfactor`.

▷ `$\delta$` $\lceil\delta\rceil$ 30–31, 64–65, 80–81, 118–119, 130–131.

▷ `$\Delta$` $\lceil\Delta\rceil$ 68–69, 120–121. In Plain TeX `${\mit\Delta}$` (`\varDelta` in $\mathcal{AMS}$-TeX) produces Δ; `${\bf\Delta}$` (`\boldsymbol\Delta` in $\mathcal{AMS}$-TeX) produces **Δ**; and `${\tt\Delta}$` produces Δ.

▷ **depth** Every box constructed by TeX is considered to have a baseline, a height, a depth and a width. The depth is the distance that the box extends below the baseline. The word is also used as a keyword: see below.

▷ **depth** 40–41, 66, 98–99, 220, 227, 229. A keyword that is used to specify the depth of an `\hrule` or a `\vrule` below the baseline. For example, '❘' was obtained by typing '`\vrule height0pt width.5pt depth3pt`'.

▷ **derivatives** 32–33, 52–55, 80–83. Usually presented as fractions. See `\nabla`, `\partial` and `\prime` for useful symbols.

▷ `$\det$` $\lceil\det\rceil$ 34–35.

▷ `$\dfrac$` $\mathcal{AMS}$ Used like `\frac`, but it automatically gives display-size fractions.

▷ `$\diagdown$` $\mathcal{AMS}$ $\lceil\diagdown\rceil$ (Class 3: relation.)

▷ **diagrams** Blank spaces may be left for pictures and diagrams. It is also possible to draw simple diagrams from within TeX. See Examples 4 and 11–13 in Chapter 2. Further, a command called `\special` allows one to import diagrams from other (typically graphical) programs, as illustrated in Example 10 in Chapter 2. The exact procedures vary from system to system, so local experts will have to be consulted on exact details. On the following page there are two further examples of PostScript pictures imported into a TeX document. The pictures were stored in files called `pict1.ps` and `pict2.ps`, and the page was generated from a separate file that summoned the two pictures.

Here are the commands used to produce the pictures on this page:

```
\input epsf
$$\epsffile{pict1.ps}$$
$$\epsffile{pict2.ps}$$
```

The commands are associated with `dvips` (see Example 10 in Chapter 2).

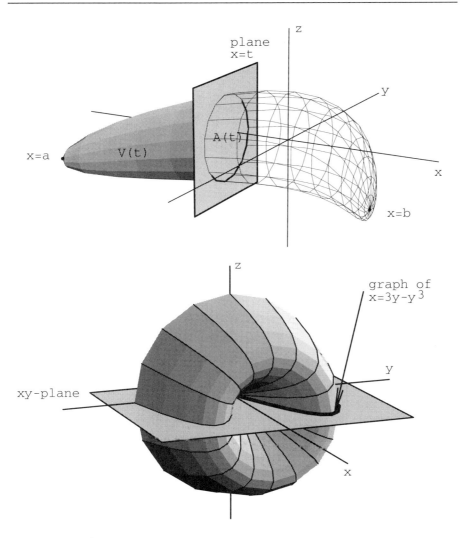

The pictures on this page were produced, and stored as Encapsulated PostScript files, with the program *Mathematica*®. They are taken from the book *An Infinite Series Approach to Calculus* by R. S. BASSEIN, *Publish or Perish* (1993), and are reprinted with the permission of the publisher.

▷ $\diagup$ 𝒜ℳ𝒮 ⌈╱⌉ (Class 3: relation.)

▷ $\diamond$ ⌈⋄⌉ (Class 2: binary operation.)

▷ $\diamondsuit$ ⌈◇⌉ (Class 0: ordinary.)

▷ **differentials** See dx, dy, etc.

▷ $\digamma$ 𝒜ℳ𝒮 ⌈𝐹⌉ (Class 0: ordinary.)

▷ $\dim$ ⌈dim⌉ 32–35, 46–47.

▷ \dimen ∗ 66, 70–71, 171, 223, 225, 227, 228, 230. Represents registers where quantities with dimension may be stored. For example, \dimen255=10pt stores the quantity 10pt in the register called \dimen255. Then, for instance, [\kern\dimen255] will give [ ].

▷ \discretionary ∗ This command underlies the way in which line break suggestions involving hyphenation may be made to TeX. As will be seen shortly, the word "hyphenation" is being used here in a general sense. The structure of the command is

   \discretionary{*A*}{*B*}{*C*}

where *A* represents the text to be used before the break, *B* the text to be used after, and *C* the text to be used if no break is taken at that point. For example, here is the Plain TeX definition of \∗:

   \def\∗{\discretionary{\thinspace\the\textfont2\char2}{}{}}

This calls for the following pre-break "text": a thin space followed by character number 2 from the font assigned to \textfont2 (see §5.5 for the meaning of this); the character is ×. No special post-break text is asked for, nor is anything special done if no break is taken (the second and third arguments are empty). See line breaks for an example.

▷ **display** See displays.

▷ $\displaybreak$ 𝒜ℳ𝒮 See \align.

▷ \displayindent ∗ A displayed formula (see displays) is usually placed with no indentation; if, however, it occurs as part of a paragraph that has a special shape (for example, hanging indentation), then the display will also be suitably indented as described below.

Suppose that the formula occurs after the nth output line (the entry under displays or under \prevgraf provides some more information on what this means). Then, immediately after reading the opening $$ signs of the display (and after setting the preceding lines in the paragraph), TeX assigns to \displayindent the value of the indentation that output line number $n + 2$ would have had, and to a parameter called

\displaywidth the length that line number $n+2$ would have had. These values are used in placing the formula in the correct place. Observe, for example, the shape of this paragraph and the placement of this formula:

$$1 + 1 = 2.$$

The input that achieved this effect was, partially,

```
\parshape 4 0in 4in .5in 3.5in 1in 3in 1.5in 2.5in \noindent
Suppose that the formula occurs after the $n$th output line (the
...
placement of this formula:
$$ 1+1=2. $$
The input that achieved this effect was, ...
```

The command \parshape, not being of special relevance to typesetting mathematics, is not discussed anywhere else in this book. What it does here should, however, be fairly obvious. The first number after the command specifies the number of lines to be shaped; the remaining quantities occur in pairs (as many pairs as the specified number) that shape, in turn, successive lines of the paragraph. The first member of each pair stipulates the left indentation to be used for the line, the second its length. Thus the first line in the paragraph above had an indention of 0 in. and a length of 4 in., the second an indentation of .5 in. and a length of 3.5 in., etc. If, as here, there are fewer pairs of sizes than lines in the paragraph, the extra lines repeat the pattern given by the last pair. And, finally, \noindent merely suppresses the normal paragraph indentation for this paragraph.

▷ $\displaylimits$ * Restores TeX's normal conventions for limits on large operators like $\sum$. See large operators.

▷ $\displaylines$ 18–20, 52–55, 62–63, 88–89, 94–95, 98–99, 102–103. Displays a list of centered equations. For example,

```
    $$\displaylines{A=B \cr C=1+D. }$$
```

gives

$$A = B$$
$$C = 1 + D.$$

The command may also be used to display an equation that is to be split over many lines (\hfill can be used to move a subformula on a line flush left or flush right, if desired).

▷ **displays** A *displayed formula* is one that is shown on a separate line from text. Formulas are displayed by sandwiching them between $$ signs in the input.

Displays in Plain TeX are automatically centered, and extra space is left above and below. The spacing of the components of a displayed formula is slightly different from that of formulas in text (see **spacing**), and some typefaces used are larger. The style is called **display style**: it can be invoked even in formulas that are not displayed, by using the command `\displaystyle`.

A question that often arises is whether the automatic centering of displays can be switched off (and replaced, say, by having the displays indented in some way). The answer is a difficult "yes": the centering of a display is deeply embedded in TeX's typesetting mechanisms, and getting around it involves work. The necessary commands are given, rather briefly, in *The TeXbook* (KNUTH [1984, 1986]); they have also been implemented in the package called `eplain`, available from several TeX archives (see the inside back cover). This solution is quite general, and it allows all the old commands (`\eqno`, `\leqno`, etc.) to still be used. A simpler way to get left-aligning displays is given on page 204 of EIJKHOUT [1992].

If one is prepared to use extra commands (not just rely on `$$...$$` to automatically give a repositioned display), then simple solutions are possible. For example,

`$$\displaylines{\indent \int_0^1 dx = 1. \hfill(1)}$$`

gives

$$\int_0^1 dx = 1. \tag{1}$$

Vertical spacing around displays can also be problematic, unless care is taken. The space left above and below displayed equations is augmented by any additional vertical space that may have been asked for just before, or just after, the display. For example, there will be extra space from `\smallskip`, `\medskip`, etc., or possibly even from a blank line that has been left just before or after the display (this will give an extra inter-paragraph space). Depending on the nature of the display, this extra space may or may not be considered desirable. If it is not, then the input file must not contain commands that call for extra space next to a display, and it must not contain blank lines next to it. If a blank line is considered useful in enhancing the readability of the input file, then the device used in Example 7 in Chapter 2 is very helpful.

The rest of this entry provides the dirty (and yet, for those whose tastes run in that direction, fulfilling) details of how TeX places displays.

When TeX typesets a **display**, it first formats the preceding part of the paragraph as though it were a full paragraph by itself. Suppose that the paragraph has n lines so far. The display is then considered to correspond to line number $n + 2$ in that paragraph (the paragraph resumes after the display with the line number assumed to be $n + 4$). The indentation of the display (`\displayindent`) and the width of the space available for the display

(\displaywidth) will obey the specifications for this line.

Now consider the end result: TEX has read the input for a displayed formula and has constructed it. (The rules for the construction of the actual formula are not being discussed here, just its placement.) It has assigned the formula a width W, has placed it shifted by an amount S from the left margin of the paragraph, has put white space of vertical size G_a above the formula and of size G_b below it, and has placed an equation number in the right place. How did TEX make these placements?

Each will be discussed in turn. It helps, first, to list all the variables involved in the calculation:

- w_0: The "natural width" of the constructed formula; i.e., the width it would have if one ignored all the elasticity (the **plus** and **minus** specifications) in the white spaces around its constituents.
- p: The value of \predisplaysize; i.e., the length of the line immediately preceding the display. The separate entry under that command lists a few qualifications.
- dw: The value of \displaywidth; i.e., the line length assigned to the line in the paragraph where the display is considered to occur. This represents the available horizontal space for the display.
- di: The value of \displayindent; i.e., the indentation assigned to the line in the paragraph where the display is considered to occur.
- eq: The width of the equation number, if there is one; zero otherwise. (Note: it is possible to have an equation number present, yet fool TEX into assigning it zero width by typing, say, \eqno\llap{$...$}.)
- eeq: The extended width of the equation number—eq plus the width of one quad in the symbols font—if there is an equation number; zero otherwise. Each part of the question can now be considered separately.

- *Dirty detail 1: the width W.*

If $w_0+eeq \leq dw$, then $W = w_0$. If $w_0+eeq > dw$, then the formula is squeezed as follows: if $eq \neq 0$ and if the white space in the formula has enough shrinkability assigned to it (the total **minus** component), the formula is compressed to a width given by $W = dw - eeq$; in all other cases, eq is set to zero and the formula is assigned a width given by $W = \min(w_0, dw)$. (When W is assigned a value here, what is really done is to put the formula in an hbox of width W; the last possibility may result in an "overfull hbox," in which case a warning message will be issued. The input for the formula will then have to be fiddled with to reduce its horizontal size, or to break the formula over two lines.)

- *Dirty detail 2: horizontal positioning.*

This corresponds to calculating the quantity S. The first choice is $S = \frac{1}{2}(dw - W)$. If $eq > 0$ and if this choice for S makes $S < 2eq$, then S is reset: if the formula begins with white space (e.g., if \quad is the first item in it) then $S = 0$; otherwise $S = \frac{1}{2}(dw - W - eq)$.

• *Dirty detail 3: vertical positioning.*

First some background: TeX provides four parameters that govern the amount of space above and below displays. The parameters along with their Plain TeX values are

```
\abovedisplayskip=12pt plus 3pt minus 9pt
\abovedisplayshortskip=0pt plus 3pt
\belowdisplayskip=12pt plus 3pt minus 9pt
\belowdisplayshortskip=7pt plus 3pt minus 4pt
```

The first and third of these represent the spaces left above and below displays, respectively. The second and fourth represent the spaces left instead when the line immediately preceding the display is short (i.e., it will not overhang the displayed formula when the formula is placed in the correct horizontal position; this is discussed more precisely below). In all of these, **plus** and **minus** refer to the stretchability and shrinkability of the space.

Before placing white space above the equation, TeX inserts a **penalty** item, which it will use later in calculating the cost of a page break at that point. The value of the penalty here is the one specified by the current value of \predisplaypenalty. (The value assigned to this in Plain TeX is 10000, effectively infinite in TeX's calculations. The extreme value is chosen because it is generally considered bad typesetting form to end a page just before a displayed formula, or—equivalently—to begin a fresh page with a formula at the top.) After the equation has been placed, along with any following white space, another penalty is inserted, this time the value of \postdisplaypenalty. (This is zero in Plain TeX, since no great harm results from a page ending with a display.)

The vertical positioning depends on what is chosen as the value of G_a (the choice for G_b then follows): \abovedisplayskip is chosen if there is a left equation number, or if $S + di \le p$; otherwise the choice is \abovedisplayshortskip. After a choice has been made for the value of G_a, a space corresponding to that value is left above the formula and one corresponding to G_b is left below it, except in two cases. The exceptions are the ones listed in case *1* (when G_a is effectively considered zero) and case *3* (when G_b is considered zero) below.

• *Dirty detail 4: placing equation numbers.*

After all of the choices listed above have been made, the equation numbers are placed. There are three cases:

1. If $eq = 0$, but there is nevertheless a left equation number (specified by \leqno), the equation number is placed by itself right after the line in the paragraph that immediately precedes the display, in a horizontal position indented by di. An infinite **penalty** is also inserted right after to prevent a page break between the equation number and the formula. The formula is then placed below, without extra vertical white space, shifted to the right by an amount $di + S$.

2. If $eq \neq 0$, the equation number and the formula are combined in a single hbox, separated by a distance (more precisely, a *kern*—see \kern) of width $dw - W - eq - S$. If the equation number is specified by \eqno, it goes in the right end of the box; if it is specified by \leqno, it goes in the left end and, further, S is then set to zero. Once this is done, the box containing the formula, the kern and the equation number is appended to the material above it (with extra white space given by G_a), shifted to the right by an amount $di + S$.

3. If $eq = 0$, but there is nevertheless a right equation number, the box containing the formula is appended just as in case *2*, but without augmenting it by adding an equation number. An infinite penalty is placed just below the formula and the equation number is placed below that, shifted to the right by $di + dw$ minus the true width of the equation number. No extra white space is added below the formula in this case.

▷ **display style** The style used automatically in mathematics displays. See styles in mathematics and \displaystyle.

▷ $\displaystyle$ * 34–35, 64–65, 74, 94–95. Invokes the spacing rules, typeface styles, etc., of the mathematics display mode (the mode that is used when expressions are entered between double $ signs), even when formulas are not being displayed; e.g., $\displaystyle \sum_1^9$ gives $\sum_1^9$, whereas $\sum_1^9$ gives $\sum_1^9$. Also see styles in mathematics.

▷ \displaywidowpenalty * "Widow" is a typesetting term (reflecting old social attitudes) that refers to the last line of a paragraph that appears by itself at the top of a new page. This is generally considered aesthetically undesirable, and there are penalties associated with such occurrences in TeX. The \displaywidowpenalty represents the penalty associated with having the line of a paragraph that immediately precedes a display occur at the top of a new page. The value in Plain TeX is 50 (compare this to 150 for a true widow line, and 10000 for starting the new page with the display itself).

▷ \displaywidth * Represents the horizontal size available for a display based on the line length of the current line, which in turn is based on the current paragraph shape. See \displayindent for more details and for an example.

▷ $\div$ $\div$ (Class 2: binary operation.)

▷ $\divideontimes$ $_{AMS}$ $*$ (Class 2: binary operation.)

▷ **document style** 6, 116, 134. A set of specifications that governs the overall style of the document: the page size, the typefaces used, the position of equation numbers, etc. The specifications are usually listed in a separate file, called a style file, which is invoked at the start of the document. Plain TeX comes with just one automatic style, so the notion of document styles is not as widely exploited there as it is in other packages (though style files do exist that alter the default format). In $\mathcal{A}_{\mathcal{M}}\mathcal{S}$-TeX, document styles are imposed through the \documentstyle command (though, again, a style declaration isn't insisted on). See AMS preprint style and Chapter 3.

▷ \documentstyle $_{\mathcal{A}_{\mathcal{M}}\mathcal{S}}$ 134. This declaration may be used at the start of an $\mathcal{A}_{\mathcal{M}}\mathcal{S}$-TeX document: it selects a style. See AMS preprint style and Chapter 3. Several things in $\mathcal{A}_{\mathcal{M}}\mathcal{S}$-TeX are style-dependent (the positions of equation numbers placed by the command \tag, for example). Further, certain commands (e.g., \proclaim) are defined only in style files, and so they cannot be used at all unless a style is chosen. Many features of $\mathcal{A}_{\mathcal{M}}\mathcal{S}$-TeX, however, work even without a choice of style.

▷ $\dot$ 126–127. $\dot a$ produces $\dot{a}$.

▷ $\Dot$ $_{\mathcal{A}_{\mathcal{M}}\mathcal{S}}$ See \accentedsymbol.

▷ $\doteq$ $\lceil \doteq \rceil$ (Class 3: relation.)

▷ $\Doteq$ $_{\mathcal{A}_{\mathcal{M}}\mathcal{S}}$ $\lceil \doteqdot \rceil$ (Class 3: relation.) Alternate name for \doteqdot.

▷ $\doteqdot$ $_{\mathcal{A}_{\mathcal{M}}\mathcal{S}}$ $\lceil \doteqdot \rceil$ (Class 3: relation.) Alternate name: \Doteq.

▷ \dotfill '[.....]' is produced by '\hbox to 1cm{[\dotfill]}'.

▷ $\dotplus$ $_{\mathcal{A}_{\mathcal{M}}\mathcal{S}}$ $\lceil \dotplus \rceil$ (Class 2: binary operation.)

▷ **dots** The standard dot-producing commands in mathematics are \ldots (for low dots) and \cdots (for centered dots). $\mathcal{A}_{\mathcal{M}}\mathcal{S}$-TeX offers a variety of further commands and a more sophisticated approach to dots in formulas. It has a \dots command that is automatically translated into one of four other commands, depending on the context in which it occurs: \dotsb between binary operations or between relations, \dotsc before commas or semicolons, \dotsi between integral signs, and \dotso in all other cases. The exact positioning of each of these four types of dots depends on the document style being used. (Thus the examples shown under each of the commands below may not work in exactly the same way in other situations.) If the automatic translation of \dots into one of the four specific types isn't considered satisfactory, then the desired type can be explicitly asked for. There is also a command called \dotsm for use between symbols that sit next to each other, when that positioning indicates multiplication.

If a new symbol is being defined, it is useful to tell TeX explicitly what sorts of dots it should use next to the symbol. The commands \DOTSB and \DOTSI are used for this purpose: they do not leave any dots themselves, but they tell TeX to use \dotsb and \dotsi, respectively, when the occasion arises. For example, if the symbol $\int_0^\infty$ frequently appears in a document, it may be useful to define an abbreviation for it. If \DOTSI is added into the definition by typing

```
\define\I{\DOTSI\int_0^\infty}
```

then the correct dots (i.e., `\dotsi`) are called upon in, for instance, `\I\dots\I`. There is a similar command called `\DOTSX` for use when the abbreviation represents a closing **delimiter** (it ensures that a little space will be left at the end between whatever dots are left by `\dots` and the delimiting symbol).

▷ `\dots` ⌜...⌟ 32–33, 40–41, 86–87, 102–103, 116–117, 130–131. Also see dots for a discussion of a special mathematical use of the command in $\mathcal{AMS}$-TeX.

▷ $\dotsb$ $\mathcal{AMS}$ 118–119, 128–129. [⋯] is produced by $[\dotsb]$. See dots.

▷ `\DOTSB` $\mathcal{AMS}$ ($\mathcal{AMS}$-TeX alert: don't `\redefine`.) See dots.

▷ $\dotsc$ $\mathcal{AMS}$ [...] is produced by $[\dotsc]$. See dots.

▷ $\dotsi$ $\mathcal{AMS}$ [⋅⋅] is produced by $[\dotsi]$. See dots.

▷ `\DOTSI` $\mathcal{AMS}$ ($\mathcal{AMS}$-TeX alert: don't `\redefine`.) See dots.

▷ $\dotsm$ $\mathcal{AMS}$ [⋯] is produced by $[\dotsm]$. See dots.

▷ $\dotso$ $\mathcal{AMS}$ [...] is produced by $[\dotso]$. See dots.

▷ `\DOTSX` $\mathcal{AMS}$ ($\mathcal{AMS}$-TeX alert: don't `\redefine`.) See dots.

▷ **double integrals** $\int\!\!\int$ gives $\iint$. Omitting the negative spaces given by `\!` would yield a less attractive result. $\mathcal{AMS}$-TeX has a similar feature built in: see `\iint`, `\iiint` and `\iiiint`.

▷ $\doublebarwedge$ $\mathcal{AMS}$ ⌜$\overline{\wedge}$⌟ (Class 2: binary operation.)

▷ $\doublecap$ $\mathcal{AMS}$ ⌜⋒⌟ (Class 2: binary operation.) Alternate name for `\Cap`.

▷ $\doublecup$ $\mathcal{AMS}$ ⌜⋓⌟ (Class 2: binary operation.) Alternate name for `\Cup`.

▷ $\downarrow$ ⌜↓⌟ 56–57, 100–101. (Class 3: relation.) Also see delimiters.

▷ $\Downarrow$ ⌜⇓⌟ (Class 3: relation.) Also see **delimiters**.

▷ `\downbracefill` Horizontal spaces may be filled with a "downbrace":

```
\hbox to 1in{\downbracefill}     gives
```

Also see `\upbracefill`, and Example 9 in Chapter 2 for a use of that command.

▷ $\downdownarrows$ $\mathcal{AMS}$ ⌜⇊⌟ (Class 3: relation.)

▷ $\downharpoonleft$ $\mathcal{AMS}$ ⌜↽⌟ (Class 3: relation.)

▷ $\downharpoonright$ $\mathcal{AMS}$ ⌜⇂⌟ (Class 3: relation.)

▷ `\dp` ∗ 66. Represents the **depth** of boxes. Also see `\ht`.

▷ **d-size** An $\mathcal{AMS}$-TeX term for display size. Also see `\dsize`.

▷ $\dsize$ $\mathcal{AMS}$ ($\mathcal{AMS}$-TeX alert: don't `\redefine`.) Forces "d-size" (display size) on an expression.

▷ **dx, dy, etc.** 82–83, 92–95, 98–99. The readability of formulas is enhanced if quantities like dx and dy are set off from preceding material by a little white space: for example, for most applications it is a lot better to have $x\,dx$ (obtained from `$x\,dx$`) than xdx (obtained from `$xdx$`)—unless one does mean for the combination to be interpreted as the product of x, d and x. Some typesetting conventions go even further and argue that the d in dx, being an operator rather than a variable, should be set in different type than that used for variables. So, for example, one would type something like `$\int x\,{\rm d}x$`, and get $\int x\,\mathrm{d}x$. This may seem extreme to many practicing mathematicians, but it is in fact the official position of at least one international standards-setting body. Others argue that in mathematics, as in other languages, context is everything, and as long as it is possible to clearly distinguish the roles that various symbols play in common formulas, there doesn't seem to be an overwhelming need for further typographical differentiation.

▷ `\edef` ∗ 168, 227–228, 230. See `\def`.

▷ `\egroup` An implicit right brace. The Plain TEX definition of the command is '`\let\egroup=}`'; see **grouping**.

▷ `\eightpoint` 𝒜𝓂𝒮 (𝒜𝓜𝒮-TEX alert: don't `\redefine`.) A command that comes with the **AMS preprint style** that allows a switch to 8-point type.

▷ `\eject` 219. Forces a page break; in most cases it is necessary to precede the command by `\vfil` or `\vfill`, to maintain proper vertical spacing. The definition is

    `\def\eject{\par\penalty-10000 }`

▷ `$\ell$` $\lceil\ell\rfloor$ (Class 0: ordinary.)

▷ `\else` ∗ 14–15, 42–43, 74–75, 216, 223, 225–230, 232. Used in conjunction with TEX's conditional commands. See `\if...` for examples.

▷ **em** 170. See **keywords** and **units**.

▷ `\empty` 168, 169, 227–228. Defined in Plain TEX by `\def\empty{}`. It is useful in situations where one wants to test if some sequence of items is empty.

▷ `$\emptyset$` $\lceil\emptyset\rfloor$ (Class 0: ordinary.)

▷ `\end` ∗ All TEX documents must end with this, either directly, or—more commonly—with some other command that has `\end` built into it.

▷ `\end...` 𝒜𝓂𝒮 A large number of structures in 𝒜𝓜𝒮-TEX are created by saying something like `\structure` at the start and then `\endstructure` at the end (for example, `\align` and `\endalign`). Information about each `\end...` command can be found under the corresponding beginning `\...` command (e.g., information about `\endalign` under `\align`, and so on). None of these `\end...` commands should be redefined.

▷ `\endgraf` 74, 222, 232. Defined by `\let\endgraf=\par`. The command is useful as an end-of-paragraph command in situations where `\par` itself might have been redefined.

▷ \endgroup ∗ 36–37, 74. See grouping.

▷ ϵ ⌈ε⌉ 14–15, 38–39, 62–63. ε produces ε.

▷ $\eqalign$ 20–21, 30–31, 38–39, 52–53, 56–57, 60–63, 70–71, 86–87, 98–99.
Aligns a set of formulas and allows an equation number to be assigned to the
entire set. For example,

    $$\eqalign{A&=B \cr x&=y-z.}\eqno7$$

gives

$$A = B$$
$$x = y - z.$$

7

More than one \eqalign can go in the same display. The alignments will go
side by side. See \aligned for a similar $\mathcal{AMS}$-TEX command.

The mechanism used in \eqalign is based on TEX's primitive command
\halign. Here, as an example of how such structures are put together, is the
definition from the Plain TEX file:

    \def\eqalign#1{\null\,\vcenter{\openup1\jot \m@th
      \ialign{\strut\hfil$\displaystyle{##}$&$\displaystyle{{}##}$\hfil
        \crcr#1\crcr}}\,}

\m@th ensures that no extra space is left around formulas, and \ialign first
appropriately initializes \everycr and \tabskip, then calls the alignment
command \halign.

▷ $\eqalignno$ 48–49, 86–87, 100–101. Aligns formulas and allows equation
numbers to be assigned to each formula (if that is desired). For example,

    $$\eqalignno{A&=B &(1)\cr x&=y-z.}$$

gives

$$A = B \tag{1}$$
$$x = y - z.$$

The Plain TEX definition of this command is more complicated than that of
\eqalign.

▷ $\eqcirc$ $\mathcal{AMS}$ ⌈≖⌉ (Class 3: relation.)

▷ $\eqno$ ∗ 92–93, 98 99. Puts equation numbers on the right.

▷ $\eqslantgtr$ $\mathcal{AMS}$ ⌈⪈⌉ (Class 3: relation.)

▷ $\eqslantless$ $\mathcal{AMS}$ ⌈⪇⌉ (Class 3: relation.)

▷ **equation alignments** See \eqalign, \eqalignno and \leqalignno. For $\mathcal{AMS}$-
TEX alignment commands, see \align and \aligned, and commands mentioned there.

▷ **equation numbers** 58–73. See \eqno, \leqno, \eqalignno and \leqalignno.
For special $\mathcal{AMS}$-TEX equation numbering devices, see \tag.

▷ $\equiv$ ⌈≡⌉ 16–17, 26–27, 102–103, 124–125. (Class 3: relation.)

▷ **errors** It is assumed that readers are familiar enough with TₑX to have seen most of the common error messages and to know how to deal with them. The most common error in mathematical typesetting is likely to be a missing $ sign. Since there is a whole host of commands that work only in math mode, such errors are usually spotted right away and are easily corrected.

▷ `$\eta$` $\lceil \eta \rceil$ 30–31, 100–101, 116–119, 130–131.

▷ `$\eth$` $\mathcal{AMS}$ $\lceil \eth \rceil$ (Class 0: ordinary.)

▷ **Euler typefaces** See AMS Euler.

▷ `\every...` ∗ A class of commands, covering `\everycr`, `\everydisplay`, `\everyhbox`, `\everyjob`, `\everymath`, `\everypar` and `\everyvbox`. Here is an example of how `\everymath` works:

The formulas "→ $1 + 1 = 2$" and "→ $\int \alpha^2 d\alpha$." were produced by typing:

```
The formulas {\everymath={\rightarrow\;} ''$1+1=2$''
and ''$\int\alpha^2d\alpha$.''} were produced by ...
```

So, `\everymath` inserts the specified tokens at the start of every formula. Braces were used to localize the effect of `\everymath`; to check that things are back to normal, try a formula again: "$\int \alpha^2 d\alpha$."

The others in this class of commands may be used in the same way to insert something after every `\cr`, at the start of every mathematics display, at the start of every `\hbox`, at the start of the job, at the start of every paragraph and at the start of every `\vbox`, respectively.

▷ **ex** 170, 222. See keywords and units.

▷ `$\exists$` $\lceil \exists \rceil$ 126–127. (Class 0: ordinary.)

▷ `$\exp$` $\lceil \exp \rceil$ 26–27, 64–65, 72–73, 82–83, 92–95.

▷ `\expandafter` ∗ 74–75, 219, 222, 225, 228, 232. Postpones the expansion of a token until the the one immediately following it has been expanded. (Expansion is the process whereby a token is replaced by others; for example, the substitution of a defined command by the replacement text in its definition.) See `\romannumeral` for an example.

▷ **exponents** See indices and spacing.

▷ `$\fallingdotseq$` $\mathcal{AMS}$ $\lceil \fallingdotseq \rceil$ (Class 3: relation.)

▷ `\fam` ∗ 162–163, 166–167, 176, 219, 220. Allows typeface changes within formulas. See § 5.5.

▷ **families** 158–169, 171–173, 176, 188, 190. The typefaces that TₑX uses to compose formulas are organized into families. The program has built-in conventions about what families it will turn to in what contexts, and other conventions can be added. For example, the Plain TₑX definition of a command like `\sum` tells TₑX to go to family such-and-such and to pick character so-and-so. See § 5.5 for a more detailed treatment of this notion.

▷ `\fi` ∗ Ends all of TₑX's conditional commands. See `\if...`.

▷ **fields** See atom.

▷ **figures** See Examples 4, 10–13 in Chapter 2.

▷ $\Finv$ $\mathcal{A_MS}$ $\ulcorner \dashv \urcorner$ (Class 0: ordinary.)

▷ \firstmark * 223, 225, 230. See \mark.

▷ \five... 112–113, 149, 150, 158–159, 166–169, 220, 228. Five-point fonts are assigned names like \fiverm, etc., in Plain TEX. See § 5.3 for a list of the 5-point fonts that are automatically loaded in Plain TEX.

▷ $\flat$ $\ulcorner \flat \urcorner$ (Class 0: ordinary.)

▷ \flushpar $\mathcal{A_MS}$ 124–127. Starts a paragraph flush left, exactly as \noindent does. (It is, in fact, defined as \par\noindent.)

▷ $\foldedtext$ $\mathcal{A_MS}$ 122–123. This is similar to \text, but it is used to place a long string of text in a formula. The text is automatically "folded"—i.e., broken into smaller lines—and set as a paragraph (without indentation). The width of this paragraph is determined automatically; it can also be explicitly specified through the command \foldedwidth. For example,

```
$$ x=\cos x \qquad \foldedtext\foldedwidth{1.5in}{for at least
one positive real number $x$.} $$
```

(where the \foldedwidth{1.5in} part is optional) produces

$$ x = \cos x \qquad \text{for at least one positive real number } x. $$

The text that is specified in \foldedtext is vertically centered. The commands \botfoldedtext and \topfoldedtext are used similarly, but they align the bottom line and the top line of the text, respectively, with the rest of the material on the line where they occur.

▷ \font * 40–41, 84–85, 114–115, 124–125, 128–129, 141–142, 149, 151, 153, 156, 166–169, 176, 210, 216, 219–221, 224, 226, 229, 231, 232. Allows access to fonts. See § 5.1.

▷ \fontdimen * 169–172, 177–211. See § 5.6.

▷ **fonts** A typographical term usually referring to collections of characters that have the same style and size. Chapter 5 discusses the topic in detail.

▷ \footline 24–27, 76–77, 225, 226, 228, 232. Allows a line of material to be placed at the bottom of every page.

▷ \footnote 16–19, 24–25, 60–63, 66–67, 134, 136, 230. Produces footnotes. There is a discussion under that topic of how the command is used in both Plain and $\mathcal{A_MS}$-TEX. Also see \vfootnote.

▷ \footnotemark $\mathcal{A_MS}$ ($\mathcal{A_MS}$-TEX alert: don't \redefine.) In the AMS preprint style \footnote internally involves two commands: \footnotemark and \footnotetext. If, for some reason, a footnote has to be issued from deep within some structure (like an \hbox within a displayed equation), then the two have to be used separately—the text in \footnote will not migrate out if it is buried deeply. For example:

```
$$n^2>n \quad\text{for all natural numbers\footnotemark\ $n$.}$$
\footnotetext{Examples of natural numbers are 1, 2 and 3.}
```

For this to work, the \footnotetext command must be issued soon after \footnotemark. If two different \footnotemarks are asked for within a formula, then the following contortions have to be gone through in order to get the correct markers on the footnotes:

```
\adjustfootnotemark{-1}%
\footnotetext{first footnote}%
\adjustfootnotemark{1}%
\footnotetext{second footnote}%
```

▷ **footnotes** 220, 230. Though the basic footnote command in both Plain and $\mathcal{AMS}$-TEX is \footnote, there are differences in how it is used and in what it does. The first part of the discussion below applies to Plain TEX, with a discussion of $\mathcal{AMS}$-TEX following.

There are several examples of Plain TEX footnotes in this book. (See the page numbers listed under \footnote.) It should be quite clear from them how the command is used.

It is possible to make changes to the default footnote style of Plain TEX fairly easily. For example, if a smaller typeface is desired, one can define a font-switching command to the smaller size and use this size in the text of footnotes. An example of how this may be done is provided by the new command \SmallFootnotes defined in the EightPt file; the file is displayed in Chapter 6.

It is also possible to change the rule that divides the footnote from the page. The default is specified by a command called \footnoterule whose Plain TEX definition is

```
\def\footnoterule{\kern-3pt \hrule width 2 true in \kern2.6pt}
```

By copying this definition into one's file, then editing it, the appearance of the rule can be changed. Finally, one way to change the indentation of a footnote is to reset the value of \parindent just before using \footnote (and then restoring it after).

In $\mathcal{AMS}$-TEX the structure of footnotes is determined by the style that is being used. If no document style is specified, but $\mathcal{AMS}$-TEX is still used (by typing \input amstex), then \footnote will be an undefined command; there will still be available, however, the command \plainfootnote: this can be used exactly like \footnote in Plain TEX.

In the AMS preprint style, one just says \footnote{stuff}; the footnotes are automatically numbered consecutively through the document. If a special footnote marker is desired, the one provided by default can be overridden. For example,

```
\footnote"**"{stuff}
```

will give a footnote with ** as the marker.

Also see \footnotemark.

▷ \footnotetext ₐₘₛ (𝒜𝑀𝒮-TₑX alert: don't \redefine.) See the discussion under \footnotemark.

▷ $\forall$ ⌜∀⌝ 126–127. (Class 0: ordinary.)

▷ \format ₐₘₛ 122–125, 128–129. (𝒜𝑀𝒮-TₑX alert: don't \redefine.) See matrices.

▷ $\frac$ ₐₘₛ 118–121. Builds fractions; e.g., $\frac{1+a}{1-a}$ gives $\frac{1+a}{1-a}$. There are also the associated commands \dfrac and \tfrac, which force the display and text styles, respectively. The definitions of these commands are

```
\def\frac#1#2{{#1\over#2}}
\def\dfrac#1#2{{\displaystyle{#1\over#2}}}
\def\tfrac#1#2{{\textstyle{#1\over#2}}}
```

Also see \fracwithdelims and \thickfrac.

▷ **fractions** See \over and also \abovewithdelims, \atopwithdelims and \overwithdelims. 𝒜𝑀𝒮-TₑX provides some special fraction constructions: see \frac, \fracwithdelims, \thickfrac and \thickfracwithdelims.

▷ $\fracwithdelims$ ₐₘₛ $\fracwithdelims[]{1+a}2$ gives $\left[\frac{1+a}{2}\right]$. The definition of the command is

```
\def\fracwithdelims#1#2#3#4{{#3\overwithdelims#1#2#4}}
```

▷ $\frak$ ₐₘₛ 76–77, 88–91, 102–103, 120–123. Gives characters from the fraktur fonts in formulas if the eufm fonts have been loaded (see § 3.3). $\frak A$ gives 𝔄. An alternate name is \goth. Here is the entire alphabet:

$$\mathfrak{A\,B\,C\,D\,E\,F\,G\,H\,I\,J\,K\,L\,M\,N\,O\,P\,Q\,R\,S\,T\,U\,V\,W\,X\,Y\,Z}$$
$$\mathfrak{a\,b\,c\,d\,e\,f\,g\,h\,i\,j\,k\,l\,m\,n\,o\,p\,q\,r\,s\,t\,u\,v\,w\,x\,y\,z}$$

These are of medium weight; bold characters are also available. Full font tables are displayed in Chapter 5.

▷ **fraktur** A traditional typeface, made available in 𝒜𝑀𝒮-TₑX distributions as part of the AMS Euler class of typefaces. See \frak.

▷ $\frown$ ⌜⌢⌝ (Class 3: relation.)

▷ \galleys ₐₘₛ An option that 𝒜𝑀𝒮-TₑX provides, which allows users to see if their document contains any overfull boxes, without actually generating any output. The command is to be typed at the top of the file (but after the \documentstyle). See \printoptions.

▷ $\Game$ ₐₘₛ ⌜⅁⌝ (Class 0: ordinary.)

▷ γ ⌜γ⌝ 22–23, 24–25, 32–33, 82–83, 130–131.

▷ Γ ⌜Γ⌝ 24–25, 82–83, 116–119, 162. In Plain TₑX ${\mit\Gamma}$ (\varGamma in 𝒜𝑀𝒮-TₑX) produces Γ; ${\bf\Gamma}$ (\boldsymbol\Gamma in 𝒜𝑀𝒮-TₑX) produces $\boldsymbol{\Gamma}$; and ${\tt\Gamma}$ produces Γ.

▷ $\gather$ ᴬᴹˢ 116–117. Used in conjunction with \endgather to produce a set of displayed formulas, not aligned in any particular way. (If one uses $$ ... $$ for each, there is often too much vertical space between the formulas.) For example,

    $$\gather A=B \\ C=1+D. \endgather $$

will give

$$A = B$$
$$C = 1 + D.$$

Any of the formulas can be numbered with \tag (as in \align). It is also possible to align a subset of formulas within \gather ... \endgather. One does this by saying {\align at the start of the subset and \endalign}\\ at the end, where the braces are strictly necessary. In between, the standard rules for \align apply.

▷ $\gathered$ ᴬᴹˢ This is used like \gather, but it produces a group of formulas that are regarded as a single unit. The group must end with \endgathered. The overall behaviour is similar to that of \aligned.

▷ $\gcd$ ⌜gcd⌝

▷ $\ge$ ⌜≥⌝ 94–97, 120–121. (Class 3: relation.) Alternate name for \geq.

▷ $\geq$ ⌜≥⌝ 38–39. (Class 3: relation.) Alternate name: \ge.

▷ $\geqq$ ᴬᴹˢ ⌜≧⌝ 90–91. (Class 3: relation.)

▷ $\geqslant$ ᴬᴹˢ ⌜⩾⌝ 78–81. (Class 3: relation.)

▷ $\gets$ ⌜←⌝ (Class 3: relation.) Alternate name for \leftarrow.

▷ $\gg$ ⌜≫⌝ (Class 3: relation.)

▷ $\ggg$ ᴬᴹˢ ⌜⋙⌝ (Class 3: relation.) Alternate name: \gggtr.

▷ $\gggtr$ ᴬᴹˢ ⌜⋙⌝ (Class 3: relation.) Alternate name for \ggg.

▷ $\gimel$ ᴬᴹˢ ⌜ℷ⌝ (Class 0: ordinary.)

▷ **glue** A construct in TEX that allows white space to have elastic properties such as stretchability and shrinkability.

▷ $\gnapprox$ ᴬᴹˢ ⌜⪊⌝ (Class 3: relation.)

▷ $\gneq$ ᴬᴹˢ ⌜⪈⌝ (Class 3: relation.)

▷ $\gneqq$ ᴬᴹˢ ⌜≩⌝ (Class 3: relation.)

▷ $\gnsim$ ᴬᴹˢ ⌜⋧⌝ (Class 3: relation.)

▷ \goodbreak 231. Indicates to TEX that it is a good place for page break. See penalties.

▷ $\goth$ ᴬᴹˢ See \frak.

▷ **graphics** See diagrams.

▷ $\grave$ à is produced by $\grave a$.

▷ $\Grave$ ᴬᴹˢ See \accentedsymbol.

▷ \graveaccent ᴬᴹˢ May be used in place of \' to get a grave text accent.

▷ **Greek letters**

α `$\alpha$`	β `$\beta$`	γ `$\gamma$`	δ `$\delta$`
ϵ `$\epsilon$`	ε `$\varepsilon$`	ζ `$\zeta$`	η `$\eta$`
θ `$\theta$`	ϑ `$\vartheta$`	ι `$\iota$`	κ `$\kappa$`
λ `$\lambda$`	μ `$\mu$`	ν `$\nu$`	ξ `$\xi$`
o `$o$`	π `$\pi$`	ϖ `$\varpi$`	ρ `$\rho$`
ϱ `$\varrho$`	σ `$\sigma$`	ς `$\varsigma$`	τ `$\tau$`
υ `$\upsilon$`	ϕ `$\phi$`	φ `$\varphi$`	χ `$\chi$`
ψ `$\psi$`	ω `$\omega$`		

Γ `$\Gamma$`	Δ `$\Delta$`	Λ `$\Lambda$`	Θ `$\Theta$`
Ξ `$\Xi$`	Π `$\Pi$`	Υ `$\Upsilon$`	Σ `$\Sigma$`
Φ `$\Phi$`	Ψ `$\Psi$`	Ω `$\Omega$`	

Slanted capitals may be obtained by prefacing the commands by `\mit` or, in $\mathcal{AMS}$-TeX, by using `\var` to start the command. For example, {`\mit\Gamma`} produces $\varGamma$, as does `\varGamma` in $\mathcal{AMS}$-TeX. Bold capitals are obtained by prefacing the commands by `\bf` (or by using `\boldsymbol` in $\mathcal{AMS}$-TeX). See AMS symbols for a few additional Greek letters.

▷ **group** See grouping.

▷ **grouping** The notion of a group is basic to how TeX applies commands. The idea is that portions of material can be corralled off from other material, and that the effects of certain commands can be kept confined within this "grouped" portion. The symbols that normally begin and end groups are { and }, respectively. The commands `\bgroup` and `\egroup` have the same effect, as do the primitive commands `\begingroup` and `\endgroup`. If the effect of a command issued within a group is required to persist outside it, it must usually be prefixed by `\global`.

▷ $\gtrapprox$ `$\gtrapprox$` $\mathcal{AMS}$ $\gtrapprox$ (Class 3: relation.)

▷ $\gtrdot$ `$\gtrdot$` $\mathcal{AMS}$ $\gtrdot$ (Class 3: relation.)

▷ $\gtreqless$ `$\gtreqless$` $\mathcal{AMS}$ $\gtreqless$ (Class 3: relation.)

▷ $\gtreqqless$ `$\gtreqqless$` $\mathcal{AMS}$ $\gtreqqless$ (Class 3: relation.)

▷ $\gtrless$ `$\gtrless$` $\mathcal{AMS}$ $\gtrless$ (Class 3: relation.)

▷ $\gtrsim$ `$\gtrsim$` $\mathcal{AMS}$ $\gtrsim$ (Class 3: relation.)

▷ $\gvertneqq$ `$\gvertneqq$` $\mathcal{AMS}$ $\gvertneqq$ (Class 3: relation.)

▷ `\halign` * 88–89, 98–99. TeX's basic alignment command.

▷ $\haswidth$ `$\haswidth$` $\mathcal{AMS}$ See commutative diagrams.

▷ $\hat$ `$\hat$` 30–31, 120–121, 167. `$\hat a$` produces $\hat{a}$.

▷ $\Hat$ `$\Hat$` $\mathcal{AMS}$ See `\accentedsymbol`.

▷ `\hataccent` $\mathcal{AMS}$ May be used in place of `\^` to give a "hat accent" in text.

▷ $\hbar$ `$\hbar$` $\hbar$ (Class 0: ordinary.) The symbol is a composite in Plain TeX, made out of a superposition of two separate characters: h and $\bar{\ }$:

`\def\hbar{{\mathchar'26\mkern-9muh}}`

The `\mkern-9mu` adjusts spacing satisfactorily in text style but not as well in other styles: consider $\hbar^{\hbar^\hbar}$.

As an alternative, the `msbm` font provides a similar single character, $\hbar$, which can be invoked in $\mathcal{AMS}$-TeX using the same command `\hbar`, provided `\undefine` has first been used (see that entry for details).

▷ `\hbox` * 30–31, 36–37, 40–41, 48–49, 52–57, 62–63, 66, 68–71, 74, 86–89, 100–105, 112–113, 124–129, 162, 216, 219, 220, 223, 224, 227, 228. ($\mathcal{AMS}$-TeX alert: don't `\redefine`.) A horizontal box. The notion of a box is fundamental to TeX's approach to typesetting; the chief use of an `\hbox` in typesetting mathematics is to allow the insertion of correctly spaced text inside formulas. For example, '`$x \hbox{ is greater than } y$`' gives 'x is greater than y'. The typeface automatically used inside the `\hbox` in such circumstances is the one that was in force when the formula began.

▷ `\hcorrection` $\mathcal{AMS}$ Shifts the position of the page left or right; for example, `\hcorrection{1cm}` shifts the page right by 1 cm. The definition is

`\def\hcorrection#1{\advance\hoffset#1\relax}`

Thus `\hcorrection` augments any previous value that `\hoffset` may have had and does not supplant it.

▷ `$\hdots$` $\mathcal{AMS}$ See matrices.

▷ `$\hdotsfor$` $\mathcal{AMS}$ See matrices.

▷ `\headline` 14–15, 24–25, 225–228, 232. Allows a line of material to be placed at the top of every page.

▷ `$\heartsuit$` $\lceil\heartsuit\rceil$ (Class 0: ordinary.)

▷ **Hebrew letters** See AMS symbols.

▷ **height** 100, 140, 148, 224. Every box constructed by TeX is considered to have a baseline, a height, a depth and a width. The height is the distance that the box extends above the baseline. The word is also used as a **keyword**: see below.

▷ **height** 40–41, 66, 98–99, 216, 219, 220, 224, 225, 227, 229. A keyword that is used to specify the height of an `\hrule` or a `\vrule` above the baseline. For example, '❙' was obtained by typing '`\vrule height5pt width.5pt depth0pt`'.

▷ `\hfil` * 11, 14–15, 20, 24–25, 34–35, 58–61, 66, 70–71, 74–77, 84–85, 88–91, 98–99, 102–105, 108–109, 219, 224, 225, 227, 228, 232. Fills a horizontal region with white space.

▷ `\hfill` * 16–19, 24–27, 52–53, 56–57, 62–65, 72–73, 94–95, 126–127. Like `\hfil`, only stronger.

▷ `\hoffset` * 225, 232. Sets the horizontal position of the page.

▷ `$\hom$` $\lceil\text{hom}\rceil$

▷ `$\hookleftarrow$` $\lceil\hookleftarrow\rceil$ (Class 3: relation.)

▷ `$\hookrightarrow$` $\lceil\hookrightarrow\rceil$ (Class 3: relation.)

▷ \hphantom   30–31, 40–41, 62–63, 86–87, 98–99, 120–121, 130–131.
\hphantom{stuff} calculates the width of stuff as it normally would, but sets
the height and depth to zero. It does not print this stuff, but leaves instead
a space of the correct width. For example, here is a space   of width
exactly equal to that of the word 'phantom'. The facility is useful when one
wants to trick TeX. Also see \phantom and \vphantom.

▷ \hrule *   216, 219, 220. Gives a horizontal rule. See rules.

▷ \hrulefill   40–41, 98–99, 227. \hbox to 1cm{[\hrulefill]} gives [___].
The command is defined by

> \def\hrulefill{\leaders\hrule\hfill}

▷ \hsize *   14–15, 42–43, 58–59, 64–66, 70–71, 74–75, 84–85, 104–105, 110–113,
216, 224–228. Represents the horizontal size of pages, or of vboxes.

▷ \hskip *   40–41, 66, 70–71, 88–89, 100–105, 108–109, 224, 225, 228, 229.
\hskip1cm gives a horizontal space of 1 cm. See \relax for an important
warning, and \quad and \qquad for applications.

▷ $\hslash$ $_{A\!M\!S}$ $\lceil \hbar \rfloor$ (Class 0: ordinary.)

▷ \hss *   225, 228. Horizontal glue that can stretch or shrink as needed.

▷ \ht *   66, 70, 224, 227. Represents the height of boxes; for example, \ht22 is
the height of box register number 22.

▷ \ialign   An appropriately initialized \halign:

> \def\ialign{\everycr={}\tabskip=0pt \halign}

Plain TeX uses alignments in its internal definitions of several commands (see
\angle and \eqalign), and \ialign is used to make sure that new settings
for \everycr and \tabskip that users may have chosen elsewhere don't make
everything go haywire.

▷ $\idotsint$ $_{A\!M\!S}$ $\lceil \int \cdots \int \rfloor$ in a line, $\lceil \int \cdots \int \rfloor$ in a display. (Class 1: large
operator.)

▷ \if... *   14–15, 40–42, 74, 98–101, 168, 216, 219, 222–232. TeX provides
a plethora of conditional commands; they all look structurally something like
this:

> \if... *condition action to take if the condition is true*
>   \else *action to take if the condition is false* \fi

Examples of conditional commands are \ifodd (tests whether a number is odd);
\ifnum (tests whether an integer is greater than, less than, or equal to another
one); \ifdim (tests whether a dimensional quantity—like a size—is greater
than, less than, or equal to another); \ifhmode, \ifvmode and \ifmmode (test
whether the current mode is horizontal, vertical and math, respectively); \ifx
(tests whether the immediately following two tokens agree: e.g., whether they
both represent the same character playing the same role, or the same TeX
primitive command, or—if they are higher level commands—whether they have

the same definition, etc.); \iftrue and \iffalse (don't test anything—they are, respectively, always true and always false—but are useful nevertheless); \ifcase (selects a case from a list of alternatives—see \or). An entire chapter could be written just on TEX's conditional commands, but it would be going too far afield to say more here. Examples of the use of these commands are scattered throughout this book—especially in Chapter 6—and they should suggest some of the ways in which conditional commands can be gainfully employed.

▷ $\iff$ $\iff$ (Class 3: relation.) This is the same symbol as the one produced by \Longleftrightarrow, but it leaves an extra thick space on each side.

▷ \ignorespaces ∗ 222, 223, 227, 230. Makes TEX ignore spaces that follow. It is often helpful to include it at the end of the definition of a new command, since it "eats up" spurious spaces that might inadvertently be introduced.

▷ $\iiiint$ $\mathcal{A}_{\mathcal{M}}S$ $\iiiint$ in a line, $\iiiint$ in a display. (Class 1: large operator.)

▷ $\iiint$ $\mathcal{A}_{\mathcal{M}}S$ $\iiint$ in a line, $\iiint$ in a display. (Class 1: large operator.)

▷ $\iint$ $\mathcal{A}_{\mathcal{M}}S$ $\iint$ in a line, $\iint$ in a display. (Class 1: large operator.)

▷ **illustrations** See diagrams.

▷ $\Im$ $\Im$ (Class 0: ordinary.)

▷ $\imath$ $\imath$ (Class 0: ordinary.)

▷ $\impliedby$ $\Longleftarrow$ $\mathcal{A}_{\mathcal{M}}S$

▷ $\implies$ $\Longrightarrow$ $\mathcal{A}_{\mathcal{M}}S$

▷ **in** See keywords and units.

▷ $\in$ $\in$ 26–31, 34–41, 44–57, 76–81, 86–89, 96–99, 102–103, 120–121. (Class 3: relation.)

▷ \indent ∗ 32–35, 52–53, 60–67, 86–91, 104–105. Leaves a space of size equal to the paragraph indentation.

▷ **indices** Lower indices are obtained with _, and upper with ^. For example, A_1 gives A_1, A^2 gives A^2, and A_a^b gives A_a^b. If the _ or ^ key is broken, \sb and \sp may be used instead. Also see spacing and subscripts.

▷ $\inf$ $\inf$ 26–27, 38–39, 78–79.

▷ ∞ ∞ 26–29, 36–39, 54–57, 78–81, 92–95, 126–127. (Class 0: ordinary.)

▷ $\injlim$ $\mathcal{A}_{\mathcal{M}}S$ $\text{inj}\lim$

▷ **Inner** See atom and \mathinner.

▷ $\innerhdotsfor$ $\mathcal{A}_{\mathcal{M}}S$ See matrices.

▷ \input * 7, 14, 74, 84, 92, 96, 102, 104, 106, 110, 114, 116, 122, 128, 133–135, 137, 151, 169, 221, 226, 229. Inserts a file. For example, a file called file.ext can be inserted by saying \input file.ext. If the extension is .tex, then it does not have to be explicitly specified.

▷ **inserts** See \midinsert, \pageinsert and \topinsert.

▷ $\int$ $\int$ in a line, $\int$ in a display. 62–65, 70–71, 82–83, 92–95, 98–99, 118–121, 157, 158, 159, 164. (Class 1: large operator.)

▷ **integrals** See \int (and \iint, etc.), \oint and large operators.

▷ $\intercal$ $\mathcal{AMS}$ $\top$ (Class 2: binary operation.)

▷ $\intertext$ $\mathcal{AMS}$ 116–119. Interjects text between the lines of an alignment given by \align. For example,

```
$$\align A&=B \\
\intertext{textual stuff}
C&=D. \endalign $$
```

will give

$$A = B$$

textual stuff

$$C = D.$$

The textual stuff can be as long as desired. It gets wrapped to additional lines automatically if it is too long to fit on one line.

▷ ι ι

▷ \it 9, 18–23, 26–29, 32–35, 38–41, 44–47, 52–53, 62–67, 72–73, 86–97, 102–103, 116–117, 126–127, 136, 139, 149–153, 163, 173, 176, 216, 220. ($\mathcal{AMS}$-TEX alert: don't \redefine.)

Italics are produced by typing {\it Italics\/}. (\/ gives an "italic correction": a small extra space that allows the last italicized character to keep from leaning into the space of the following upright character.) See §5.5 for a full discussion of how commands like \it work.

$\mathcal{AMS}$-TEX does not permit the command in formulas; \italic must be used instead.

▷ $\italic$ $\mathcal{AMS}$ 126–127. Gives text italic letters in a formula, rather as \bold gives boldface letters. The italic letters will not, however, shrink in indices.

▷ **italics** 9, 38, 75, 140–142, 144, 149–152, 156, 158, 159, 166, 167, 169, 173, 174, 176, 177, 178, 180, 184, 186, 196, 204, 206, 210, 212, 213, 219, 234. See \it.

▷ \item 126–127. Makes itemized lists.

▷ \itemitem 126–127. Gives a doubly indented itemized list.

▷ $\jmath$ $\jmath$ (Class 0: ordinary.)

▷ $\joinrel$ 98–99. See special effects.

▷ \jot 124–125. (*AMS*-TeX alert: don't \redefine.) A parameter typically used to change the spacing of multiline displays. Its value in Plain TeX is 3pt. It may be reset to 300 points, for example, by typing '\jot=300pt'. See \openup.

▷ κ $\lceil \kappa \rceil$ 130–131.

▷ $\ker$ $\lceil \ker \rceil$

▷ \kern * 20–21, 216, 219, 220, 224, 228. Leaves a horizontal space in horizontal mode and a vertical space in vertical mode.

▷ **keywords** These are words that modify, or continue, the instructions in certain commands, but only when used in conjunction with them. The improper use of keywords can be the source of many frustrating—and sometimes mystifying—errors. Here is a full list of all the keywords and the exact contexts in which they perform special functions:

- bp, cc, cm, dd, in, mm, pc, pt, sp: these are all units of size. Every size specification given to TeX must be in terms of one of these units: e.g., \hskip 1cm, \parindent 20pt, \vsize 40pc, etc. Away from this context, TeX will just typeset the characters as they appear: e.g., '20pt' in the input, away from a size command, will give '20pt' in the output. Also see units.

- em, ex: font-dependent units. See individual entries for each, and units.

- at, scaled: used in conjunction with \font to specify font sizes. See §5.1. These are a little tricky, because such specifications are optional; small mistakes in usage will not trigger any warnings. The instruction \font\BigRm=cmr10 at 50pt will assign the name \BigRm to the 10-point Computer Modern roman font scaled to 50-point size. On the other hand, \font\BigRm=cmr10 .at 50pt will assign the name to the regular 10-point roman font at its normal size, and will then typeset '.at 50pt' in the output.

- by: used (if desired) in conjunction with \multiply, \divide or \advance to multiply, divide or "advance" the values stored in TeX's registers. For example, when a page is composed and "shipped out" by Plain TeX, it carries out a sequence of instructions that may be summarized as \advance\pageno by 1: this adds 1 to the contents of the standard page number register.

- depth, height, width: used to specify the depth, height and width of a rule. As with at and scaled, their use is optional, and the warning given there applies here as well. See \relax for other situations where care must be taken.

- fil: a special unit used to specify infinitely stretchable or shrinkable glue. For example, TeX adds on some glue at the end of every paragraph: the amount is the value assigned to \parfillskip. In Plain TeX it is

`0pt plus 1fil`; i.e., no white space is necessarily left, but the space can stretch to fill as much of the last line as needed.

- `l`: added on after `fil` (to make `fill` or `filll`) to give stronger degrees of stretchability or shrinkability: a `fill` will overpower a `fil`, pretty much as `\hfill` overpowers `\hfil`. Since `l` is a keyword by itself, `fil l` means the same thing as `fill`. This means that one can occasionally get into a mess if TeX interprets some following `l` as a keyword extension to `fil`: e.g., in '`the \hskip1pt plus 1fil lower bound`'; the solution is to say `\relax` after the `fil`.

- `minus`, `plus`: used to specify the stretchability and shrinkability of glue. Also see `\relax`.

- `mu`: the math unit of glue. A special unit is useful because of the different typeface sizes that are routinely used in the same formula. The unit adjusts itself automatically by being tied to the size of the em from family 2 in the current style: 18 mu equal 1 em. It is used to specify sizes of horizontal spaces in formulas through the command `\mskip`. For example, the negative thin space command in mathematics, `\!`, is defined in Plain TeX by '`\mskip -3mu`'. The other spacing commands, `\,`, `\>` and `\;`, are of size '`3mu`', '`4mu plus2mu minus4mu`' and '`5mu plus5mu`', respectively. See `\medmuskip`, `\thickmuskip` and `\thinmuskip`.

- `spread`, `to`: used to specify the size of a horizontal or vertical box.

- `true`: used to specify magnification-independent sizes.

▷ `\l` ⌈l⌉ 122–123. In $\mathcal{AMS}$-TeX the command plays a second role when formatting **matrices** with the command `\format`; this is discussed under that topic. (And in LA$\mathcal{MS}$-TeX it plays a still further role when labelling arrows in **commutative diagrams**; see pages 128–131.)

▷ `\L` ⌈L⌉ The command plays a further role in LA$\mathcal{MS}$-TeX (see `\l`).

▷ `$\lambda$` ⌈λ⌉ 16–17, 28–29, 32–33, 38–39, 48–49, 78–83, 94–95, 120–121, 126–127.

▷ `$\Lambda$` ⌈Λ⌉ 94–95. In Plain TeX `${\mit\Lambda}$` (`\varLambda` in $\mathcal{AMS}$-TeX) gives $\mathit{\Lambda}$; `${\bf\Lambda}$` (`\boldsymbol\Lambda` in $\mathcal{AMS}$-TeX) gives $\boldsymbol{\Lambda}$; and `${\tt\Lambda}$` gives $\mathtt{\Lambda}$.

▷ `$\land$` ⌈∧⌉ (Class 2: binary operation.) Alternate name for `\wedge`.

▷ `$\langle$` ⌈⟨⌉ 42, 120–121, 165. (Class 4: opening.) Also see **delimiters**.

▷ `\language` * 85. An addition to TeX from version 3.0 on; it allows the use of different hyphenation rules for different languages.

▷ **large operators** These are operators that come in large sizes. They are considered a separate class (see **classes**) and they thereby follow their own spacing rules. See **spacing**. The larger member of each pair shown below is used automatically for displayed equations, or when `\displaystyle` has been explicitly invoked.

large operators

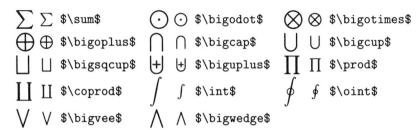

$\sum$	$\sum$ `$\sum$`	$\odot$ `$\bigodot$`	$\otimes$ `$\bigotimes$`
$\bigoplus$ `$\bigoplus$`	$\bigcap$ `$\bigcap$`	$\bigcup$ `$\bigcup$`
$\bigsqcup$ `$\bigsqcup$`	$\biguplus$ `$\biguplus$`	$\prod$ `$\prod$`
$\coprod$ `$\coprod$`	$\int$ `$\int$`	$\oint$ `$\oint$`
$\bigvee$ `$\bigvee$`	$\bigwedge$ `$\bigwedge$`	

There is also a small integral sign, $\int$, produced by `$\smallint$` (but in the "ordinary" class, not the "large operator" one). $\mathcal{A}\mathcal{M}\mathcal{S}$-TeX makes available the commands \iint, \iiint, \iiiint and \idotsint, useful for properly spaced multiple integrals. Their effects are shown in separate entries.

Limits may be introduced on integrals, sums, products, etc. These are entered in the input as superscripts and subscripts. With two exceptions, the limits will be automatically set above and below the symbol in **display style** and to its right otherwise. For example, \sum_i will appear as $\sum_i$ if used as part of a formula in a line, and as

$$\sum_i$$

if used in display. The exceptions are the two integral signs $\int$ and $\oint$, obtained from `$\int$` and `$\oint$`, respectively. They have limits placed on the right, even in displays. All these positions for limits can be changed by using the commands \limits (to move limits to the top and bottom), \nolimits (to move limits to the side) and \displaylimits (to restore the default conventions). For example,

```
$$\int\limits_0^1 \qquad \sum\nolimits_1^n \qquad
    \sum\nolimits\displaylimits_1^n \qquad
    \sum\limits\displaylimits_1^n $$
```

produces

$$\int\limits_0^1 \qquad \sum\nolimits_1^n \qquad \sum_1^n \qquad \sum_1^n$$

Multiline limits can also be introduced on large operators:

```
$$ \sum_{{\scriptstyle 0<i<5 \atop \scriptstyle 0<j<7}
        \atop \scriptstyle 0<k<9}. $$
```

gives

$$\sum_{\substack{0<i<5 \\ 0<j<7 \\ 0<k<9}}.$$

$\mathcal{A}\mathcal{M}\mathcal{S}$-TeX has a built-in mechanism for such multiline limits; for example,

```
$$ \sum \Sb 0<i<5 \\ 0<j<7 \endSb A_{ij}. $$
```
gives

$$\sum_{\substack{0<i<5 \\ 0<j<7}} A_{ij}.$$

For upper limits, \Sp is used.

▷ \last... ∗ A class of commands encompassing \lastbox, \lastkern, \lastpenalty and \lastskip. They refer, respectively, to the last box (\hbox or \vbox) made, to the last kern taken (see \kern), the last penalty inserted, and to the last skip.

▷ LaTeX 1, 4, 5, 6, 11, 12, 13, 112, 114–115, 137, 138, 148, 173, 174, 208–209, 219, 237, 238. A widely used package of commands built out of the primitive commands of TeX. The package makes it very easy to carry out certain standard typesetting tasks like, for example, double-column output. LaTeX commands are not always compatible with those of Plain TeX, so one has to be careful if one is switching between the two. More significantly, as is briefly discussed in Chapter 1, LaTeX adopts a rather different conceptual approach to typesetting, encouraging logical rather than visual formatting.

A few aspects of font selection in LaTeX are discussed in § 5.7.

▷ $\lbrace$ ⌈{⌉ (Class 4: opening.) Alternate name: \{. Also see delimiters.

▷ $\lbrack$ ⌈[⌉ 78–79. (Class 4: opening.) Also see delimiters.

▷ $\lceil$ ⌈⌈⌉ (Class 4: opening.) Also see delimiters.

▷ $\lcfrac$ $\mathcal{A_MS}$ See continued fractions.

▷ $\ldots$ ⌈...⌉ 26–35, 38–39, 46–47, 62–65, 76–83 86–91, 94–95, 136. A constructed symbol, assigned the "Inner" atomic type. See atom, spacing and \mathinner.

▷ $\le$ ⌈≤⌉ 40–41, 94–97, 118–121. (Class 3: relation.) Alternate name for \leq.

▷ \leaders ∗ 66. Used for putting together repeated material; see, for example, \hrulefill.

▷ $\left$ ∗ 16–17, 26–27, 32–33, 40–41, 56–57, 62–65, 70–71, 76, 78–79, 88–89, 98–103, 116–119, 122–123, 128–129, 157, 163–165. ($\mathcal{A_MS}$-TeX alert: don't \redefine.)

Creates left delimiters (i.e., brackets, parentheses, etc.) that will have the correct size for a large formula. The \left command is to be used in displayed equations, and always in conjunction with \right: TeX will expect \left and \right to occur in pairs in each equation, and for \left to precede \right. The actual symbols used need not occur in balanced pairs: \left(may be paired with \right], for example, or even with \right[if an unusual effect is wanted. If just a single delimiter is needed, then '\left.' or '\right.' must be typed for the missing one. For example,

`$\leftarrow$`

```
$$ \left. {d \sin x\over dx}\right|_{x=0} = 1
\qquad
1 \left/ {a\over b} \right.
\qquad
\left. 3\int_0^1 x^2\,dx = x^3 \right[_{x=0}^{x=1} = 1. $$
```

gives

$$\left. \frac{d\sin x}{dx}\right|_{x=0} = 1 \qquad 1 \left/ \frac{a}{b} \right. \qquad 3\int_0^1 x^2\,dx = x^3 \left[_{x=0}^{x=1}\right. = 1.$$

Observe that `\left` and `\right` must sandwich the correct subformulas in order to get the correct heights for the delimiters.

▷ `$\leftarrow$` $\lceil\leftarrow\rfloor$ 66, 112–113. (**Class 3:** relation.) Alternate name: `\gets`.

▷ `$\Leftarrow$` $\lceil\Leftarrow\rfloor$ (**Class 3:** relation.)

▷ `\leftarrowfill` '$\longleftarrow$' is produced by '`\hbox to 1cm{\leftarrowfill}`'.

▷ `$\leftarrowtail$` $\mathcal{AMS}$ $\lceil\leftarrowtail\rfloor$ (**Class 3:** relation.)

▷ `$\leftharpoondown$` $\lceil\leftharpoondown\rfloor$ (**Class 3:** relation.)

▷ `$\leftharpoonup$` $\lceil\leftharpoonup\rfloor$ (**Class 3:** relation.)

▷ `$\leftleftarrows$` $\mathcal{AMS}$ $\lceil\leftleftarrows\rfloor$ (**Class 3:** relation.)

▷ `\leftline` 12, 66, 74, 104–105, 216, 223, 225, 227. `\leftline{stuff}` puts stuff on the left. If used in mid-paragraph, the preceding line must be explicitly broken.

▷ `$\leftrightarrow$` $\lceil\leftrightarrow\rfloor$ (**Class 3:** relation.)

▷ `$\Leftrightarrow$` $\lceil\Leftrightarrow\rfloor$ (**Class 3:** relation.)

▷ `$\leftrightarrows$` $\mathcal{AMS}$ $\lceil\leftrightarrows\rfloor$ (**Class 3:** relation.)

▷ `$\leftrightharpoons$` $\mathcal{AMS}$ $\lceil\leftrightharpoons\rfloor$ (**Class 3:** relation.)

▷ `$\leftrightsquigarrow$` $\mathcal{AMS}$ $\lceil\leftrightsquigarrow\rfloor$ (**Class 3:** relation.)

▷ `$\leftroot$` $\mathcal{AMS}$ Used along with `\root` to position the root horizontally.

▷ `\leftskip` * 225, 231. Moves the left margin (for text) in or out.

▷ `$\leftthreetimes$` $\mathcal{AMS}$ $\lceil\leftthreetimes\rfloor$ (**Class 2:** binary operation.)

▷ `$\leq$` $\lceil\leq\rfloor$ 26–31, 34–39, 56–57, 60–61. (**Class 3:** relation.) Alternate name: `\le`.

▷ `$\leqalignno$` 26–27, 48–49. Aligns equations and places equation numbers to the left. For example,

```
$$\leqalignno{A&=B&(1)\cr C&=0\cr x&=D+y.&(2)\cr}$$
```

gives

$$
\begin{aligned}
(1) \qquad\qquad\qquad A &= B \\
C &= 0 \\
(2) \qquad\qquad\qquad x &= D + y.
\end{aligned}
$$

▷ $\leqno$ *   16–17, 20–21, 26–27, 30–31, 34–35, 38–39, 60–65, 68–69, 72–73. Works like \eqno, but it places equation numbers to the left.

▷ $\leqq$ $\mathcal{A_MS}$ $\lceil \leqq \rfloor$ 90–91. (Class 3: relation.)

▷ $\leqslant$ $\mathcal{A_MS}$ $\lceil \leqslant \rfloor$ 78–81. (Class 3: relation.)

▷ $\lessapprox$ $\mathcal{A_MS}$ $\lceil \lessapprox \rfloor$ (Class 3: relation.)

▷ $\lessdot$ $\mathcal{A_MS}$ $\lceil \lessdot \rfloor$ (Class 3: relation.)

▷ $\lesseqgtr$ $\mathcal{A_MS}$ $\lceil \lesseqgtr \rfloor$ (Class 3: relation.)

▷ $\lesseqqgtr$ $\mathcal{A_MS}$ $\lceil \lesseqqgtr \rfloor$ (Class 3: relation.)

▷ $\lessgtr$ $\mathcal{A_MS}$ $\lceil \lessgtr \rfloor$ (Class 3: relation.)

▷ $\lesssim$ $\mathcal{A_MS}$ $\lceil \lesssim \rfloor$ (Class 3: relation.)

▷ \let *   153, 216, 220, 221, 223, 226, 228, 231. Allows a command to be set equal to the current meaning of another command, or to some other token. See \bgroup for an example.

▷ $\lfloor$ $\lceil \lfloor \rfloor$ (Class 4: opening.) Also see delimiters.

▷ $\lg$ $\lceil \lg \rfloor$

▷ $\lgroup$ See delimiters.

▷ $\lim$ $\lceil \lim \rfloor$ 26–27, 80–81, 92–93.

▷ $\liminf$ $\lceil \liminf \rfloor$

▷ limits See large operators.

▷ $\limits$ *   26–27, 38–39, 62–63, 70–71, 78–81, 90–91. ($\mathcal{A_MS}$-TEX alert: don't \redefine.) Forces limits to appear above and below large operators.

▷ \LimitsOn... $\mathcal{A_MS}$ The AMS preprint style allows global settings of whether limits (see, for example, large operators) are to be placed on integrals, names of functions (log, exp, etc.) and sums. The ... above must be replaced by Ints, Names and Sums, respectively, to finish the command name. The opposite effect is achieved by saying '\NoLimitsOn...'.

▷ $\limsup$ $\lceil \limsup \rfloor$

▷ \line 26–27, 34–35, 70–71, 90–91, 104–105, 108 109, 224, 225, 227, 228. \line{stuff} puts stuff on a single line. One usually also needs to include something like \hfil to explicitly ask for white space to fill the rest of the line.

▷ line breaks TEX decides where line breaks should occur after combining elaborate sets of penalties and demerits, and then computing the breakpoints that cost the least. See the discussion of penalties. Line breaks may be forced or prevented by typing (when in horizontal mode) \break or \nobreak. They may be suggested by typing \allowbreak. $\mathcal{A_MS}$-TEX offers several further commands that allow control over line breaks. They are listed at the end.

Formulas in text, sandwiched between single $ signs, will be broken by
TeX only at a few places: after a relation symbol like $=$ or $<$, or (as a less
preferred choice) after an operation symbol like $+$ or $-$. In all these cases, the
break occurs only if the symbol is not buried within an explicit group (i.e., it
is not between { and }). Thus, line breaks may be prohibited by hiding parts
of the formula, or all of it, between { and }—though one must be careful that
one isn't also interfering with the normal spacing rules that TeX follows for
mathematics (see spacing).

Awkward line breaks between text and formulas (or even within straight
text) may be prevented by using ties, i.e., by using the command character ~.
This gives a single space at which the line will not be broken. Ties are useful
in several places: when referring to names of entities (one types 'Theorem~3'
or 'Section~B', for example), when referring to variables in a sentence (for
example, `height~$h$`), or when attaching units to numbers (for example,
`lines 12~points apart`). In general, a tie should be used whenever a line
break between words would be awkward.

In a formula with implicit multiplications, line breaks may be suggested by
using * in places where multiplications are implied; e.g., consider $x = (a - b) \times (c - d)(u - v)$. The input for this was:

```
... consider $x=(a-b)\*(c-d)\*(u-v)$. The input ...
```

Notice that TeX ignores * unless it is at a good place for a line break; at
such a place it is replaced by '$\times$'. See `\discretionary` for the definition of the
command.

Displayed formulas, typed between double $ signs, are never broken by
TeX: they have to be explicitly divided into subformulas. In some situations
it is better to display an explicitly broken equation using `\displaylines` (or
`\gather` in $\mathcal{AMS}$-TeX) rather than with a pair of $$ signs for each line of the
equation (the second approach may leave too much space between the lines).

In addition to the mechanisms discussed here, $\mathcal{AMS}$-TeX offers a few extra
ones (some are just refinements of the ones discussed above). See
`\allowlinebreak, \allowmathbreak, \linebreak, \mathbreak,`
`\nolinebreak,` and `\nomathbreak.`

▷ **line spacing**   Controlled by setting `\baselineskip`. The default value in
Plain TeX is 12pt. In addition to this, there are two other parameters that
control interline spacing. The `\lineskiplimit` is related to how close the top
of one line can get to the bottom of the one above. Lines that are closer
than the chosen value of `\lineskiplimit` are then separated by the value
of `\lineskip` (the commands are discussed in separate entries above). See
`\offinterlineskip` to find out how to suppress normal interline spacing. See
also `\openup`.

▷ `\linebreak` $\mathcal{AMS}$ May be placed in a line to force a line break (and is thus stronger than \allowlinebreak). The \nolinebreak command has the opposite effect.

▷ \lineskip * 100–101, 154, 216, 224. Specifies the gap to be left between two lines if the value of \baselineskip that one is using brings them too close. See \lineskiplimit to see what "too close" means. Plain TeX sets the value of \lineskip to be 1pt. See line spacing.

▷ \lineskiplimit * 100–101, 154. The parameter that determines how close two adjacent lines must be for extra space (see \lineskip) to be inserted. The Plain TeX value is 0pt. See line spacing.

▷ $\ll$ $\lceil \ll \rceil$ (Class 3: relation.)

▷ \llap 14–15, 18–21, 48–49, 58–59, 68–69, 100–101, 124–125, 224, 225, 228, 232. Allows material to overlap other material to the left. (In particular, the overlap could be with the left margin.) For example, a\llap/ produces ⱥ. Also see \rlap.

▷ $\llcorner$ ᴬᴹˢ $\lceil \llcorner \rceil$ (Class 4: opening.)

▷ $\Lleftarrow$ ᴬᴹˢ $\lceil \Lleftarrow \rceil$ (Class 3: relation.)

▷ $\lll$ ᴬᴹˢ $\lceil \lll \rceil$ (Class 3: relation.) Alternate name: \llless.

▷ $\llless$ ᴬᴹˢ $\lceil \lll \rceil$ (Class 3: relation.) Alternate name for \lll.

▷ $\lmoustache$ 165. See delimiters.

▷ $\ln$ $\lceil \ln \rceil$ 118–119.

▷ $\lnapprox$ ᴬᴹˢ $\lceil \lnapprox \rceil$ (Class 3: relation.)

▷ $\lneq$ ᴬᴹˢ $\lceil \lneq \rceil$ (Class 3: relation.)

▷ $\lneqq$ ᴬᴹˢ $\lceil \lneqq \rceil$ (Class 3: relation.)

▷ $\lnot$ $\lceil \lnot \rceil$ (Class 0: ordinary.) Alternate name for \neg.

▷ $\lnsim$ ᴬᴹˢ $\lceil \lnsim \rceil$ (Class 3: relation.)

▷ \load... ᴬᴹˢ 120–121, 135, 168–169. A class of commands that allows users to load fonts selectively. The complete list is \loadbold (bold symbols), \loadeufb (bold fraktur), \loadeufm (medium fraktur), \loadeurb (bold AMS Euler), \loadeurm (medium AMS Euler), \loadeusb (bold script), \loadeufm (medium script), \loadmsam (AMS math symbols, A) and \loadmsbm (AMS math symbols, B). A table for each of these fonts is displayed in Chapter 5, where a description of the last two \load... commands is also given (§ 5.5).

▷ $\log$ $\lceil \log \rceil$ 80–83.

▷ $\longleftarrow$ $\lceil \longleftarrow \rceil$ 70–71. (Class 3: relation.)

▷ $\Longleftarrow$ $\lceil \Longleftarrow \rceil$ (Class 3: relation.)

▷ $\longleftrightarrow$ $\lceil \longleftrightarrow \rceil$ (Class 3: relation.)

▷ $\Longleftrightarrow$ $\lceil \Longleftrightarrow \rceil$ (Class 3: relation.)

▷ $\longmapsto$ $\lceil \longmapsto \rceil$ (Class 3: relation.)

▷ $\longrightarrow$ $\lceil \longrightarrow \rceil$ 56–57, 70–71, 98–99. (Class 3: relation.)

▷ $\Longrightarrow$ $\lceil \Longrightarrow \rceil$ (Class 3: relation.)

▷ \m@th 13. A command that switches off extra space around formulas:

    \def\m@th{\mathsurround=0pt }

The use of @ in command names is a standard device in large packages of commands to ensure that internally used commands in the package are not inadvertently redefined by users. (The category code of @ is changed at the start of the package to allow it to be used in a command name, then reset at the end.)

Marks pieces of input so that they can be extracted later, after the page is set; \firstmark represents the first piece of marked text on the page, \botmark the last piece, and \topmark the last piece on the previous page.

Invokes characters that represent accents. Each use of the command requires a specification of the position of the character in a particular family, and an assignment to it of a class. Here, for example, is the Plain TeX definition of \ddot:

    \def\ddot{\mathaccent"707F }

This method of calling upon characters is discussed in detail (in connection with \mathchar) in §5.5. Though \mathaccent is used in the same way as \mathchar, characters summoned by the two commands are treated differently when TeX positions them in a formula.

▷ \mathbin * TₑX needs to know the "atomic type" of the constituents of formulas so that it can space them correctly: see **atom** and **spacing**. Single characters are assigned **classes**, which then also determine their atomic type, when they are summoned by commands like \mathchar, \mathchardef, \delimiter, \mathaccent or \radical. More complex structures can also be assigned a class as shown below.

The declaration \mathbin assigns to its argument the **class** "binary operation"; e.g., the definition of \bmod in Plain TₑX uses \mathbin as follows:

```
\def\bmod{\mskip-\medmuskip \mkern5mu
    \mathbin{\rm mod} \penalty900 \mkern5mu \mskip-medmuskip}
```

This definition ensures that "mod" will be set in roman type, and that the space around it will be appropriate for a binary operation. The \mskip and \mkern further adjust spacing, and the \penalty discourages a line break.

▷ $\mathbreak$ 𝒜ℳ𝒮 Forces a break in a formula within text. Also see \allowmathbreak and, for an opposite effect, \nomathbreak.

▷ \mathchar * 161, 164, 165. Summons characters from specified font families and assigns them each a specified class. See §5.5 for a full discussion.

▷ \mathchardef * 161, 162, 164, 166, 167, 168. See §5.5.

▷ \mathchoice * There may sometimes be a need for a single command to have different effects in display, text, script and scriptscript styles (see **styles in mathematics**). \mathchoice accommodates this need by allowing the specification of four subformulas: the first is automatically picked when the current style is display style, the second when it is text, and so on. For example,

```
\def\Ex {\mathchoice{A}{B}{C}{D}}
$$ \Ex \qquad {\textstyle \Ex}^{\Ex^{\Ex}}. $$
```

has this effect

$$A \qquad B^{C^D}.$$

Since TₑX actually privately typesets each of the four subformulas (here just A, B, C and D) when it encounters a \mathchoice, then picks the appropriate one as needed, this command isn't efficient and should be used sparingly. Also, further work is sometimes needed to ensure that everything does indeed work correctly in all styles: see \mathpalette.

▷ \mathclose * See \mathbin for a general comment. The declaration \mathclose assigns to its argument the **class** called "closing." For example, the Plain TₑX definitions of \bigr, \Bigr, etc., are

```
\def\bigr{\mathclose\big}      \def\Bigr{\mathclose\Big}
```

and so on.

▷ \mathcode * 160–162, 166. Assigns a code number to input characters, so that TₑX will know how to interpret them in the output. It is the appropriate assignment of \mathcode numbers that makes, say, an input '+' in a formula lead to a '+' in the output. By reassigning the codes, one can change the output at will (while leaving the input untouched). See § 5.5.

▷ `\mathinner` ∗ See `\mathbin` for a general comment. The declaration `\mathinner` assigns its argument the atomic type called "Inner." For example, the Plain TeX definitions of `\ldots` and `\cdots` are

```
\mathchardef\ldotp="613A \mathchardef\cdotp="6201
% The line above calls on single characters, each a 'dot'.
\def\ldots{\mathinner{\ldotp\ldotp\ldotp}}
\def\cdots{\mathinner{\cdotp\cdotp\cdotp}}
```

▷ `\mathop` ∗ 76–77, 90–91. See `\mathbin` for a general comment. The declaration `\mathop` assigns to its argument the class called "large operator." As examples, here are the Plain TeX definitions of `\arccos` and `\sup`:

```
\def\arccos{\mathop{\rm arccos}\nolimits}
\def\sup{\mathop{\rm sup}}
```

The use of `\nolimits` makes the placement of limits for some of these functions different from others. See roman type for special functions.

▷ `\mathopen` ∗ See `\mathbin` for a general comment. The declaration `\mathopen` assigns to its argument the class called "opening." For example, the Plain TeX definitions of `\bigl`, `\Bigl`, etc., are

```
\def\bigl{\mathopen\big}      \def\Bigl{\mathopen\Big}
```

and so on.

▷ `\mathord` ∗ 30–31, 34–35, 124–125. See `\mathbin` for a general comment. `\mathord` assigns to its argument the class called "ordinary."

▷ `\mathpalette` First see `\mathchoice`. The `\mathpalette` command is designed to work with `\mathchoice` in order to ensure that the correct style is indeed used in different contexts. For example, some constructions of composite symbols in Plain TeX involve placing individual characters above each other, or on top of each other. Such placements are conveniently achieved using the alignment command `\halign`, or commands based on it. The command `\halign` is, however, unacceptable to TeX in math mode, so it has to be placed in a `\vbox`: the upshot of all this is that TeX is no longer in math mode within the alignment, much less employing any particular math style. `\mathpalette` works by explicitly reintroducing the correct style in each of the four branches of `\mathchoice`. Here is its definition:

```
\def\mathpalette#1#2{\mathchoice{#1\displaystyle{#2}}
  {#1\textstyle{#2}}{#1\scriptstyle{#2}}{#1\scriptscriptstyle{#2}}}
```

For example, consider a first attempt at producing a "less than/greater than" symbol:

```
\def\LG{\mathrel{\mathchoice{\AbvGT<}{\AbvGT<}{\AbvGT<}{\AbvGT<}}}
\def\AbvGT #1{\lower2pt\vbox{\baselineskip0pt \lineskip-.5pt
        \halign{$##$\cr #1 \crcr > \cr}}}
```

The command \AbvGT takes one argument and places it above a "greater than" symbol, via an \halign. The other commands in its definition are merely spacing and positioning adjustments. Then the command \LG places a "less than" symbol above the "greater than," and makes TeX regard the whole construction as a relation (through \mathrel). Despite the use of \mathchoice the new symbol does not shrink when needed; for instance, the input

```
$$A\LG B \qquad \sum_{A\LG B}. $$
```

produces

$$A \lessgtr B \qquad \sum_{A \lessgtr B} .$$

(The new symbol has not shrunk in the "subscript.") What is to be done? \mathpalette can be brought to the rescue as follows:

```
\def\LG {\mathrel{\mathpalette\AbvGT < }}
\def\AbvGT #1#2{\lower2pt\vbox{\baselineskip0pt \lineskip-.5pt
        \halign{$#1 ##$\cr #2\crcr >\cr}}}
```

Following the thread of the argument assignments in the commands above is instructive: \mathpalette acts on two arguments; they are \AbvGT and < in the input shown above. Each gets inserted in four places in the replacement text of \mathpalette. For example, the first branch of the \mathchoice there now looks like '\AbvGT \displaystyle {<}'. Finally, \AbvGT, which itself acts on two arguments, takes the command \displaystyle as its first argument and the symbol < as its second; the definition of the command shows where the two get inserted.

With these new definitions, the very same input that was used for the display above will produce

$$A \lessgtr B \qquad \sum_{A \lessgtr B} .$$

Another use of \mathpalette is shown under special effects. It should be noted that the actual alignment command that is used in Plain TeX when constructing such composite symbols is often a controlled version of \halign; see, for example, \ialign and \oalign.

▷ \mathpunct ∗ See \mathbin for a general comment. \mathpunct assigns to its argument the class called "punctuation."

▷ \mathrel ∗ 20–21, 76–77. See \mathbin for a general comment. \mathrel assigns to its argument the class called "relation." Here are some samples of its use, from the Plain TeX file:

```
\def\models{\mathrel|\joinrel=}
\def\bigm{\mathrel\big}
```

▷ **$\mathstrut$** 128–129. Assigns a standard vertical size to formulas. For example, the input for

$$\sqrt{a} + \sqrt{b} \qquad \text{and} \qquad \sqrt{a} + \sqrt{b}.$$

is

```
$$\sqrt{a} + \sqrt{b} \qquad \hbox{and} \qquad
\sqrt{\mathstrut a} + \sqrt{\mathstrut b}.$$
```

The Plain TEX definition of the command is '\vphantom('.

▷ **\mathsurround** ∗ A parameter that determines how much space TEX will leave before and after a formula. In Plain TEX the value is 0pt. For example, here are the effects of two different values for \mathsurround:

2 points: $a < b$ and $b < c \Rightarrow a < c$.

5 points: $a < b$ and $b < c \Rightarrow a < c$.

The input for the two is

```
{\mathsurround=2pt 2 points: $a<b$ and $b<c$ $\Rightarrow$ $a<c$.}
\hfil\break
{\mathsurround=5pt 5 points: $a<b$ and $b<c$ $\Rightarrow$ $a<c$.}
```

▷ **matrices** ($\mathcal{A}_{\mathcal{M}}\mathcal{S}$-TEX) $\mathcal{A}_{\mathcal{M}}\mathcal{S}$-TEX provides a number of commands that simplify the making of matrices. Some have the same names as the matrix commands of Plain TEX. The $\mathcal{A}_{\mathcal{M}}\mathcal{S}$-TEX matrix commands are, however, used differently from the ones in Plain TEX. For example, \pmatrix is used in $\mathcal{A}_{\mathcal{M}}\mathcal{S}$-TEX in this way:

```
$$ M=\pmatrix a & b \\ c & d \endpmatrix. $$
```

The output from this is

$$M = \begin{pmatrix} a & b \\ c & d \end{pmatrix}.$$

There are several other similar commands: \bmatrix gives brackets, \vmatrix gives vertical lines, and \Vmatrix gives double vertical lines. These matrix commands must end with \endbmatrix, \endvmatrix and \endVmatrix, respectively. For more general **delimiters**, there is the \matrix command:

```
$$\left\lfloor\matrix a & b \\ c & d \endmatrix\right\rfloor $$
```

gives

$$\begin{bmatrix} a & b \\ c & d \end{bmatrix}$$

and

`$$ \matrix x & y \\ z & \sqrt{x^2+y^2+z^2} \endmatrix. $$`

gives

$$ \begin{matrix} x & y \\ z & \sqrt{x^2+y^2+z^2} \end{matrix} . $$

(To get just one invisible delimiter, say on the right, one uses '`\right.`'.) If the vertical spacing seems awkward, it can be adjusted as discussed at the end of the entry.

If, for some reason, one wants a matrix in the middle of text, say $\begin{bmatrix} 1 & 0 \\ 0 & -1 \end{bmatrix}$, there is a command called `\smallmatrix`. The matrix here, for example, is obtained from this input:

`... say $\left[\smallmatrix 1&0\\ 0&-1 \endsmallmatrix\right]$, ...`

The entries in all of these matrices are centered and columns separated by a `\quad`; this can be changed by specifying some other format:

`$$ \bmatrix \format\l&\quad\c\quad&\r\\`
`    a+b&x+y+z&500.25\\ u&v&20.12 \endbmatrix. $$`

This gives

$$ \begin{bmatrix} a+b & x+y+z & 500.25 \\ u & v & 20.12 \end{bmatrix} . $$

The `\format` line determines the number of columns. In this context, `\l` gives a column of entries set flush left, `\c` a centered column and `\r` a column of entries set flush right. Spacing commands, like `\quad` here, are needed to prevent the columns from touching. (The `\format` line is like the preamble line in text alignments that are constructed using `\halign`.) As with other alignments in TEX, repetitive preambles can be specified. If the format line says

`\format &\qquad\r\\`

then all entries will be set flush right with a `\qquad` worth of space on the left. If it says

`\format \c\quad&\r\quad&&\l\quad \\`

then entries in the first column will be set centered, ones in the second column set flush right, and all subsequent ones set flush left. A `\quad` worth of space will also be left in all the specified places.

Finally, one can put in lots and lots of dots:

`$$ \vmatrix A & \hdots & B \\ \vdots & \ddots & \vdots \\`
`    C & \hdots & D \endvmatrix. $$`

gives

$$ \begin{vmatrix} A & \dots & B \\ \vdots & \ddots & \vdots \\ C & \dots & D \end{vmatrix} . $$

The dots may also span columns:

```
$$ \vmatrix A & \hdots & B \\ \hdotsfor3 \\
   \spacehdots .5\for3 \\ C & \hdots & D \endvmatrix. $$
```

gives

$$
\begin{vmatrix}
A & \dots & B \\
\cdots\cdots\cdots\cdots \\
\cdots\cdots\cdots\cdots \\
C & \dots & D
\end{vmatrix}.
$$

The number after \hdotsfor specifies how many columns should be filled with dots, as does the number after \for. \spacehdots allows the dot spacing to be specified. The value 1.5 would give the default spacing of \hdotsfor. As is seen above, smaller values give closer spacing. One normally would not use both \hdotsfor and \spacehdots in the same matrix.

If the dots are to begin after the first column, they will also impinge on the intercolumn space to the left. This can be avoided:

```
$$ \pmatrix A & B & \hdots & C \\
   D &\innerhdotsfor2 \after \quad& E \\
   F & G & \hdots & H \endpmatrix. $$
```

gives

$$
\begin{pmatrix}
A & B & \dots & C \\
D & \dots\dots & E \\
F & G & \dots & H
\end{pmatrix}.
$$

A \quad is asked for before the dots, because that is the normal intercolumn space. The \after command is not a general purpose one: it is used just in this context. Any space may be specified with it. The spacing of the dots can be adjusted by saying \spaceinnerhdots in place of \innerhdots, and using it just like \spacehdots above.

The vertical spacing between pairs of lines in a matrix can be adjusted by using \vspace. See that entry for a discussion. If all the lines of a matrix are to be spread apart by say, 5 pt, one can say \spreadmatrixlines 5pt within the display (but before the matrix commands).

▷ **matrices** (Plain TEX) $\mathcal{AMS}$-TEX's matrix commands are discussed above. Here is how matrices are made in Plain TEX.

A matrix with parentheses is obtained by using \pmatrix. For example,

```
$$ \pmatrix{a&b\cr c&d\cr}. $$
```

gives

$$
\begin{pmatrix}
a & b \\
c & d
\end{pmatrix}.
$$

Otherwise, \matrix allows a specification of delimiters:

```
$$ \matrix{a&b\cr c&d\cr}$$
```

gives

$$
\begin{matrix}
a & b \\
c & d
\end{matrix}
$$

and

$$ \left[\matrix{a&b\cr c&d\cr}\right] . $$

gives

$$
\left[\begin{matrix} a & b \\ c & d \end{matrix}\right].
$$

Plain TₑX also provides a command called \bordermatrix which produces entities like

$$
\begin{matrix}
 & \triangle & \heartsuit & \wp \\
\Im & & & \\
\ell & \begin{pmatrix} a & b & c \\ d & e & f \\ 1 & \mu & -73 \end{pmatrix} \\
\aleph & & &
\end{matrix}
$$

if given

$$\bordermatrix{&\triangle&\heartsuit&\wp\cr
\Im&a&b&c\cr \ell&d&e&f\cr \aleph&1&\mu&-73\cr}$$

as input. In addition, it is possible to produce matrices with invisible left or right delimiters by typing '\left.' or '\right.' when using \matrix.

It is possible to span columns with dots in this way:

$$ \pmatrix{A & \cdots & B \cr
    \multispan3\dotfill \cr
    C & \cdots & D}. $$

This gives

$$
\begin{pmatrix}
A & \cdots & B \\
\hdotsfor{3} \\
C & \cdots & D
\end{pmatrix}.
$$

Plain TₑX goes to some lengths to ensure that the spacing in a matrix will be uniform. It is therefore difficult to adjust from the outside. Examples 2 and 8 in Chapter 2 show a couple of tricks that can be used to tamper with the normal spacing. Another trick that allows control over vertical spacing is to use an invisible vertical line of the desired height. For example, if one defines

    \def\Prop #1{\vrule width0pt depth0pt height#1pt}

then

$$ \pmatrix{A&B\cr \Prop{40}C&D}. $$

will give

$$
\begin{pmatrix}
A & B \\
\\
\\
C & D
\end{pmatrix}.
$$

▷ $\mkern$ ∗ 76–77. Creates rigid spaces inside math formulas; the size must be given in mu (see keywords). For example, $[\mkern20mu]^{[\mkern20mu]}$ gives [ ][ ].

▷ mm See keywords and units.

▷ $\mod$ 𝒜ℳ𝒮 $7\equiv 2 \mod{5}$, gives $7 \equiv 2 \mod 5$. Compare this with \pmod and \pod.

▷ $\models$ ⌜⊨⌝ (Class 3: relation.)

▷ **modes** As TEX processes a file, it is, at any given point, in one of six possible modes. The mode it is in will determine how it responds to certain commands (like \break). The modes are

  • *Horizontal*: When stringing things together horizontally, towards building a paragraph. Once TEX starts putting together material for a paragraph, it stays in horizontal mode until the paragraph ends.
  • *Restricted horizontal*: When stringing things together horizontally, towards building an \hbox.
  • *Vertical*: When putting things together vertically, towards building a page. TEX is in this mode when between paragraphs.
  • *Internal vertical*: When putting things together vertically, towards building a \vbox.
  • *Display math*: When putting together a formula that will be placed on a new line (i.e., when the input is entered between $$ signs). After the formula is built, the interrupted paragraph resumes.
  • *Math*: When putting together a formula to be placed in a paragraph (i.e., when the input is entered between $ signs).

▷ \moveleft ∗ 66. Moves boxes to the left; it may only be used in vertical mode.

▷ \moveright ∗ Moves boxes to the right; it may only be used in vertical mode.

▷ $\mp$ ⌜∓⌝ 72–73. (Class 2: binary operation.)

▷ msam 135, 168, 192–193. The name of the first font of extra math symbols that 𝒜ℳ𝒮-TEX supplies. Chapter 5 shows a font table. Also see \load....

▷ msbm 135, 169, 194–195. The name of the second font of extra math symbols that 𝒜ℳ𝒮-TEX supplies. Chapter 5 shows a font table. Also see \load....

▷ $\mskip$ ∗ The mathematics mode analogue of \hskip. The space to be skipped must be specified in terms of the math unit called mu (see keywords). Though \hskip also works in formulas, it cannot be used with mu; thus, \mskip is often preferable (because mu automatically adjusts itself to the style being used in the formula).

▷ mu 76–77. See keywords.

▷ μ ⌜μ⌝ 92–95, 120–121, 130–131.

▷ **multiline displays** See \displaylines, \gather and \multline.

▷ $\multimap$ 𝒜ℳ𝒮 ⌜⊸⌝ (Class 3: relation.)

▷ \multispan 40–41, 102–103. Permits entries to span many columns of an alignment.

▷ $\multline$ *AMS* 118–119. The following input

```
$$\multline Z=a-b+c-d+e-f \\ g-h+i-j+k-l+m-n \\
\shoveright{o-p+q-r+s-t} \\ +u-v+w-x+y-z.\endmultline $$
```

produces

$$
Z = a - b + c - d + e - f \\
g - h + i - j + k - l + m - n \\
o - p + q - r + s - t \\
+ u - v + w - x + y - z.
$$

The first and last lines are positioned automatically as shown, and lines in-between normally centered. The \shoveright command does precisely that; there is also \shoveleft if the opposite effect is needed.

▷ $\multlinegap$ *AMS* As the example under \multline shows, that command leaves a little space at the left and right margins. The space can be changed, say to 1 pt, by typing \multlinegap{1pt} within the display (but before \multline). The effect of this is confined to that particular display. The command \nomultlinegap is defined to be \multlinegap{0pt}.

▷ $\MultLineGap$ *AMS* Like \multlinegap, but it has a global effect.

▷ \muskip ∗ Represents **registers** where gobs of math **glue** may be stored. For example, \muskip0=5mu plus 3mu stores the specified (elastic) size of math glue in the register \muskip0. Then, $[\mskip\muskip0]$ will create a space of that size: [].

▷ ∇ ⌈∇⌉ 52–55, 59, 116–117. (Class 0: ordinary.)

▷ $\natural$ ⌈♮⌉ (Class 0: ordinary.)

▷ $\ncong$ *AMS* ⌈≇⌉ (Class 3: relation.)

▷ $\ne$ ⌈≠⌉ 20–21, 28–29, 32–33, 40–41, 48–49, 62–65, 76–77, 98–99, 102–103. (Class 3: relation.) Alternate name for \not= (as is \neq).

▷ $\nearrow$ ⌈↗⌉ (Class 3: relation.)

▷ $\neg$ ⌈¬⌉ (Class 0: ordinary.) Alternate name: \lnot.

▷ **negating stuff** The command \not, with the help of small spacing adjustments, can be used to negate many things. Some negations have their own names (\neq) or their own special definitions (\notin). See **relations**.

▷ \negmedspace *AMS* Gives a negative medium space: [\negmedspace] gives |.

▷ \negthickspace *AMS* Gives a negative thick space: [\negthickspace] gives].

▷ \negthinspace 44–45. Gives a negative thin space: [\negthinspace] gives ‖.

▷ $\neq$ ⌜≠⌟ 32–33, 48–49. (Class 3: relation.) Alternate name for \not= (as is \ne).

▷ \new... 66, 74, 159, 166–168, 216, 222, 225–228, 230–232. Used in allocating new registers: for example, \newbox, \newcount, \newdimen, \newmuskip, \newskip and \newtoks assign new \box, \count, \dimen, \muskip, \skip and \toks registers, respectively. Also used in allocating new families for typesetting mathematics (\newfam). There are further variants, but these are not discussed or used anywhere in the present book.

▷ \newline *AMS* 126–127. Starts a new line; it works like \hfill\break, but it first checks to see if the command is being issued at an appropriate spot.

▷ \newpage *AMS* Starts a new page, rather like \vfill\break does. But, unlike that combination, this command can only be issued between paragraphs, not within one.

▷ \newsymbol *AMS* 135. First see § 3.3 for a discussion of fonts in $\mathcal{AMS}$-TEX. If only a few symbols are needed from the symbols fonts, they can be selectively added through the command \newsymbol. To use it, one looks up the code number of the symbol, say 107B, then types something like

    \newsymbol\complement 107B

The name \complement will then call up the new symbol. Though any name can be used, it is best to stick to standard ones. A complete list of code numbers and standard names is given under **AMS symbols**. Also see \undefine for some related information.

▷ $\nexists$ *AMS* ⌜∄⌟ (Class 0: ordinary.)

▷ $\ngeq$ *AMS* ⌜≱⌟ (Class 3: relation.)

▷ $\ngeqq$ *AMS* ⌜≩⌟ (Class 3: relation.)

▷ $\ngeqslant$ *AMS* ⌜≱⌟ (Class 3: relation.)

▷ $\ngtr$ *AMS* ⌜≯⌟ (Class 3: relation.)

▷ $\ni$ ⌜∋⌟ (Class 3: relation.) Alternate name: \owns.

▷ $\nleftarrow$ *AMS* ⌜↚⌟ (Class 3: relation.)

▷ $\nLeftarrow$ *AMS* ⌜⇍⌟ (Class 3: relation.)

▷ $\nleftrightarrow$ *AMS* ⌜↮⌟ (Class 3: relation.)

▷ $\nLeftrightarrow$ *AMS* ⌜⇎⌟ (Class 3: relation.)

▷ $\nleq$ *AMS* ⌜≰⌟ (Class 3: relation.)

▷ $\nleqq$ *AMS* ⌜≨⌟ (Class 3: relation.)

▷ $\nleqslant$ *AMS* ⌜≰⌟ (Class 3: relation.)

▷ $\nless$ *AMS* ⌜≮⌟ (Class 3: relation.)

▷ $\nmid$ *AMS* ⌜∤⌟ (Class 3: relation.)

▷ \noalign * 16–17, 40–41, 62–65, 86–87, 94–95, 98–99. Allows the insertion of nonaligned material into an alignment.

▷ \NoBlackBoxes *A M S* Switches off TEX's normal black box indicator next to an overfull box. Also see \BlackBoxes.

▷ \nobreak 223, 227, 231. Prevents a line break in horizontal mode and a page break in vertical mode. (See modes for more information about that topic.) The definition is given under penalties.

▷ \noexpand * 223. Suppresses expansion of a token. See \expandafter for a short description of expansion.

▷ \noindent * 16–23, 26–37, 52–53, 66, 92–95, 98–99, 106–109, 222–224, 227, 230, 232. Suppresses paragraph indentation at the point where it occurs.

▷ \nointerlineskip 226–228. A "one-shot" command that suppresses the normal interline spacing for the following line. It is based on a trick: it gets TEX to think that the depth of the previous line had a large negative value, thereby effectively counteracting the normal mechanism (see \lineskip and \lineskiplimit) that prevents two lines from touching. The definition is {\prevdepth=-10000pt }.

▷ $\nolimits$ * 38–39, 76. (*A M S*-TEX alert: don't \redefine.) Forces limits to appear to the right of large operators.

▷ \NoLimitsOn... *A M S* See \LimitsOn....

▷ \nolinebreak *A M S* Forbids a line break in text (see \linebreak).

▷ \nomathbreak *A M S* Forbids a line break in a formula (see \mathbreak).

▷ $\nomultlinegap$ *A M S* See \multlinegap.

▷ \nonscript * Consider, as examples, the following definitions:

```
\def\SpaceOne{\;}
\def\SpaceTwo{\nonscript\;}
```

Then, $a \SpaceOne b^{c \SpaceOne d}$ will produce $a\ b^{c\ d}$, whereas $a \SpaceTwo b^{c \SpaceTwo d}$ will produce $a\ b^{cd}$.

As the examples show, the use of \nonscript before a spacing command suppresses the extra spacing in indices. More precisely, the extra space will not be left in either of the script or scriptscript styles. See spacing.

▷ \nopagebreak *A M S* Forbids a page break (see \pagebreak). If typed within a displayed formula, it blocks a page break just after.

▷ \nopagenumbers 229. Suppresses Plain TEX's normal page numbering. If the AMS preprint style is being used, then \NoPageNumbers must be used instead.

▷ \normalbaselines 100–101, 153–154, 220 Sets various line spacing parameters to "normal values"; see section § 5.4.

▷ $\not$ ⌐/⌐ 98–99. (*A M S*-TEX alert: don't \redefine.) It is mostly used as a prefix to Plain TEX's relation commands in order to negate them. See relations.

▷ $\not=$ ⌐$\neq$⌐ (Class 3: relation.) Alternate names are \ne and \neq.

▷ $\notin$ ⌜∉⌟ 76–77. (Class 3: relation.)

▷ $\nparallel$ ᴬᴹˢ ⌜∦⌟ (Class 3: relation.)

▷ $\nprec$ ᴬᴹˢ ⌜⊀⌟ (Class 3: relation.)

▷ $\npreceq$ ᴬᴹˢ ⌜⋠⌟ (Class 3: relation.)

▷ $\nrightarrow$ ᴬᴹˢ ⌜↛⌟ (Class 3: relation.)

▷ $\nRightarrow$ ᴬᴹˢ ⌜⇏⌟ (Class 3: relation.)

▷ $\nshortmid$ ᴬᴹˢ ⌜∤⌟ (Class 3: relation.)

▷ $\nshortparallel$ ᴬᴹˢ ⌜∦⌟ (Class 3: relation.)

▷ $\nsim$ ᴬᴹˢ ⌜≁⌟ (Class 3: relation.)

▷ $\nsubseteq$ ᴬᴹˢ ⌜⊈⌟ (Class 3: relation.)

▷ $\nsubseteqq$ ᴬᴹˢ ⌜⊈⌟ (Class 3: relation.)

▷ $\nsucc$ ᴬᴹˢ ⌜⊁⌟ (Class 3: relation.)

▷ $\nsucceq$ ᴬᴹˢ ⌜⋡⌟ (Class 3: relation.)

▷ $\nsupseteq$ ᴬᴹˢ ⌜⊉⌟ (Class 3: relation.)

▷ $\nsupseteqq$ ᴬᴹˢ ⌜⊉⌟ (Class 3: relation.)

▷ $\ntriangleleft$ ᴬᴹˢ ⌜⋪⌟ (Class 3: relation.)

▷ $\ntrianglelefteq$ ᴬᴹˢ ⌜⋬⌟ (Class 3: relation.)

▷ $\ntriangleright$ ᴬᴹˢ ⌜⋫⌟ (Class 3: relation.)

▷ $\ntrianglerighteq$ ᴬᴹˢ ⌜⋭⌟ (Class 3: relation.)

▷ ν ⌜ν⌟ 26–27, 94–95, 102–103, 118–121.

▷ **nucleus** See atom.

▷ \null 228. An abbreviation for \hbox{}.

▷ \nulldelimiterspace ∗ Commands like \left. and \right. create invisible, or null, delimiters. They do result, however, in a little extra space being left, of width given by a parameter called \nulldelimiterspace. The Plain TeX value is 1.2pt.

▷ $\nvdash$ ᴬᴹˢ ⌜⊬⌟ (Class 3: relation.)

▷ $\nvDash$ ᴬᴹˢ ⌜⊭⌟ (Class 3: relation.)

▷ $\nVdash$ ᴬᴹˢ ⌜⊮⌟ (Class 3: relation.)

▷ $\nVDash$ ᴬᴹˢ ⌜⊯⌟ (Class 3: relation.)

▷ $\nwarrow$ ⌜↖⌟ (Class 3: relation.)

▷ \o ⌜ø⌟

▷ \O ⌜Ø⌟

▷ \oalign First see \ialign; \oalign is an extension to that command: it adjusts line spacing so that characters are placed over each other. The Plain TeX definition is

```
\def\oalign#1{\leavevmode\vtop{\baselineskip0pt \lineskip.25ex
    \ialign{##\crcr#1\crcr}}} % put characters over each other
```

There is an extension to `\oalign` called `\ooalign`, used for superposing characters:

```
\def\ooalign{\lineskiplimit-\maxdimen \oalign}
```

where `\maxdimen` stands for the largest legal dimensional quantity.

▷ $\odot$ $\lceil \odot \rceil$ (Class 2: binary operation.)

▷ `\oe` $\lceil œ \rceil$

▷ `\OE` $\lceil Œ \rceil$

▷ $\of$ $\mathcal{A}_{\mathcal{M}}\!S$ See `\root`.

▷ `\offinterlineskip` 70–71, 98–99. Switches off the normal spacing between lines.

▷ $\oint$ $\lceil \oint \rceil$ in a line, $\lceil \oint \rceil$ in a display. (Class 1: large operator.)

▷ `\oldnos` $\mathcal{A}_{\mathcal{M}}\!S$ `\oldnos{123}` will give 123.

▷ `\oldstyle` 136. {`\oldstyle` 0123456789} will produce 0123456789. The command is undefined in $\mathcal{A}_{\mathcal{M}}\!S$-TEX, but `\oldnos` is available instead.

▷ ω $\lceil \omega \rceil$ 20–21, 82–83, 92–95, 100–101, 126–127, 166–167.

▷ Ω $\lceil \Omega \rceil$ 80–81, 120–121. In Plain TEX ${\mit\Omega}$ (\varOmega in $\mathcal{A}_{\mathcal{M}}\!S$-TEX) gives $\mathit{\Omega}$; ${\bf\Omega}$ (\boldsymbol\Omega in $\mathcal{A}_{\mathcal{M}}\!S$-TEX) gives $\boldsymbol{\Omega}$; and ${\tt\Omega}$ gives Ω.

▷ $\omicron$ Not available as a separate command in Plain TEX; o gives o, which is pretty close.

▷ $\ominus$ $\lceil \ominus \rceil$ (Class 2: binary operation.)

▷ **Op** See atom and `\mathop`.

▷ **Open** See atom and `\mathopen`.

▷ `\openup` Opens up vertical spacing; `\openup2pt`, for example, will add 2pt to all of TEX's line-spacing parameters. Plain TEX also provides a "unit" called `\jot` in terms of which `\openup` may be specified; one `\jot` equals 3pt. Multiline equation displays in Plain TEX have a built-in default of 1\jot for `\openup`. The size of `\jot` may be reset (see the separate entry for it); this will change the line spacing for all such displays. For example, the input for

$$x = y$$
$$y \approx z$$

is

```
$$\openup-\jot  % SPACING REDUCED BY A 'JOT'
\displaylines{x=y\cr y\approx z\cr}$$
```

▷ **operations** See binary operations.

▷ `\operatorname` $\mathcal{A}_{\mathcal{M}}\!S$ 118–119. Typing

```
\define\Trace{\operatorname{Trace}}
```

will allow `\Trace` to be used exactly as `\sin`, `\log`, etc., are used. For example, `$\Trace^2(M)$` will give $\mathrm{Trace}^2(M)$.

▷ `\operatornamewithlimits` $\mathcal{A}_{\mathcal{M}}\mathcal{S}$ Typing

```
\define\Union{\operatornamewithlimits{Union}}
```

will make `$$\Union_{i\in N} A_i.$$` give

$$\operatorname*{Union}_{i \in N} A_i.$$

▷ `$\oplus$` $\lceil \oplus \rceil$ 32–33, 130–131. (Class 2: binary operation.)

▷ `\or` ∗ 168. Used with `\ifcase`. For example, if one types `\count255=2` then '`\ifcase\count255 A\or B\or C\or D\else E\fi`' gives 'C'.

▷ **Ord** See atom and `\mathord`.

▷ `$\oslash$` $\lceil \oslash \rceil$ (Class 2: binary operation.)

▷ `$\otimes$` $\lceil \otimes \rceil$ 130–131. (Class 2: binary operation.)

▷ `\outer` ∗ 222, 227. Used as a prefix to `\def`. It forbids the command being defined from appearing in a number of situations: in an argument, in the parameter text or the replacement text of another definition, in the preamble to an alignment, or in conditional text. It is useful to have commands that are designated as "outer" in one's document: errors that might otherwise go undetected for a while (the missing closing brace in a new definition, for example) will be discovered sooner. See `\proclaim` for an example.

▷ `\output` ∗ 226, 232. The routine that composes a page before TEX ships it off to the dvi file. This is where material that pertains to the entire page—a footline, a headline, a "frame" (a ruled box) around the page, for example—may be added. The standard output routine of Plain TEX does several things: it places a headline at the top of a `\vbox` representing the current page, and a footline at its bottom. In between, it puts another `\vbox` that contains the actual page contents. It also advances the page number and ships the page out. Examples of output routines are shown in Chapter 6.

▷ **Over** See atom.

▷ `$\over$` ∗ 16–17, 22–23, 44–45, 54–57, 59, 62–65, 68–73, 82–83, 92–95, 98–99, 157. ($\mathcal{A}_{\mathcal{M}}\mathcal{S}$-TEX alert: don't `\redefine`.) Used in making fractions: `${a+b \over c-d}$` gives $\frac{a+b}{c-d}$.

▷ `$\overarrow$` $\mathcal{A}_{\mathcal{M}}\mathcal{S}$ Alternate name for `\overrightarrow`.

▷ `$\overbrace$` ($\mathcal{A}_{\mathcal{M}}\mathcal{S}$-TEX alert: don't `\redefine`.) Covers an expression with a brace; for example, the input

$$A=\overbrace{a\times \cdots \times a}^{k\;\rm factors}.$$

gives

$$A = \overbrace{a \times \cdots \times a}^{k \text{ factors}}.$$

▷ **overlaps** \rlap and \llap create right and left overlaps, respectively.

▷ $\overleftarrow$ $\overleftarrow{\text{Long arrows may be put on top of stuff.}}$ Here's how it was done:

    $\overleftarrow{\hbox{Long arrows may be put on top of stuff.}}$

▷ $\overleftrightarrow$ $\mathcal{A}_{\mathcal{M}}\mathcal{S}$ Works like \overleftarrow.

▷ $\overline$ 92–93, 126–127. $\overline{a^2 = c}$ is produced by $\overline{a^2=c}$.

▷ $\overrightarrow$ Alternate name in $\mathcal{A}_{\mathcal{M}}\mathcal{S}$-TeX: \overarrow. Works like \overleftarrow.

▷ $\overset$ $\mathcal{A}_{\mathcal{M}}\mathcal{S}$ $\overset a\to A$, gives $\overset{a}{A}$. See also \underset.

▷ $\oversetbrace$ $\mathcal{A}_{\mathcal{M}}\mathcal{S}$ The input

    $\oversetbrace \text{$n$ factors}\to {A\times\dots\times A}$

will give $\overbrace{A \times \ldots \times A}^{n \text{ factors}}$.

▷ $\overwithdelims$ ∗ 171–172. One of three primitive "generalized fraction" commands (see \abovewithdelims and \atopwithdelims). The input

    $${A \overwithdelims ( ) B}.$$

produces

$$\left(\frac{A}{B}\right).$$

The thickness of the fraction bar is the default size being used for fractions.

▷ $\owns$ $\lceil \ni \rceil$ (Class 3: relation.) Alternate name for \ni.

▷ \P $\lceil \P \rceil$ Part of the cmsy Computer Modern symbols font, but is available outside math mode. Also see \dag.

▷ **page breaks** These may be forced by saying \vfil\break, or \vfill\break (in the appropriate places). $\mathcal{A}_{\mathcal{M}}\mathcal{S}$-TeX offers more sophisticated page break mechanisms; see the discussions under \newpage, \nopagebreak and \pagebreak. Page-breaking in systems of displayed formulas is discussed under \align.

Here is how some Plain TeX structures handle page breaks: alignments formed by using \eqalign do not permit page breaks; those formed by \eqalignno and \leqalignno automatically do (i.e., the page may be broken, if necessary, in the normal manner after any line in the alignment; the alignment then continues on the next page). Systems of formulas displayed with \displaylines also admit page breaks in this way.

In $\mathcal{A}\mathcal{M}\mathcal{S}$-TEX, neither \align nor \aligned normally permits page breaks, but this can be overridden for \align (as discussed under that command). \gather permits normal page breaks between lines; \gathered does not.

▷ **page numbers** Page numbers are provided automatically by Plain TEX, centered at the bottom of the page. This page numbering may be switched off by typing '\nopagenumbers'; starting values may be set by resetting \pageno. Page numbers in roman numerals may be obtained by using a negative starting value for \pageno. In $\mathcal{A}\mathcal{M}\mathcal{S}$-TEX the page-numbering style is determined by the document style in use. If the AMS preprint style is being used, then page numbering is switched off by saying \NoPageNumbers.

More sophisticated page numbering commands can be constructed fairly easily. The files listed in Chapter 6 display examples of such commands.

▷ \pagebreak Used pretty much exactly like \nopagebreak, but with the opposite effect. When used between paragraphs, it ends the page without pushing everything towards the top (unlike \newpage); when used in a paragraph, it ends the page after the current line; and when used in a mathematics display, it ends the page right after the display (this is the correct way to force a page break after a display if no extra white space is wanted at the bottom).

▷ \pageheight $\mathcal{A}\mathcal{M}\mathcal{S}$ A page height of, say, 6 in. is set by \pageheight{6in}. The definition of the command is

```
\def\pageheight#1{\vsize#1\relax}
```

▷ \pageinsert Inserts a full page's worth of material. Also see \midinsert and \topinsert.

▷ \pageno 14–39, 44–45, 68–73, 76–77, 86–87, 228, 229, 231, 232. A name assigned to the register \count0, since that is where Plain TEX stores the value of the current page number. See page numbers.

▷ \pagewidth $\mathcal{A}\mathcal{M}\mathcal{S}$ A page width of, say, 4 in. is set by \pagewidth{4in}. The definition of the command is

```
\def\pagewidth#1{\hsize#1\relax
  \captionwidth@\hsize\advance\captionwidth@-1.5in}
```

Thus, the command also fixes the \captionwidth (readers who are puzzled by the @ signs will find an explanation in §1.8, or under \m@th) to be 1.5 in. less than the horizontal size. See \botcaption.

▷ \par * 26–29, 32–39, 56, 66, 70–71, 74–75, 126–127, 146, 222, 223, 230–232. ($\mathcal{A}\mathcal{M}\mathcal{S}$-TEX alert: don't \redefine.) TEX's paragraph-ending command: the program inserts it at the end of every paragraph, even if the paragraph was ended in some other way (for example, by a blank line). The \par command is also built into several other commands (\item, for example).

Plain TeX provides the equivalent command \endgraf, defined by saying \let\endgraf=\par. This makes it possible to redefine \par temporarily so as to achieve some special effect automatically at the end of every paragraph. *This must be done with care, even in Plain TeX, and usually only for small portions of the document.* The automatic effects can be switched off whenever desired by using '\def\par{\endgraf}' (thus resetting \par).

▷ $\parallel$ ⌈‖⌋ 88–91. (Class 3: relation.)

▷ **parentheses** See delimiters.

▷ \parfillskip * 219, 224, 231. The glue automatically added at the end of every paragraph. Plain TeX makes this assignment:

> \parfillskip=0pt plus 1fil

(see keywords). A whole host of special effects can be obtained by assigning special values to \parfillskip: 0pt will make all subsequent paragraphs end flush right (if they are not long enough, a complaint will ensue from TeX); 1cm plus 1fil will make them all end at least 1 cm from the right margin, etc.

▷ \parindent * 14–15, 58–59, 74–75, 84–85, 216, 222, 224–228, 231. Sets the paragraph indentation.

▷ \parskip * 74–75, 222, 225, 228. Sets the space between paragraphs.

▷ ∂ ⌈∂⌋ 62–63, 68–69, 82–83, 116–121. (Class 0: ordinary.)

▷ pc See keywords and units.

▷ **penalties** TeX approaches the task of cutting up an input paragraph into lines, or of slicing off a piece from a list of formatted paragraphs to make a page, as an optimization question. The cost of different breaks is computed by evaluating an elaborate set of penalties and demerits. Good or bad break points can be suggested to TeX by explicitly inserting penalties. Here, for example, are some Plain TeX definitions:

```
\def\break{\penalty-10000 }
\def\nobreak{\penalty10000 }
\def~{\penalty10000\ }
\def\allowbreak{\penalty0 }
\def\goodbreak{\par\penalty-500 }
```

and so on. A value of 10000 is effectively infinite in TeX's internal calculations, which is why \break and \nobreak, for example, respectively force and prohibit breaks. TeX also comes with a number of built-in penalty parameters that control breaks in a variety of circumstances. See the discussions of the commands \binoppenalty, \relpenalty, \displaywidowpenalty, \postdisplaypenalty and \predisplaypenalty.

▷ **penalty** See penalties.

▷ \penalty * 222. See penalties.

▷ **periods** A little extra space is normally left after a period. It can be squelched by typing \ just after (i.e., \ followed by a single space). $\mathcal{AMS}$-TEX allows the command \. to be used in place of the period, in order to get the same result. On the other hand, TEX does not leave extra space after a period that just follows a capital letter (since this usually occurs after initials and other abbreviations, and not often at the end of a sentence.) To get an extra space, one can type something like '\null.' instead of just the period. Again, $\mathcal{AMS}$-TEX has a built in mechanism: @. gives an end-of-sentence period; the package also offers @, , @; , @: , @! and @? for similar use.

▷ $\perp$ $\lceil\perp\rfloor$ (Class 3: relation.)

▷ \phantom calculates the height, depth and width of stuff as it normally would, but does not print this stuff; it leaves instead a space of the correct dimensions. See \hphantom and \vphantom.

▷ ϕ $\lceil\phi\rfloor$ 32–33, 94–95, 100–103, 118–119. φ produces φ.

▷ Φ $\lceil\Phi\rfloor$ 18–19, 42. In Plain TEX ${\mit\Phi}$ (\varPhi in $\mathcal{AMS}$-TEX) produces Φ; ${\bf\Phi}$ (\boldsymbol\Phi in $\mathcal{AMS}$-TEX) produces $\boldsymbol\Phi$; and ${\tt\Phi}$ produces Φ.

▷ π $\lceil\pi\rfloor$ 26–27, 46–47, 54–55, 60–65, 70–73, 82–83, 102–103, 120–121, 128–129, 157, 162. ϖ produces ϖ.

▷ Π $\lceil\Pi\rfloor$ In Plain TEX ${\mit\Pi}$ (\varPi in $\mathcal{AMS}$-TEX) produces Π; ${\bf\Pi}$ (\boldsymbol\Pi in $\mathcal{AMS}$-TEX) produces $\boldsymbol\Pi$; and ${\tt\Pi}$ produces Π.

▷ **pictures** See diagrams.

▷ $\pitchfork$ $\mathcal{AMS}$ $\lceil\pitchfork\rfloor$ (Class 3: relation.)

▷ **Plain TEX** 5–7, 9, 11, 13, 14, 38, 41, 59, 75, 85, 95, 100, 117, 123, 136, 142, 146–151, 153–154, 156–159, 161–167, 172–173, 178–191. A package of commands built out of the primitive commands of TEX. The package is so universally available, wherever TEX is, that when people say "TEX" they almost always mean "Plain TEX." Any command listed in this Glossary that is not marked by an * or by $\mathcal{AMS}$ is a Plain TEX command—it is just an abbreviation for a collection of primitive ones. Appendix B of *The TEXbook* (KNUTH [1984, 1986]) contains the definitions, in terms of primitives, of essentially all Plain TEX commands.

The default format that is normally thought of as TEX's is really a format provided by Plain TEX. It is suitable for the production of preprints and technical reports. For other kinds of documents, the format must be altered or an entirely different package used.

▷ \plainfootnote $\mathcal{AMS}$ See footnotes.

▷ \plainproclaim $\mathcal{AMS}$ See \proclaim.

▷ **plus** 58, 74, 84, 219, 220, 222, 223, 231. See keywords.

▷ $\pm$ $\lceil\pm\rfloor$ 16–17, 64–65, 72–73, 94–95. (Class 2: binary operation.)

▷ $\pmatrix$ 40–41, 122–123 ($\mathcal{A}_{\mathcal{M}}S$), 136. See matrices.

▷ $\pmod$ 136. $7\equiv 2 \pmod{5}$, gives $7 \equiv 2 \pmod 5$. Compare this with \bmod. The Plain TeX definition is

```
\def\pmod #1{\allowbreak \mkern18mu ({\rm mod}\,\,#1)}
```

$\mathcal{A}_{\mathcal{M}}S$-TeX uses a more sophisticated definition that adjusts the spacing automatically if the command is used in a display. It also provides variants through \mod and \pod.

▷ $\pmb$ $\mathcal{A}_{\mathcal{M}}S$ Gives pretty measly bold characters (to be used when true bold characters are not available) by slightly shifting and superposing ones of normal weight: $\pmb{x}\pmb{+}\pmb\gamma$ gives $\boldsymbol{x + \gamma}$, for example.

▷ $\pod$ $\mathcal{A}_{\mathcal{M}}S$ $7\equiv 2 \pod{5}$, gives $7 \equiv 2 \;(5)$. Compare this with \mod and \pmod.

▷ **point** See units.

▷ \postdisplaypenalty * 230. The penalty imposed by TeX for a page break right after a display. Plain TeX effectively assigns the value 0 to this parameter (i.e., no penalty is imposed). This can be changed if desired. Also see penalties.

▷ $\Pr$ $\ulcorner\mathrm{Pr}\urcorner$

▷ $\prec$ $\ulcorner\prec\urcorner$ (Class 3: relation.)

▷ $\precapprox$ $\mathcal{A}_{\mathcal{M}}S$ $\ulcorner\precapprox\urcorner$ (Class 3: relation.)

▷ $\preccurlyeq$ $\mathcal{A}_{\mathcal{M}}S$ $\ulcorner\preccurlyeq\urcorner$ (Class 3: relation.)

▷ $\preceq$ $\ulcorner\preceq\urcorner$ (Class 3: relation.)

▷ $\precnapprox$ $\mathcal{A}_{\mathcal{M}}S$ $\ulcorner\precnapprox\urcorner$ (Class 3: relation.)

▷ $\precneqq$ $\mathcal{A}_{\mathcal{M}}S$ $\ulcorner\precneqq\urcorner$ (Class 3: relation.)

▷ $\precnsim$ $\mathcal{A}_{\mathcal{M}}S$ $\ulcorner\precnsim\urcorner$ (Class 3: relation.)

▷ $\precsim$ $\mathcal{A}_{\mathcal{M}}S$ $\ulcorner\precsim\urcorner$ (Class 3: relation.)

▷ \predefine $\mathcal{A}_{\mathcal{M}}S$ See \redefine.

▷ \predisplaypenalty * The penalty imposed by TeX for a page break just before a display. The Plain TeX assignment is \predisplaypenalty=10000, since it is considered bad form in the best typesetting circles to start a page with a display. This can be changed if desired. Also see penalties.

▷ \predisplaysize * First see \prevgraf for a discussion of what happens in a paragraph prior to the typesetting of a displayed formula. The \predisplaysize is a parameter that represents, roughly, the length of the line just preceding the display. It is calculated as follows: if there is no line of text preceding the display in that paragraph, the parameter is set equal to the smallest legal dimension (given by $-$\maxdimen). If there is a line, then the parameter is set to equal the length of that line (more precisely, the position of the right edge of the rightmost box inside the hbox that represents that line), plus whatever indentation the line may have, plus a length of size \qquad (i.e.,

2em in the current font). If, however, the spaces in the line preceding the display had been allowed to stretch or shrink (so that the line does not have its natural length), then \predisplaysize is set to \maxdimen.

▷ $\pretend$ *A_MS* Used in making **commutative diagrams**.

▷ \pretolerance * 36–37, 221, 222, 225, 228, 231. TEX's tolerance for white space before it tries hyphenation. The Plain TEX value is 100 (on a scale where 10000 is effectively infinite). Also see \tolerance.

▷ **pretty-printing** An awkward name for printing nicely formatted **computer programs**. See \cleartabs.

▷ \prevdepth * The depth of the most recently constructed box added in vertical **mode**. See \nointerlineskip.

▷ \prevgraf * The number of lines in the most recently constructed paragraph. When TEX typesets a **display**, it first formats the preceding part of the paragraph as though it were a full paragraph by itself. Suppose that the resulting value of \prevgraf is n. This number is used in a few places: see, for example, \displayindent. When the interrupted paragraph continues after the display, \prevgraf is changed to $n + 3$: i.e., the display is considered to occupy 3 lines of the paragraph, whatever its actual vertical size.

▷ $\prime$ ⌐′⌐ (Class 0: ordinary.)

▷ **primitive commands** The basic commands of TEX, out of which other commands may be built: for example, the commands of packages like Plain TEX and *A_MS*-TEX. This book discusses both the primitive, indecomposable commands of TEX as well as the composite commands of other packages. To distinguish between the two in this Glossary, the primitive commands are labelled by an * when they are listed (but not when they are discussed).

▷ \printoptions *A_MS* *A_MS*-TEX permits three processing options: a syntax check (with no output pages produced), a "galleys" run that reports overfull boxes (but again produces no output), and a full run that makes pages. If the command \printoptions is put in after the \documentstyle declaration, then each time the file is processed the program will ask which option is to be used for that run. Also see \galleys and \syntax.

▷ \proclaim 38, 134, 136. A Plain TEX illustration of a how a high-level command may be defined to take care automatically of several small formatting details. *A_MS*-TEX has a command of the same name, but it functions differently; this is discussed at the end of the entry. The definition is

```
\outer\def\proclaim #1. #2\par{\medbreak
\noindent{\bf#1.\enspace}{\sl#2\par}%
\ifdim\lastskip<\medskipamount
    \removelastskip\penalty55\medskip\fi}
```

The command is useful in typesetting theorems and other special pronouncements. For example,

$\prod$

> \proclaim Theorem A. Life is sweet.
>
> \noindent Proof: Suppose it is not. We show that a contradiction ensues...

gives

Theorem A. *Life is sweet.*

Proof: Suppose it is not. We show that a contradiction ensues...

Thus, \proclaim accepts everything up to the first period followed by a space, '. ', as its first argument and sets it in boldface. Everything after that until the end of the paragraph (signalled by a blank line, say, or an explicit \par command) is set in slanted type. The other commands in its definition adjust vertical spacing, suggest good and bad breaks, etc. It is easy to model similar commands on \proclaim and to adjust them to suit one's private needs.

In $\mathcal{AMS}$-TeX, the definition of \proclaim is left to style files. If no document style is selected, the command \proclaim will be undefined. (The package, however, provides at all times the command \plainproclaim—exactly Plain TeX's \proclaim, but under a new name.)

▷ $\prod$ $\prod$ in a line, $\prod$ in a display. 78–79. (Class 1: large operator.)

▷ $\projlim$ $\mathcal{AMS}$ $\mathrm{proj\,lim}$

▷ $\propto$ $\propto$ (Class 3: relation.)

▷ ψ ψ 24–27, 102–103.

▷ Ψ Ψ 18–19, 30–31. In Plain TeX ${\mit\Psi}$ (\varPsi in $\mathcal{AMS}$-TeX) gives Ψ; ${\bf\Psi}$ (\boldsymbol\Psi in $\mathcal{AMS}$-TeX) produces **Ψ**; and ${\tt\Psi}$ produces Ψ.

▷ pt 14, 16–17, 20–21, 30–31 40–41, 46–47, 54–55, 58, 66, 70–71, 74, 84, 98–103, 110–115, 122–129, 142, 153, 171, 172, 216, 219, 220, 222–229, 231. See keywords and units.

▷ **Punct** See atom and \mathpunct.

▷ **punctuation** Standard punctuation characters can be used in the mathematics mode, with one exception: '$a:b$' gives '$a : b$' (the spacing is more appropriate for a relation, which is how TeX interprets ':' in this mode). If a colon is needed as punctuation, the command \colon must be used; for example, '$a\colon b$' gives '$a\colon b$'. When typing a formula in text (i.e., one that is not to be displayed on a separate line), one should usually put punctuation outside the $ signs. For example, to get '$a = x$, y, or z.', one types '$a=x$, y, or z.' instead of '$a=x, y,$ or $z.$'. The spacing will be more uniform the first way and it will offer TeX more line break possibilities. If some of these possibilities are to be discouraged, that may be done with a tie. (For example, one types 'or~z.'; see line breaks for a discussion.)

In displayed formulas (between double $ signs), the punctuation must be put in as part of the formula. If the formula contains text, then the text and associated punctuation should all go in an \hbox.

▷ \qquad 24–27, 38–39, 48–49, 52–55, 62–63, 68–69, 72–74, 86–87, 102–105, 124–125. Leaves horizontal space the size of 2em in the font currently being used. The definition is

> \def\qquad{\hskip2em\relax}

For example, a mathematical formula embedded in some boldface text will use a different size for \qquad than one in the middle of regular roman: {\bf $a[\qquad]b$} will produce $a[\qquad]b$, whereas $a[\qquad]b$ will produce $a[\qquad]b$. A similar difference holds when \qquad is used in straight text.

▷ \quad 10, 18–19, 24–27, 32–33, 38–39, 40–41, 46–49, 52–53, 56–57, 62–63, 68–69, 74, 86–89, 92–97, 100–103, 120–121, 126–129, 143, 162, 170, 223, 232. ($\mathcal{AMS}$-TEX alert: don't \redefine.) Leaves horizontal space the size of 1em in the font currently being used. The definition is

> \def\quad{\hskip1em\relax}

The comments made above under \qquad apply here as well.

▷ \r $\mathcal{AMS}$ 122–125. Used with \format in formatting matrices (see that topic).

▷ **Rad** See atom.

▷ \radical * 166. ($\mathcal{AMS}$-TEX alert: don't \redefine.) Used rather like \delimiter. For example, here is how Plain TEX defines \sqrt:

> \def\sqrt{\radical"270370 }

See the discussion of \delimiter in §5.5 to see what this means. Though \radical and \delimiter are used in the same way, TEX will position characters differently depending on whether it regards them as delimiters or radicals, so the two are not interchangeable.

▷ \raggedright 36–37, 222, 225. Switches off right justification.

▷ \raise * 66, 219, 223. TEX^{lifts} a box 4 pt if it sees \raise4pt\hbox{lifts} in horizontal mode.

▷ $\rangle$ $)$ 42, 120–121, 165. (Class 5: closing.) Also see delimiters.

▷ $\rbrace$ $\}$ 62–65. (Class 5: closing.) Alternate name: \}. Also see delimiters.

▷ $\rbrack$ $]$ 79–80. (Class 5: closing.) Also see delimiters.

▷ $\rceil$ $]$ (Class 5: closing.) Also see delimiters.

▷ $\rcfrac$ $\mathcal{AMS}$ See continued fractions.

▷ $\Re$ $\Re$ (Class 0: ordinary.)

▷ \redefine $\mathcal{A}_{\mathcal{M}}\mathcal{S}$ When defining a command with \define, $\mathcal{A}_{\mathcal{M}}\mathcal{S}$-TEX checks to see if the name is already in use and ignores the new definition if it is. (It also issues a warning.) If the new name is wanted anyway, \redefine can be used instead, because it does no checking (the command is the same as the primitive command \def). The old meaning of the chosen name may be preserved, if desired, by using a command called \predefine. For example, if the abbreviation \L is desired for \Longleftrightarrow, but its old meaning (it normally gives the Polish suppressed-L, Ł) is also to be stored somewhere, the following lines are typed:

```
\predefine\Psl{\L}
\redefine\L{\Longleftrightarrow}
```

The name \Psl is arbitrary; any legitimate name can be used instead. It is usually a good idea to first use \define and then to fall back on \redefine only if necessary.

▷ **registers** TEX can store a variety of types of information in "registers." There are different classes of registers available for storing different types of information. This book illustrates several of these. Registers used to store pure whole numbers are called \count registers, those used to store quantities with dimension are called \dimen registers, those that store material in boxes are called \box registers (these are used a little differently from the others), those that store lists of **tokens** are called \toks registers, those that store **glue** specifications are called \skip registers and those that store glue for use in formulas (i.e., "muglue"—see **keywords**) are called \muskip registers. They are all briefly discussed in individual entries.

▷ **Rel** See **atom** and \mathrel.

▷ **relations**

< `$<$`	> `$>$`	= `$=$`
⊥ `$\perp$`	≤ `$\leq$`	≥ `$\geq$`
≡ `$\equiv$`	≐ `$\doteq$`	≺ `$\prec$`
⪯ `$\preceq$`	≻ `$\succ$`	⪰ `$\succeq$`
∼ `$\sim$`	≃ `$\simeq$`	≪ `$\ll$`
≫ `$\gg$`	⊂ `$\subset$`	⊆ `$\subseteq$`
⊃ `$\supset$`	⊇ `$\supseteq$`	⊑ `$\sqsubseteq$`
⊒ `$\sqsupseteq$`	≍ `$\asymp$`	≈ `$\approx$`
≅ `$\cong$`	⋈ `$\bowtie$`	⊨ `$\models$`
⊢ `$\vdash$`	⊣ `$\dashv$`	⌣ `$\smile$`
⌢ `$\frown$`	| `$\mid$`	‖ `$\parallel$`
∈ `$\in$`	∋ `$\ni$`	∝ `$\propto$`

These relations can be negated by prefixing the commands by \not; for example, $\not\perp$ gives ⊥̸. This does not always give pleasing results: $\not\parallel$ gives ‖̸. In some cases, TEX comes equipped with an alternative: $\notin$ gives $\notin$; $\neq$ and $\ne$ give $\neq$. In other cases, one has to make adjustments oneself; here is one crude way: $\not\kern1.33pt\parallel$ gives $\nparallel$. There are some alternate names here as well: $\le$ and $\ge$ have the same effect as $\leq$ and $\geq$; $\owns$ is equivalent to $\ni$.

Other relation symbols that are available are

←	`$\leftarrow$`	⟵	`$\longleftarrow$`
→	`$\rightarrow$`	⟶	`$\longrightarrow$`
⇐	`$\Leftarrow$`	⟸	`$\Longleftarrow$`
⇒	`$\Rightarrow$`	⟹	`$\Longrightarrow$`
↑	`$\uparrow$`	↓	`$\downarrow$`
⇑	`$\Uparrow$`	⇓	`$\Downarrow$`
↔	`$\leftrightarrow$`	⟷	`$\longleftrightarrow$`
⇔	`$\Leftrightarrow$`	⟺	`$\Longleftrightarrow$`
↕	`$\updownarrow$`	⇕	`$\Updownarrow$`
↦	`$\mapsto$`	⟼	`$\longmapsto$`
↩	`$\hookleftarrow$`	↪	`$\hookrightarrow$`
↗	`$\nearrow$`	↘	`$\searrow$`
↙	`$\swarrow$`	↖	`$\nwarrow$`
↼	`$\leftharpoonup$`	⇀	`$\rightharpoonup$`
↽	`$\leftharpoondown$`	⇁	`$\rightharpoondown$`
⇌	`$\rightleftharpoons$`		

All the vertical arrows also come in larger sizes; these sizes are obtained by the same commands that are used for large delimiters. Alternate names here are $\to$ and $\gets$ for $\rightarrow$ and $\leftarrow$, respectively. $\iff$ produces $\iff$ just as $\Longleftrightarrow$ does, but with more space on either side. In $\mathcal{AMS}$-TEX \implies and \impliedby produce the same arrows as do \Longrightarrow and \Longleftarrow, respectively, but also with more space.

See AMS symbols for displays of additional relation symbols and arrows.

▷ **\relax** ∗ 40–41, 100–103, 218, 219, 222, 227, 228, 230, 231. This pretty much means "do nothing." It is a surprisingly useful command. For example, after a spacing command like \hskip 1in, TEX is on the lookout for a plus or a minus (see keywords). If no elasticity is intended in that space, it is important to say \relax just after. Otherwise words like "plush" that immediately follow will be misinterpreted, and TEX will complain of a missing number (it expects a number after plus). Similar precautions must be taken after typing \hrule or \vrule (since TEX is on the lookout here for height, width and depth).

▷ **\relpenalty** ∗ The penalty assigned to breaking a line at a relation (like =) in a math formula. The Plain TEX assignment is \relpenalty=500; this can be changed if it is desired to make line breaks at relations more, or less, attractive. See line breaks and penalties.

▷ \repeat See \loop.

▷ \ResetBuffer $\mathcal{A}_{\mathcal{M}}\mathcal{S}$ See \buffer.

▷ $\restriction$ $\mathcal{A}_{\mathcal{M}}\mathcal{S}$ $\lceil\upharpoonright\rfloor$ (Class 3: relation.) Also called \upharpoonright.

▷ $\rfloor$ $\lceil\rfloor\rfloor$ (Class 5: closing.) Also see delimiters.

▷ $\rgroup$ See delimiters.

▷ ρ $\lceil\rho\rfloor$ 18–19, 30–31, 36–37, 72–73, 82–83, 90–91, 94–95, 102–103, 130–131. ϱ produces ϱ.

▷ $\right$ ∗ 16–17, 26–27, 32–33, 40–41, 56–57, 62–65, 70–71, 76, 78–79, 88–89, 98–103, 116–119, 122–123, 128–129, 157, 163–165. ($\mathcal{A}_{\mathcal{M}}\mathcal{S}$-TEX alert: don't \redefine.) Makes a right delimiter of the correct size in a displayed mathematical expression. It can be used only in conjunction with $\left$; see that entry for a fuller discussion.

▷ $\rightarrow$ $\lceil\rightarrow\rfloor$ 66. (Class 3: relation.) Alternate name: \to.

▷ $\Rightarrow$ $\lceil\Rightarrow\rfloor$ (Class 3: relation.)

▷ \rightarrowfill '——→' is made by '\hbox to1cm{\rightarrowfill}'.

▷ $\rightarrowtail$ $\mathcal{A}_{\mathcal{M}}\mathcal{S}$ $\lceil\rightarrowtail\rfloor$ (Class 3: relation.)

▷ $\rightharpoondown$ $\lceil\rightharpoondown\rfloor$ (Class 3: relation.)

▷ $\rightharpoonup$ $\lceil\rightharpoonup\rfloor$ (Class 3: relation.)

▷ $\rightleftarrows$ $\mathcal{A}_{\mathcal{M}}\mathcal{S}$ $\lceil\rightleftarrows\rfloor$ (Class 3: relation.)

▷ $\rightleftharpoons$ $\lceil\rightleftharpoons\rfloor$ (Class 3: relation.) This is a composite symbol in Plain TEX; the msam symbols font provides a similar single character (which will work correctly in subscripts and superscripts), $\rightleftharpoons$; it can be invoked in $\mathcal{A}_{\mathcal{M}}\mathcal{S}$-TEX after using \undefine.

▷ \rightline 216, 225, 228, 231. \rightline{stuff} puts stuff on the right. If used in mid-paragraph, the preceding line must be explicitly broken.

▷ $\rightrightarrows$ $\mathcal{A}_{\mathcal{M}}\mathcal{S}$ $\lceil\rightrightarrows\rfloor$ (Class 3: relation.)

▷ \rightskip ∗ 219, 225, 231. Moves the right margin (for text) in or out.

▷ $\rightsquigarrow$ $\mathcal{A}_{\mathcal{M}}\mathcal{S}$ $\lceil\rightsquigarrow\rfloor$ (Class 3: relation.)

▷ $\rightthreetimes$ $\mathcal{A}_{\mathcal{M}}\mathcal{S}$ $\lceil\rightthreetimes\rfloor$ (Class 2: binary operation.)

▷ $\risingdotseq$ $\mathcal{A}_{\mathcal{M}}\mathcal{S}$ $\lceil\risingdotseq\rfloor$ (Class 3: relation.)

▷ \rlap 14, 18–19, 26–27, 34–35, 58, 66, 68–69, 86–87, 90–91, 100–101, 225, 232. Overlaps material on the right (for instance, to place stuff in the right margin). For example, '∅' is produced by '\rlap{/}0'.

▷ \rm 26–27, 32–41, 48–49, 52–53, 75, 76, 78–83, 112–113, 136, 142, 149–151, 159, 163, 174, 176, 178–179, 220. ($\mathcal{A}_{\mathcal{M}}\mathcal{S}$-TEX alert: don't \redefine.) Allows users to force roman type (e.g., for a word in a formula). See § 5.3 and § 5.5 for the definition and a full discussion. $\mathcal{A}_{\mathcal{M}}\mathcal{S}$-TEX does not obey the command in formulas: \roman must be used instead.

▷ $\rmoustache$ 164–165. See delimiters.

▷ $\roman$ $\mathcal{A}\mathcal{M}\mathcal{S}$ 75, 118–119, 126–127. Produces roman characters in a formula, in the same way that \bold produces bold characters.

▷ **roman text** See \rm and \hbox. Also see Chapter 5 for a full discussion of all typeface issues.

▷ \romannumeral * 225, 232. Converts a number to roman numerals. For example, the value currently stored in the page number register (\pageno) may be expressed as CCCXXIX in these numerals. This was obtained by typing

    \uppercase\expandafter{\romannumeral\pageno}

where \expandafter works a trick that makes \uppercase act on the expanded \romannumeral command (i.e., *after* the page number has been converted to a roman numeral). If lowercase numerals are satisfactory, \romannumeral\pageno is sufficient.

▷ **roman type for special functions** Functions like sin, cos and log, as well as a few other expressions, are usually set in roman type, even in formulas. This can be done in TeX by using:

\arccos	\arcsin	\arctan	\arg	\cos	\cosh	\cot	\coth
\csc	\deg	\det	\dim	\exp	\gcd	\hom	\inf
\ker	\lg	\lim	\liminf	\limsup	\ln	\log	\max
\min	\Pr	\sec	\sin	\sinh	\sup	\tan	\tanh

It is possible to put expressions above and below some of these, when displaying equations. For example,

$$\lim_{x\to\infty} \qquad \overset{\heartsuit}{\max}$$

The commands that achieve this are the ones normally used for superscripts and subscripts. The expressions that will have such indices placed above or below are \det, \gcd, \inf, \lim, \liminf, \limsup, \max, \min, \Pr and \sup. For the others, TeX will use the normal placements for indices:

$$\sin^2 x + \cos^{-2} x \neq 1 \quad \text{in general.}$$

The input for this is

    $$\sin^2x + \cos^{-2}x \neq 1 \quad {\rm in\ general.}$$

See \mathop for sample definitions that reveal why there are these differences in placement.

$\mathcal{A}\mathcal{M}\mathcal{S}$-TeX offers some additional functions:

\injlim	\projlim	\varliminf	\varlimsup
\varinjlim	\varprojlim		

It also offers commands like \operatorname that allow users to define names of this type for additional functions that they might use.

There are two varieties of "mod" to cover the two ways it is often used: between variables, or at the end of a formula. They are produced by `$\bmod$` and `$\pmod$`, respectively; e.g., `$(a\bmod b)$` gives $(a \bmod b)$, and `$\pmod{n}$` gives $(\bmod\ n)$. $\mathcal{A}\mathcal{M}\mathcal{S}$-TeX also offers `\mod` and `\pod` as variations on `\pmod`.

▷ `$\root$`   The command '`$\root n\of{x+1}$`' gives '$\sqrt[n]{x+1}$'. In $\mathcal{A}\mathcal{M}\mathcal{S}$-TeX one can additionally use `\uproot{`i`}` and `\leftroot{`j`}` right after `\root` in order to move the position of the root n up by i units and to the left by j units (negative values move it in the opposite direction). The units used internally in computing these shifts involve the math unit called `mu` (see **keywords**), and they allow very small adjustments to be made.

▷ **roots**   These may be obtained by using `\root` and `\sqrt`.

▷ `\rq` $\mathcal{A}\mathcal{M}\mathcal{S}$ ($\mathcal{A}\mathcal{M}\mathcal{S}$-TeX alert: don't `\redefine`.) May be used in place of '.

▷ `$\Rrightarrow$` $\mathcal{A}\mathcal{M}\mathcal{S}$ $\lceil \Rrightarrow \rceil$ (Class 3: relation.)

▷ `$\Rsh$` $\mathcal{A}\mathcal{M}\mathcal{S}$ $\lceil \Rsh \rceil$ (Class 3: relation.)

▷ `$\rtimes$` $\mathcal{A}\mathcal{M}\mathcal{S}$ $\lceil \rtimes \rceil$ (Class 2: binary operation.)

▷ **rules**   40–41, 113, 139.   A printer's term for drawn lines like _____. TeX provides two kinds: an `\hrule` (for rules drawn in vertical **mode**) and a `\vrule` (for horizontal **mode**). The `height`, `depth` and `width` may be specified, if desired. See separate entries for each, and **keywords**. For example, the input for the rule above was

      `... like \vrule height.2pt depth.2pt width2cm.`

Also see `\hrulefill`.

▷ `\S` $\lceil \S \rceil$ 26–27, 36–37, 58, 62–63, 84, 90–91, 219.   Part of the `cmsy` Computer Modern symbols font, but is available outside math mode. Also see `\dag`.

▷ **sans serif**   A style of type without little protrusions ("serifs") at the ends of the strokes that make up characters.

▷ `$\sb$`   May be used instead of _ to get subscripts (in case _ is unavailable): `$A\sb{x+1}$` gives A_{x+1}.

▷ `$\Sb$` $\mathcal{A}\mathcal{M}\mathcal{S}$ Used for multiline lower limits under **large operators**; see that topic.

▷ `\sc`   This is not a command either in Plain TeX or in $\mathcal{A}\mathcal{M}\mathcal{S}$-TeX, but it is commonly used in other packages (LaTeX, for example) to switch to the **capitals and small capitals** typeface. Also see `\smc`.

▷ `scaled` 42, 124–125, 128–129, 142, 153, 221, 224, 226.   See **keywords**.

▷ **script**   Script alphabets are often used in mathematics when distinctive characters are needed to denote special things. None of the sets of faces that come with standard TeX packages offers a really curly script face. The standard available faces are the script fonts that come with the AMS Euler typefaces, and the calligraphic letters that come with the Computer Modern faces. Curly letters may, however, be obtained from other sources; see page 96 for a short discussion.

▷ **script style** 164–165. The style used automatically in indices and in fractions set in text (i.e., in **text** style). See **styles in mathematics** and \scriptstyle.

▷ \scriptfont * 158–159, 163, 165–168, 175, 220, 221. Specifies which font is to be used for indices, etc. See §5.5 and **styles in mathematics**.

▷ **scriptscript style** 165. The size and style used automatically in indices on indices, etc. See **styles in mathematics** and \scriptscriptstyle.

▷ \scriptscriptfont * 158–159, 165, 166–169, 175, 220. Specifies which font is to be used for indices on indices, etc. See §5.5 and **styles in mathematics**.

▷ $\scriptscriptstyle$ * 112–113. Explicitly invokes scriptscript style. Also see **styles in mathematics**.

▷ \scriptspace * Represents the extra space after a subscript or a superscript. Plain TeX gives it a value through this declaration: '\scriptspace=0.5pt'.

▷ $\scriptstyle$ * 36–37, 56–57, 66, 70–71, 94–95, 100–101, 114–115, 219. Explicitly invokes script style. Also see **styles in mathematics**.

▷ $\searrow$ ⌐↘⌐ 56–57. (Class 3: relation.)

▷ $\sec$ ⌐sec⌐

▷ \setbox * 66, 70–71, 74, 216, 220, 224, 227. Puts stuff into box **registers**. See \copy for an example.

▷ $\setminus$ ⌐\⌐ (Class 2: binary operation.)

▷ \settabs 46–47. This offers an easy way to align material. For example,

```
{\settabs 3\columns
\+ One&Two&Three\cr
\+ &Four\cr}
```

produces

One	Two	Three
	Four	

▷ \seven... 56–57, 114–115, 149, 157, 158, 166, 222, 232. Seven-point fonts are assigned names like \sevenrm, etc., in Plain TeX. See §5.3 for a list of the 7-point fonts that are automatically loaded in Plain TeX.

▷ \sfcode * See \spacefactor.

▷ $\sharp$ ⌐♯⌐ (Class 0: ordinary.)

▷ $\shave$ *AMS* (*AMS*-TeX alert: don't \redefine.) Shaves off space from the top and bottom of large operators with limits so that **delimiters** around them, or radicals over them, are not too tall (the limits introduce extra space). For example,

\shipout

$$ \left(\shave{\sum_1^n} x_i \right). $$

gives

$$ \left(\sum_1^n x_i \right). $$

Space can be selectively shaved off the top or bottom by using \topshave or \botshave instead.

▷ \shipout * 229. The command that actually ships a page out into the ".dvi" world. See the \output routines in various files in Chapter 6.

▷ $\shortmid$ 𝒜𝓂𝒮 ⌈∣⌋ (Class 3: relation.)

▷ $\shortparallel$ 𝒜𝓂𝒮 ⌈∥⌋ (Class 3: relation.)

▷ $\shoveleft$ 𝒜𝓂𝒮 See \multline.

▷ $\shoveright$ 𝒜𝓂𝒮 See \multline.

▷ \sideset 𝒜𝓂𝒮 Allows the construction of symbols like

$$ {\sum}' \qquad {}^*\bigcup \qquad {}_\flat^\sharp{\prod}_\natural . $$

The input for this, including the definitions, was

```
\define\SumPrime{\sideset \and^{\prime} \to\sum}
\define\AstBigcup{\sideset^\ast\and \to\bigcup}
\define\MrMusical{\sideset_\flat^\sharp\and_\natural \to\prod}
$$ \SumPrime \qquad \AstBigcup \qquad \MrMusical. $$
```

Superscripts and subscripts may be added to these symbols in the usual way; for example,

```
$$ \SumPrime_{i=1}^n. $$
```

gives

$$ \sum_{i=1}^n{}'. $$

▷ σ ⌈σ⌋ 38–41, 98–99, 120–121, 126–127. ς produces ς.

▷ Σ ⌈Σ⌋ In Plain TeX ${\mit\Sigma}$ (\varSigma in 𝒜𝓂𝒮-TeX) produces Σ; ${\bf\Sigma}$ (\boldsymbol\Sigma in 𝒜𝓂𝒮-TeX) produces $\mathbf{\Sigma}$; and ${\tt\Sigma}$ produces Σ.

▷ $\sim$ ⌈∼⌋ 86–91, 124–127. (Class 3: relation.)

▷ $\simeq$ ⌈≃⌋ (Class 3: relation.)

▷ $\sin$ ⌈sin⌋ 60–65, 70–71, 82–83.

▷ $\sinh$ ⌈sinh⌋

▷ $\skew$ Makes adjustments of accents of accents. See spacing.

▷ \skewchar * 166–167, 219. When TeX positions an accent above a character, it makes a small spacing adjustment that depends on the font from which the character is taken. Information about the size of the adjustment is taken from the size data of a designated character in the font, named the \skewchar. Here are the Plain TeX choices of \skewchar for its math fonts:

```
\skewchar\teni='177   \skewchar\seveni='177   \skewchar\fivei='177
\skewchar\tensy='60   \skewchar\sevensy='60   \skewchar\fivesy='60
```

The first of these means, for example, that character number *'177* of the font \teni (the 10-point math italic font) is the designated \skewchar for the font.

Every new font is initially automatically assigned a default \skewchar when it is loaded. The default value is represented by \defaultskewchar, given the value -1 in Plain TeX (so that no spacing adjustment will be made for characters from those fonts whose \skewchar has not explicitly been changed).

▷ \skip * 230. Represents registers where gobs of glue may be stored. For example, '\skip0=15pt plus 3pt minus1pt' stores the specified (elastic) size of glue in the register \skip0. Then, '[\hskip\skip0]' will give '[]'.

▷ \sl 9, 44–45, 52–55, 92–93, 136, 139, 149, 150, 153, 159, 162, 163, 220. (𝒜ℳ𝒮-TeX alert: don't \redefine.) *Slanted type* results from saying {\sl Slanted type\/}; see \it for a discussion of \/. Also see §5.3 and §5.5 for a full discussion of how commands like \sl work. 𝒜ℳ𝒮-TeX does not permit the command in formulas: \slanted must be used instead.

▷ $\slanted$ 𝒜ℳ𝒮 Produces slanted characters in a formula, in the same way that \bold produces bold characters.

▷ \slash ⌈/⌋ Line breaks are permitted at this slash, as opposed to the one produced by just typing /.

▷ **small capitals** See capitals and small capitals, \sc and \smc.

▷ \smallbreak 32–33. Suggests a good page break point to TeX. If the program does not take the suggestion, it will leave a \smallskip worth of space there (unless it immediately follows a vertical space of a size already bigger than a \smallskip).

▷ $\smallfrown$ 𝒜ℳ𝒮 ⌈⌢⌋ (Class 3: relation.)

▷ $\smallint$ ⌈∫⌋ 120–121. (Class 0: ordinary.)

▷ $\smallmatrix$ 𝒜ℳ𝒮 Used for matrices within a line of text.

▷ \smallpagebreak 𝒜ℳ𝒮 126–127. Functions like Plain TeX's \smallbreak, but it also checks to see if the command is being issued in an appropriate place (i.e., in vertical mode).

▷ $\smallsetminus$ 𝒜ℳ𝒮 ⌈∖⌋ (Class 2: binary operation.)

▷ \smallskip 26–29, 32–33, 38–41, 46–47, 52–53, 62–63, 70–71, 90–91, 94–99, 114–115, 228. Gives a vertical space of a certain standard size. The relevant definitions are

$\smallsmile$

```
\newskip\smallskipamount % A new 'skip' register
\smallskipamount=3pt plus 1pt minus 1pt
\def\smallskip{\vskip\smallskipamount}
```

▷ $\smallsmile$ $\mathcal{A_M S}$ $\lceil\smile\rceil$ (Class 3: relation.)

▷ \smash 40–41, 66, 100–103, 128–129. \smash{stuff} will print stuff but will assign it zero height and depth in internal calculations. The \smash command is the opposite of \vphantom and is often used in conjunction with it to fool TeX. For example, \smash used by itself here causes the "overline" to be set very low (since TeX now thinks the expression has zero height): $\overline{\smash{\int_0^1 dx=1}}$. The input for this is

```
    ... $\overline{\smash{\int_0^1dx=1}}$. The input ...
```

More useful examples may be found on the pages listed at the top of the entry.

▷ $\smile$ $\lceil\smile\rceil$ (Class 3: relation.)

▷ \smc $\mathcal{A_M S}$ Produces capitals and small capitals in the AMS preprint style; other styles may not recognize the command.

▷ \snug $\mathcal{A_M S}$ Removes the extra space normally left after a formula, in situations where a snug fit is needed. For example, 'n\snug-bein' gives 'n-bein'.

▷ sp See keywords and units.

▷ $\sp$ Gives superscripts; see \sb.

▷ $\Sp$ $\mathcal{A_M S}$ Used for multiline upper limits over large operators; see that topic.

▷ \space Defined by \def\space{ }; i.e., it gives a blank space.

▷ \spacefactor * As TeX strings together horizontal material to make the one long line that it will later chop into pieces (and make a paragraph), it uses a quantity called the space factor—denoted here by f—to determine the size of an interword space. The normal value for f is 1000: this corresponds to using the standard interword glue from the current font. This value of 1000 is automatically assigned to f at the start, or whenever a math formula or a noncharacter box has been added to the list of composed horizontal material. If the value of f is different from 1000, then the interword space is calculated like this:

1. If $f \geq 2000$, the "extra space" designated in the font is added to the normal interword space.
2. The "interword stretch" for the font is then multiplied by $f/1000$, and the "interword shrink" by $1000/f$.

The notions of "interword space," "extra space," etc., are explained in §5.6. The space factor f takes on a value other than 1000 in one of two ways. It can be reset explicitly by typing '\spacefactor=*new value*'. The new value remains in effect until the next automatic resetting (as mentioned above).

The space factor value can also be made to change automatically in response to certain input by assigning appropriate "space factor codes" to characters. These codes are numbers that are attached to characters (pretty much as **character codes** and **category codes** are) and which are used in place of f: if the space factor code of a character is g, then g replaces f unless (a) $g = 0$, or (b) $f < 1000 < g$ (in which case f is set to 1000 again). Space factor codes are assigned to characters through the command `\sfcode`. Here are sample Plain TeX assignments:

```
\sfcode`\)=0    \sfcode`\'=0    \sfcode`\]=0
```

Additional assignments (like `\sfcode`\.=3000`) are also usually in effect, unless TeX is under the influence of the fixed-spacing command `\frenchspacing`. Most characters have an `\sfcode` of 1000, except for the 26 uppercase letters, which are assigned the value 999. (The non-standard value nullifies—see (b) above—the effect of the large space factor code of a punctuation mark, like a period, if it occurs right after an uppercase letter. That is why the spaces after initials in a name are not excessively large.)

▷ `$\spacehdots$` *AMS* (*AMS*-TeX alert: don't `\redefine`.) Allows the adjustment of the spacing of dots; see **matrices**.

▷ `$\spaceinnerhdots$` *AMS* (*AMS*-TeX alert: don't `\redefine`.) Allows the adjustment of the spacing of certain dots; see **matrices**.

▷ **spacing**   10, 20, 30, 156, 157, 159–162, 165, 169–172.   The discussion here will be centered on mathematics spacing. Spacing in math **mode** differs in very many ways from text spacing. For example, '(1,0)+(-1,1/2)=(0,1/2)' produces '(1,0)+(-1,1/2)=(0,1/2)', whereas '`$(1,0)+(-1,1/2)=(0,1/2)$`' produces '$(1,0) + (-1,1/2) = (0,1/2)$'.

The first part of the discussion will concentrate on how spacing adjustments may be made; the second part on explaining the underlying principles.

• *Adjusting spacing*:
The commands

$$\backslash!, \qquad \backslash,, \qquad \backslash>, \qquad \backslash;, \qquad \backslash\text{quad} \quad \text{and} \quad \backslash\text{qquad}$$

allow considerable control over spacing. The first gives a negative space, the others positive spaces of increasing lengths. The first four commands are discussed under **keywords**; `\quad` and `\qquad` are discussed in individual entries.

It is also possible to fool TeX by using grouping symbols; for example, '`$a,b$`' gives 'a,b', whereas '`$a{,}b$`' gives 'a,b'; or, '`$a:b$`' gives '$a : b$', but '`$a{:}b$`' gives '$a:b$'. Other ways to control horizontal spacing are discussed under `\mskip`.

The `\phantom` command and its variants allow further control over spacing in formulas. In particular, `\vphantom` in conjunction with `\smash` provides a powerful way to position symbols where they are wanted. The command `\phantom` may also be used to position indices. For example,

spacing

$R_{abc}{}^d$ is created by `$R_{abc}^{\phantom{abc}d}$`,

and

$R^{ab}{}_{cd}$ is created by `$R_{\phantom{ab}cd}^{ab}$`.

These effects can also be obtained somewhat more easily by fooling TEX; for example, the input for

$$R^{ab}{}_{cd},\ R^{abc}{}_d \text{ and } R^a{}_b{}^c{}_d.$$

is

`$$R^{ab}{}_{cd},\; R^{abc}{}_d\; {\rm and}\; R^a{}_b{}^c{}_d.$$`

Observe that the vertical positioning of indices in this display style is slightly different from that used in the text style expressions above the displayed line. Vertical positioning of indices (or of any material, even outside the mathematics mode) can be controlled using `\raise` and `\lower`. The positioning of accents on accents also requires some adjustments. These may be made using the `\skew` command. The amount of skew can be controlled by specifying a number; for example, the input for $\hat{\hat I}$, $\hat{\hat I}$, $\hat{\hat I}$ and $\hat{\hat I}$ is

`... the input for $\hat{\hat I}$, $\skew3\hat{\hat I}$,`
`$\skew5\hat{\hat I}$ and $\skew9\hat{\hat I}$ ...`

Also see `\rlap`, `\llap`, punctuation and `\mathsurround` for a little more information about spacing adjustments.

• *Principles of spacing*:

The space between two adjacent atoms in a formula is determined by their types. TEX chooses different spacings according to whether it thinks the atoms are ordinary (Ord), large operators (Op), binary operations (Bin), relations (Rel), opening (Open), closing (Close), punctuation (Punct), or subformulas sandwiched by delimiters (Inner). The other five types of atoms (Over, Under, Acc, Rad and Vcent) are treated as type Ord as far as spacing is concerned. The size of the spaces can be zero, thin, medium or thick. The last three of these are, respectively, precisely the spaces corresponding to `\thinmuskip`, `\medmuskip` and `\thickmuskip` (i.e., the spaces left by `\,`, `\>` and `\;`).

TEX's spacing choices are shown in this chart:

		Ord	Op	Bin	Rel	Open	Close	Punct	Inner
	Ord	zero	thin	(med.)	(thick)	zero	zero	zero	(thin)
	Op	thin	thin	*	(thick)	zero	zero	zero	(thin)
	Bin	(med.)	(med.)	*	*	(med.)	*	*	(med.)
Left	Rel	(thick)	(thick)	*	zero	(thick)	zero	zero	(thick)
atom	Open	zero	zero	*	zero	zero	zero	zero	zero
	Close	zero	thin	(med.)	(thick)	zero	zero	zero	(thin)
	Punct	(thin)	(thin)	*	(thin)	(thin)	(thin)	(thin)	(thin)
	Inner	(thin)	thin	(med.)	(thick)	(thin)	zero	(thin)	(thin)

(Right atom — column headers; Left atom — row headers)

Parentheses are used to indicate that the space is inserted in text and display styles, but not in script or scriptscript. An asterisk ($*$) indicates that the particular juxtaposition of atomic types does not arise in practice, because of an internal conversion that TEX makes when it reads a formula: a Bin atom is converted to type Ord if it is the first atom in a formula; or if it is preceded by any of these types: Bin, Op, Rel, Open, Punct; or if it is followed by any of these: Close, Punct, Rel.

▷ `$\spadesuit$` ⌈♠⌉ (Class 0: ordinary.)

▷ `$\spcheck$` $\mathcal{AMS}$ See accents.

▷ `\special` $*$ 104–105, 107, 115. Instructs the printer to do special things, like importing pictures from other programs. See diagrams.

▷ **special effects** It is possible to put symbols on other symbols. For example,

$$\overset{\mu,\nu}{\longmapsto} \qquad \texttt{\$\{\textbackslash buildrel\textbackslash mu,\textbackslash nu\textbackslash over\textbackslash longmapsto\}\$}$$

$$\overset{\text{def}}{\propto} \qquad \texttt{\$\{\textbackslash buildrel\textbackslash rm def\textbackslash over\textbackslash propto\}\$}$$

$$\overset{\circ}{g}_{\mu\nu} \qquad \texttt{\$\{\textbackslash buildrel\textbackslash circ\textbackslash over g\}\_\{\textbackslash mu\textbackslash nu\}\$}$$

$\mathcal{AMS}$-TEX offers even more control through `\overset` and `\underset`.

Another command, `$\joinrel$`, allows relation symbols to be joined. This is how several of the relation symbols available in TEX are put together. For example, the symbol ⟷ is given by the command `$\longleftrightarrow$`, defined by

    \def\longleftrightarrow{\leftarrow\joinrel\rightarrow}

In other words, the symbol is created by joining ← and →.

Next, here is a useful new command, `\Stacksymbols`:

```
\def\Stacksymbols #1#2#3#4{\def\theguybelow{#2}
    \def\verticalposition{\lower#3pt}
    \def\spacingwithinsymbol{\baselineskip0pt\lineskip#4pt}
    \mathrel{\mathpalette\intermediary#1}}
\def\intermediary#1#2{\verticalposition\vbox{\spacingwithinsymbol
        \everycr={}\tabskip0pt
        \halign{$\mathsurround0pt#1\hfil##\hfil$\crcr#2\crcr
                \theguybelow\crcr}}}
```

The command allows the construction of many composite symbols. It is set up so that the symbols will automatically become smaller if used in indices, and it allows the fine-tuning of spacing. For example,

$$a \qquad \underset{\tilde{}}{g} \qquad \overset{g}{\tilde{}} \qquad b$$

is obtained from this input:

```
$$a\qquad
\Stacksymbols{g}{\tilde{}}{8}{1}\qquad
\Stacksymbols{g}{\tilde{}}{-1}{4}\qquad b$$
```

The first two arguments in \Stacksymbols are the two symbols that are to go one over the other. The next argument controls the overall vertical position, and the last the spacing within the composite symbol.

Finally, it is possible to fill a horizontal space entirely with certain types of material: see \hrulefill, \dotfill, \leftarrowfill and \rightarrowfill. Other space-filling patterns can be obtained by using \leaders.

▷ **special symbols (text)**

œ	\oe	æ	\ae	å	\aa	ø	\o	ł \l
Œ	\OE	Æ	\AE	Å	\AA	Ø	\O	Ł \L
†	\dag	‡	\ddag	§	\S	¶	\P	ß \ss
ı	\i	ȷ	\j	...	\dots	©	\copyright	

There is also a £ symbol, obtained from {\it\$}. Additional text symbols are shown under **AMS symbols**. Symbols only for use in mathematics are shown either under symbol type (**Greek letters**, etc.) or under **symbols**.

▷ $\sphat$ $\mathcal{A_{M}S}$ See **accents**.

▷ $\sphericalangle$ $\mathcal{A_{M}S}$ ⌜◁⌟ (Class 0: ordinary.)

▷ $\split$ $\mathcal{A_{M}S}$ A displayed formula that is divided over many lines is conventionally assigned an equation number differently depending on whether it is numbered on the left (the number is then placed to the left of the first line) or on the right (the number is placed to the right of the last line of the display). Since the \tag mechanism places equation numbers automatically on the left or the right, based on the **document** style being used, it is helpful if the vertical position of the equation number also shifts automatically. The \split command makes this happen. For example,

$$\split A&=B+C \\ &=b+c-d-e. \endsplit\tag12 $$

(note that \tag is placed after \endsplit) produces

$$(12) \qquad \begin{aligned} A &= B + C \\ &= b + c - d - e. \end{aligned}$$

If some other document style is later imposed that places equation numbers on the left, the equation number above will automatically also move to the second line. It is also possible to use \split inside another alignment and have everything align nicely:

$$\align A&=B \tag1 \\ \split C&=abcd \\ &=\frac{x-y}{z-t} \endsplit\tag2 \\ G_i&=0. \endalign $$

produces

$$(1) \qquad A = B$$
$$(2) \qquad C = abcd$$
$$= \frac{x-y}{z-t}$$
$$G_i = 0.$$

▷ $\spreadlines$ ₳ₘₛ The command is permitted only within a display, and it spreads out (vertically) the lines of certain multiline structures in that display. The structures affected by it are those built by \align, \aligned and \gather. For example, saying \spreadlines3pt after the opening $$ signs in a display will add 3 pt to the interline spacing. It is helpful, for the sake of the overall consistency of the document, if the size of the extra space is given in terms of \jot (see the entry under that command for more details). Also see \vspace.

▷ $\spreadmatrixlines$ ₳ₘₛ Similar to \spreadlines, but used to spread the lines of a matrix or of a structure built by \cases.

▷ $\spvec$ ₳ₘₛ See accents.

▷ $\sqcap$ ⌐⊓⌐ 68–69. (Class 2: binary operation.)

▷ $\sqcup$ ⌐⊔⌐ 68–69. (Class 2: binary operation.)

▷ $\sqrt$ 16–17, 72–73, 100–101, 104–105, 118–121, 166, 167. $\sqrt{a^2+b^2}$ gives $\sqrt{a^2+b^2}$. Also see \radical.

▷ $\sqsubset$ ₳ₘₛ ⌐⊏⌐ (Class 3: relation.)

▷ $\sqsubseteq$ ⌐⊑⌐ (Class 3: relation.)

▷ $\sqsupset$ ₳ₘₛ ⌐⊐⌐ (Class 3: relation.)

▷ $\sqsupseteq$ ⌐⊒⌐ (Class 3: relation.)

▷ $\square$ ₳ₘₛ ⌐□⌐68. (Class 0: ordinary.) Also see box operator.

▷ \ss ⌐ß⌐84–85, 90–91, 155.

▷ s-size An ₳ₘₛ-TₑX term for "script size." Also see \ssize.

▷ $\ssize$ ₳ₘₛ (₳ₘₛ-TₑX alert: don't \redefine.) Forces s-size (script size) on an expression. For example, $A^{\frac12}$ gives $A^{\frac12}$, but the input $\ssize A^{\frac12}$ gives $A^{\frac12}$.

▷ ss-size An ₳ₘₛ-TₑX term for "scriptscript size." Also see \sssize.

▷ $\sssize$ ₳ₘₛ 124–125. Forces ss-size (scriptscript size) on an expression.

▷ $\star$ ⌐⋆⌐ (Class 2: binary operation.)

▷ \string * 106–107, 122–123. Causes the next token to appear verbatim: {\tt\string\bye} produces \bye.

▷ \strut 74. Props apart lines that might otherwise be too close together. The definition is

```
\newbox\strutbox
\setbox\strutbox=\hbox{\vrule height8.5pt depth3.5pt width0pt}
\def\strut{\relax\ifmmode\copy\strutbox\else\unhcopy\strutbox\fi}
```

If a switch is made to a typeface size other than 10 points, then `\strutbox` also needs suitable modification. This is illustrated in § 6.2.

▷ `\strutbox` 220. See `\strut`.

▷ **style** See styles in mathematics.

▷ **style file** See document style.

▷ **styles in mathematics** 164, 171–172. Mathematical expressions are typeset in one of eight styles: four basic ones and four variations. The four basic styles are display, text, script and scriptscript. Each has an associated "cramped" style, where the exponents are positioned slightly lower than in the normal style. The style determines the typeface used and influences the positioning of indices.

Formulas set in text (entered between $ signs) appear in text style; those that are displayed (entered between $$ signs) appear in display style. Indices and fractions in the formula can cause temporary style changes, as can accents, the use of `sqrt` and `\overline`. The rules are given below, using the following abbreviations: D (display), T (text), S (script), SS (scriptscript), D', T', S' and SS' (the last four are the cramped styles).

- *Indices*:

Original style	Superscript	Subscript
D, T	S	S'
D', T'	S'	S'
S, SS	SS	SS'
S', SS'	SS'	SS'

- *Fractions*:

Original style	Numerator	Denominator
D	T	T'
D'	T'	T'
T	S	S'
T'	S'	S'
S, SS	SS	SS'
S', SS'	SS'	SS'

- *Other changes*:

Accents, `\sqrt` and `\overline` cause a switch (if needed) to cramped styles.

The typefaces used in the different styles are these: text size (i.e., `\textfont`) in styles D, D', T and T'; script size (i.e., `\scriptfont`) in styles S and S'; scriptscript size in styles SS and SS'. The exact positioning of indices, etc., is determined by the style, the values of various `\fontdimen` parameters, and a set of carefully laid out rules. These rules are all explicitly given in Appendix G of *The TEXbook* (KNUTH [1984, 1986]).

▷ **subscripts** See indices and spacing. Also see §5.6 for a discussion of height adjustment.

▷ $\subset$ ⌐$\subset$⌐ 26–29, 36–37, 56–57, 76–79, 86–87. (Class 3: relation.)

▷ $\Subset$ *AMS* ⌐$\Subset$⌐ (Class 3: relation.)

▷ $\subseteq$ ⌐$\subseteq$⌐ 96–97, 100–101. (Class 3: relation.)

▷ $\subseteqq$ *AMS* ⌐$\subseteqq$⌐ (Class 3: relation.)

▷ $\subsetneq$ *AMS* ⌐$\subsetneq$⌐ (Class 3: relation.)

▷ $\subsetneqq$ *AMS* ⌐$\subsetneqq$⌐ (Class 3: relation.)

▷ $\succ$ ⌐$\succ$⌐ (Class 3: relation.)

▷ $\succapprox$ *AMS* ⌐$\succapprox$⌐ (Class 3: relation.)

▷ $\succcurlyeq$ *AMS* ⌐$\succcurlyeq$⌐ (Class 3: relation.)

▷ $\succeq$ ⌐$\succeq$⌐ (Class 3: relation.)

▷ $\succnapprox$ *AMS* ⌐$\succnapprox$⌐ (Class 3: relation.)

▷ $\succneqq$ *AMS* ⌐$\succneqq$⌐ (Class 3: relation.)

▷ $\succnsim$ *AMS* ⌐$\succnsim$⌐ (Class 3: relation.)

▷ $\succsim$ *AMS* ⌐$\succsim$⌐ (Class 3: relation.)

▷ $\sum$ ⌐$\sum$⌐ in a line, ⌐$\sum$⌐ in a display. 26–33, 38–39, 62–63, 68–71, 80–81, 90–95, 98–99, 102–103, 116–119, 164. (Class 1: large operator.) It is sometimes necessary to set a symbol, like a prime, against a summation symbol; for example:

$$\sum{}'$$

AMS-TEX has a general mechanism for this (see \sideset), but in Plain TEX it takes a little work. Because of the normal positioning of limits on a sum, getting the prime in the right place is tricky; typing $$\sum'.$$, for example, gives

$$\sum{}'.$$

One way out is to define \sump by typing

    \def\sump{\mathop{{\sum}'}}

(this takes advantage of the fact that {\sum} is not regarded as a large operator, even though \sum is). The weakness here is that limits may now not be centered to the satisfaction of everybody: $$\sump_{i=1}^N$$ will give

$$\sum_{i=1}^{N}{}'$$

for example, and so some further spacing adjustments may be required.

▷ **summation** See \sum.

▷ $\sup$ ⌈sup⌋ 26–27, 78–79, 118–119.

▷ **superscripts** See indices and spacing.

▷ $\supset$ ⌈⊃⌉ 76–77, 86–87, 124–125. (Class 3: relation.)

▷ $\Supset$ *AMS* ⌈⋑⌉ (Class 3: relation.)

▷ $\supseteq$ ⌈⊇⌉ (Class 3: relation.)

▷ $\supseteqq$ *AMS* ⌈⊇⌉ (Class 3: relation.)

▷ $\supsetneq$ *AMS* ⌈⊋⌉ (Class 3: relation.)

▷ $\supsetneqq$ *AMS* ⌈⊋⌉ (Class 3: relation.)

▷ $\surd$ ⌈√⌉ (Class 0: ordinary.)

▷ $\swarrow$ ⌈↙⌉ (Class 3: relation.)

▷ **symbols**

ℵ	$\aleph$	∀	$\forall$	∃	$\exists$
′	$\prime$	ℏ	$\hbar$	∅	$\emptyset$
∇	∇	¬	$\neg$	√	$\surd$
ı	$\imath$	ȷ	$\jmath$	ℓ	ℓ
♭	$\flat$	♯	$\sharp$	♮	$\natural$
⊤	$\top$	⊥	$\bot$	\	$\backslash$
ℜ	$\Re$	ℑ	$\Im$	∂	∂
∞	∞	∠	$\angle$	△	$\triangle$
♣	$\clubsuit$	◇	$\diamondsuit$	♡	$\heartsuit$
♠	$\spadesuit$	…	$\ldots$	⋯	$\cdots$
℘	$\wp$	·	$\cdotp$	.	$\ldotp$
⋮	$\vdots$	⋰	$\ddots$		

Also see AMS symbols for several other special symbols.

▷ $\syntax$ *AMS* If this is put in after the $\documentstyle$ declaration, $\mathcal{AMS}$-TeX will only check the syntax of the document and not produce any pages. See $\printoptions$.

▷ **tabbing** See $\cleartabs$ and $\settabs$.

▷ $\tabskip$ ∗ Sets the gap between columns in an alignment. The normal value is just zero, but that can be changed. For example, $\tabskip=3cm$ will push all columns (in an alignment made with $\halign$) apart by 3 cm.

▷ $\tag$ *AMS* 116–119. ($\mathcal{AMS}$-TeX alert: don't $\redefine$.) This is $\mathcal{AMS}$-TeX's command for placing equation numbers. The position of the equation number (i.e., right or left) depends on the **document style** being used. Here are several examples of the use of $\tag$:

    $$ A=B. \tag1 $$

gives

$$ (1) \qquad\qquad\qquad\qquad A = B. $$

```
$$ \align A&=x+y-z \tag4a \\ B&=xyz. \tag4b \endalign $$
```

gives

$$ A = x + y - z \tag{4a}$$
$$ B = xyz. \tag{4b}$$

```
$$ \aligned X&=A+B^2 \\ Y&=0. \endaligned \tag3--2 $$
```

gives

$$ X = A + B^2 \tag{3-2}$$
$$ Y = 0.$$

```
$$ \gather \Gamma=\alpha^2 \tag$A$ \\
   \Phi=\rho-\pi \tag$A\ast$ \\ \Psi=0. \endgather $$
```

gives

$$ \Gamma = \alpha^2 \tag{A}$$
$$ \Phi = \rho - \pi \tag{A*}$$
$$ \Psi = 0.$$

```
$$ \xalignat 2 A&=B & C&<D \tag"\bf[1]" \\
   F&=a^2+b^2 & G&=3x-2y+4z \endxalignat $$
```

gives

$$ \begin{array}{ll} A = B & C < D \\ F = a^2 + b^2 & G = 3x - 2y + 4z \end{array} \tag{[1]}$$

These examples illustrate the different contexts and the different ways in which \tag can be used. The style of the equation number (in addition to its position) is determined by the document style being used. That style can be overridden by using double quotes, as in the last example. It is also possible to refer to equation numbers in text and have the standard style automatically applied: \thetag{12} produces (12). Tags are treated as text; if there is frequent need to treat them as mathematics (as in the A, $A*$ example above), the command \TagsAsMath automatically changes the style globally. The original style can be returned to by saying \TagsAsText.

The AMS preprint style puts equation numbers on the left, but provides the commands \TagsOnRight and \TagsOnLeft, which can be used to switch positions. It is a good idea to select one position at the start, make the appropriate declaration (if needed), then stick to it. The style file also provides \CenteredTagsOnSplits to center tags (vertically) on split formulas, and \TopOrBottomTagsOnSplits to return to the default (see \split).

▷ \TagsAs... 𝒜ℳ𝒮 See \tag.

▷ \TagsOn... 𝒜ℳ𝒮 See \tag.

▷ $\tan$ ⌈tan⌋

▷ $\tanh$ ⌈tanh⌋

▷ τ ⌈τ⌋ 15, 70–71, 110–111, 118–121.

▷ $\tbinom$ 𝒜ℳ𝒮 Like \binom but it automatically uses text size.

▷ \ten... 149–150, 154–156, 158–159, 161, 163, 165–171, 178–191, 220, 222, 223, 227. (𝒜ℳ𝒮-TEX alert: don't \redefine.)

　　Ten-point fonts are assigned names like \tenrm, etc, in Plain TEX. See § 5.3 for a list of the 10-point fonts that are automatically loaded in Plain TEX.

▷ \tenpoint 𝒜ℳ𝒮 (𝒜ℳ𝒮-TEX alert: don't \redefine.) A command that comes with the AMS preprint style that allows an explicit request to be made for 10-point type (as opposed to only having it as the default size). This allows a switch back to 10 points if, for some reason, a different size is needed for part of the document.

▷ \TeX ⌈TEX⌋ The definition of the command is

```
\def\TeX{T\kern-.1667em \lower.5ex\hbox{E}\kern-.125em X}
```

The positions of the characters are chosen to be pleasing for the roman fonts, but they are not always satisfactory in others: for example, TEX looks a little awkward. Small adjustments can be made as needed, however, without too much trouble. Using the command in math mode gives *TEX* (the 'E' is in an \hbox and thus insulated from the math italics).

▷ $\text$ 𝒜ℳ𝒮 120–125. Allows the insertion of up to one line of text in a formula. For example,

```
$$ x^2 \ge 0 \quad\text{if $x$ is a real number} $$
```

gives

$$ x^2 \geq 0 \quad \text{if } x \text{ is a real number} $$

and

```
$$ Y^2=0 \quad\text{if and only if}\quad Y=0. $$
```

gives

$$ Y^2 = 0 \quad \text{if and only if} \quad Y = 0. $$

See \foldedtext for a description of putting longer pieces of text into equations.

▷ **text style** The style used automatically in the main part (i.e., not in indices and so forth) of formulas in text. See styles in mathematics and \textstyle

▷ \textfont * 158–159, 163, 165, 166–169, 172, 220, 221. The font assigned to expressions that are to appear in text and display styles. See § 5.5 and styles in mathematics.

▷ \textindent 40–41. Places material in the paragraph indentation.

▷ $\textstyle$ ∗ 16–17, 26–27, 34–35, 38–39, 62–63, 70–73, 78–83, 90–91. Produces the style normally used for mathematical expressions set in text. The command is used to override automatic style choices (for example, in a displayed equation). Also see **styles in mathematics**.

▷ **tfm** 141, 142, 148, 155, 158, 164–166, 170, 175, 176, 188. See §5.1.

▷ $\tfrac$ 𝒜ℳ𝒮 118–119. Used like \frac, but it automatically gives text-size fractions.

▷ \the ∗ 58, 74, 84, 171, 216, 222, 223, 225–228, 230–232. Extracts the contents of **registers**.

▷ $\therefore$ 𝒜ℳ𝒮 ⌈∴⌉ (Class 3: relation.)

▷ θ ⌈θ⌉ 24–27, 34–35, 100–101. ϑ produces ϑ.

▷ Θ ⌈Θ⌉ In Plain TeX ${\mit\Theta}$ (\varTheta in 𝒜ℳ𝒮-TeX) produces Θ; ${\bf\Theta}$ (\boldsymbol\Theta in 𝒜ℳ𝒮-TeX) produces **Θ**; and ${\tt\Theta}$ produces θ.

▷ \thetag 𝒜ℳ𝒮 See \tag.

▷ $\thickapprox$ 𝒜ℳ𝒮 ⌈≈⌉ (Class 3: relation.)

▷ $\thickfrac$ 𝒜ℳ𝒮 $\thickfrac{a+b}{c-d}$ gives $\frac{a+b}{c-d}$. The thickness of the fraction line can be varied at will; for example:

$\thickfrac\thickness{2.5}{a+b}{c-d}$ gives $\frac{a+b}{c-d}$.

The number after \thickness represents a multiple of the normal thickness of the fraction rule.

▷ $\thickfracwithdelims$ 𝒜ℳ𝒮 Has the same relationship to \thickfrac that \fracwithdelims does to \frac.

▷ $\thickmuskip$ ∗ One of three math glue-specifiers that are built into TeX. Plain TeX makes the following specifications:

```
\thickmuskip=5mu plus 5mu
\def\;{\mskip\thickmuskip}
```

The quantity **mu** is discussed under **keywords**.

▷ $\thickness$ 𝒜ℳ𝒮 See \thickfrac.

▷ $\thicksim$ 𝒜ℳ𝒮 ⌈∼⌉ (Class 3: relation.)

▷ \thickspace 𝒜ℳ𝒮 (𝒜ℳ𝒮-TeX alert: don't \redefine.) Gives a thick space; for example, [\thickspace] gives [ ]; '\;' may be used as an abbreviation. The command may also be used within a formula.

▷ $\thinmuskip$ ∗ One of three math glue-specifiers that are built into TeX. Plain TeX makes the following specifications:

```
\thinmuskip=3mu
\def\,{\mskip\thinmuskip}
```

The quantity **mu** is discussed under **keywords**.

▷ \thinspace 14–15, 58–59, 216, 219. Gives a thin space; for example, [\thinspace] gives []. In $\mathcal{AMS}$-TeX, '\,' may be used as an abbreviation in math mode and outside.

▷ \tie $\mathcal{AMS}$ May be used in place of ~ to give ties.

▷ **ties** 10. Blocks line breaks.

▷ \$\tilde\$ 82–83, 94–95. \$\tilde a\$ produces $\tilde{a}$.

▷ \$\Tilde\$ $\mathcal{AMS}$ See \accentedsymbol.

▷ \tildeaccent $\mathcal{AMS}$ Alternate name for \~.

▷ \$\times\$ $\lceil\times\rfloor$ 34–35, 40–41, 54–57, 76–79, 122–123, 128–129. (Class 2: binary operation.)

▷ **to** 66, 70–71, 74, 104–105, 216, 224, 226, 228. See keywords.

▷ \$\to\$ $\lceil\to\rfloor$ 26–31, 34–35, 38–39, 46–47, 50–51, 54–57, 78–81, 92–95. (Class 3: relation.) Alternate name for \rightarrow. The command also does special duty in $\mathcal{AMS}$-TeX when used in conjunction with \overset, \oversetbrace, \sideset, \underset and \undersetbrace.

▷ **tokens** The basic unit of TeX currency. When TeX reads a file it first converts the input into *tokens*—units that are not further broken down at that stage. These tokens are then passed on to lower stages of the program for further processing. A token is either a single character, with an associated category code, or the name of a command. For example, the input let \$a={1\over 2}\$ gets converted to the following sequence of tokens (category codes of characters are shown here as subscripts):

$$l_{11}\quad e_{11}\quad t_{11}\quad \sqcup_{10}\quad \$_3\quad a_{11}\quad =_{12}\quad \{_1\quad 1_{12}\quad \backslash over\quad 2_{12}\quad \}_2\quad \$_3$$

▷ \toks * Represents registers where lists of tokens may be stored. For example, \toks255={\sl A list} stores \sl A list in the register called \toks255. Then, {\the\toks255} gives *A list*.

▷ \tolerance * TeX's tolerance for white space, allowing hyphenation. The Plain TeX value is 200. Also see \pretolerance.

▷ \$\top\$ $\lceil\top\rfloor$ (Class 0: ordinary.)

▷ \$\topaligned\$ $\mathcal{AMS}$ 120–121. See \aligned.

▷ \topcaption $\mathcal{AMS}$ See \botcaption.

▷ \$\topfoldedtext\$ $\mathcal{AMS}$ See \foldedtext.

▷ \topglue 24–25, 231. Leaves space at the top of a page.

▷ \topinsert Like \midinsert, except that it first tries to insert material at the top of the current page; if there is no room, it places the material at the top of the next available page.

▷ \topmark * 230. See \mark.

▷ \TopOrBottomTagsOnSplits $\mathcal{AMS}$ See \tag.

▷ \$\topshave\$ $\mathcal{AMS}$ See \shave.

▷ \topsmash 𝒜ℳ𝒮 (𝒜ℳ𝒮-TEX alert: don't \redefine.) Like \smash, but it only smashes the height of its argument (i.e., assigns it zero height).

▷ $\triangle$ ⌈△⌉ (Class 0: ordinary.)

▷ $\triangledown$ 𝒜ℳ𝒮 ⌈▽⌉ (Class 0: ordinary.)

▷ $\triangleleft$ ⌈◁⌉ 96–97. (Class 2: binary operation.)

▷ $\trianglelefteq$ 𝒜ℳ𝒮 ⌈⊴⌉ (Class 3: relation.)

▷ $\triangleq$ 𝒜ℳ𝒮 ⌈≜⌉ (Class 3: relation.)

▷ $\triangleright$ ⌈▷⌉ (Class 2: binary operation.)

▷ $\trianglerighteq$ 𝒜ℳ𝒮 ⌈⊵⌉ (Class 3: relation.)

▷ true 220. See keywords.

▷ t-size An 𝒜ℳ𝒮-TEX term for text size. Also see \tsize.

▷ $\tsize$ 𝒜ℳ𝒮 Forces t-size (text size) on an expression.

▷ \tt 9, 84–85, 106–109, 122–123, 149, 163, 184–185, 216, 218, 220, 223. Typewriter-like type is obtained by typing {\tt Typewriter-like}. This is the only fixed-width face (i.e., each character has the same width) that automatically comes with Plain TEX, and it is suitable for various kinds of verbatim reproduction of the contents of files. See §5.5 for a full discussion of how commands like \tt work.

▷ $\twoheadleftarrow$ 𝒜ℳ𝒮 ⌈↞⌉ (Class 3: relation.)

▷ $\twoheadrightarrow$ 𝒜ℳ𝒮 ⌈↠⌉ (Class 3: relation.)

▷ typefaces The topic is discussed in detail Chapter 5. Here are a few very general comments on typefaces in formulas. The standard typeface for mathematical expressions is called *math italic*. There are also typefaces for mathematical symbols. The typefaces come in different sizes: text style for ordinary expressions that are in the middle of text; display style for expressions that are displayed on a separate line; and script style and scriptscript style for indices. TEX picks styles for the different terms in an expression on its own, but users can also issue instructions to override these choices by typing \textstyle, \scriptstyle, etc. A calligraphic typeface (see \cal) is also available. Packages like 𝒜ℳ𝒮-TEX offer styles beyond these (see, for example, \Bbb).

▷ $\ulcorner$ 𝒜ℳ𝒮 ⌈⌜⌝⌋167. (Class 4: opening.)

▷ \undefine 𝒜ℳ𝒮 If \newsymbol is being used to assign a name that is already in service, the old name must first be undefined. For example, if the angle symbol that comes with the 𝒜ℳ𝒮-TEX fonts is preferred over Plain TEX's symbol (see \angle), one types

      \undefine\angle
      \newsymbol\angle 105C

and the deed is done.

▷ Under See atom.

▷ $\underarrow$ 𝒜ℳ𝒮 Alternate name for \underrightarrow.

▷ $\underbrace$ Covers an expression with a brace; for example, the input $$A=\underbrace{a\times \cdots \times a}_{k\;\rm factors}.$$ gives

$$A = \underbrace{a \times \cdots \times a}_{k \text{ factors}}.$$

▷ $\underleftarrow$ 𝒜ℳ𝒮 $\underleftarrow{\text{Long arrows may be put under stuff.}}$ Here's how this was done:

$\underleftarrow{\hbox{Long arrows may be put under stuff.}}$

▷ $\underleftrightarrow$ 𝒜ℳ𝒮 Used like \underleftarrow.

▷ $\underline$ * (𝒜ℳ𝒮-TeX alert: don't \redefine.) Underlines formulas; e.g., $\underline{e^{i\pi}+1=0}$ yields $\underline{e^{i\pi} + 1 = 0}$.

▷ $\underrightarrow$ 𝒜ℳ𝒮 (𝒜ℳ𝒮-TeX alert: don't \redefine.) Used like \underleftarrow.

▷ \underscore 𝒜ℳ𝒮 Alternate name for _.

▷ $\underset$ 𝒜ℳ𝒮 $\underset\Lambda\to {ab}$ gives $\underset{\Lambda}{ab}$.

▷ $\undersetbrace$ 𝒜ℳ𝒮 The input

$\undersetbrace \text{$n$ factors}\to {A\times\dots\times A}$

will give $\underbrace{A \times \ldots \times A}_{n \text{ factors}}$.

▷ \unhbox * 74, 216, 224. Extracts the contents of a box **register** and strips off the outer level of hbox boxing.

▷ **units** Spacing commands like \hskip have to be given in units that TeX can understand. Here is a complete list:

inch (in)	
centimeter (cm)	2.54 cm = 1 in
point (pt)	72.27 pt = 1 in
pica (pc)	1 pc = 12 pt
big point (bp)	72 bp = 1 in
millimeter (mm)	10 mm = 1 cm
didot point (dd)	1157 dd = 1238 pt
cicero (cc)	1 cc = 12 dd
scaled point (sp)	65536 sp = 1 pt

TeX converts all units to sp and computes sizes using this unit. Since 1 sp $\approx 5 \times 10^{-7}$ cm, it is possible to position material in an extremely fine manner. TeX will not admit sizes $\geq 2^{30}$ sp (≈ 575 cm). Also see em and ex.

If the document has an overall magnification, sizes must be specified with care. If, say, it is needed that 3cm be the size of some quantity, one must type '3 true cm' in order to escape the effect of the magnification.

▷ \unskip * 227. Removes the immediately preceding horizontal **glue** (if any).

▷ \unvbox * 216. Extracts the contents of a box **register** and strips off the outer level of vbox boxing.

▷ $\uparrow$ ⌜↑⌝ (Class 3: relation.) Also see **delimiters**.

▷ $\Uparrow$ ⌜⇑⌝ (Class 3: relation.) Also see **delimiters**.

▷ \upbracefill 102–103. Horizontal spaces may be filled with an "upbrace": \hbox to 1in{\upbracefill} gives ⏜ Example 9 in Chapter 2 gives another illustration. Also see \downbracefill.

▷ $\updownarrow$ ⌜↕⌝ (Class 3: relation.) Also see **delimiters**.

▷ $\Updownarrow$ ⌜⇕⌝ (Class 3: relation.) Also see **delimiters**.

▷ $\upharpoonleft$ $\mathcal{AMS}$ ⌜↿⌝ (Class 3: relation.)

▷ $\upharpoonright$ $\mathcal{AMS}$ ⌜↾⌝ (Class 3: relation.) Also called \restriction.

▷ $\uplus$ ⌜⊎⌝ (Class 2: binary operation.)

▷ \uppercase * 221, 228, 232. '\uppercase{rd}' gives 'RD'.

▷ $\uproot$ $\mathcal{AMS}$ Used along with \root to position the root vertically.

▷ υ ⌜υ⌝

▷ Υ ⌜Υ⌝ In Plain TeX ${\mit\Upsilon}$ (\varUpsilon in $\mathcal{AMS}$-TeX) gives Υ; ${\bf\Upsilon}$ (\boldsymbol\Upsilon in $\mathcal{AMS}$-TeX) gives $\boldsymbol{\Upsilon}$; and ${\tt\Upsilon}$ gives ϒ.

▷ $\upuparrows$ $\mathcal{AMS}$ ⌜⇈⌝ (Class 3: relation.)

▷ $\urcorner$ $\mathcal{AMS}$ ⌜⌝⌝168. (Class 5: closing.)

▷ \UseAMSsymbols $\mathcal{AMS}$ 116, 133, 135, 169. This allows access to all the names of the new symbols that come in the AMS symbols fonts, saving users from defining each individually through \newsymbol. The **AMS preprint style** has the command built in.

▷ $\varDelta$ $\mathcal{AMS}$ ⌜Δ⌝ See \Delta.

▷ ε ⌜ε⌝ 94–95. ϵ produces ϵ.

▷ $\varGamma$ $\mathcal{AMS}$ ⌜Γ⌝ See \Gamma.

▷ $\varinjlim$ $\mathcal{AMS}$ ⌜$\varinjlim$⌝

▷ $\varkappa$ $\mathcal{AMS}$ ⌜$\varkappa$⌝ (Class 0: ordinary.)

▷ $\varLambda$ $\mathcal{AMS}$ ⌜Λ⌝ See \Lambda.

▷ $\varliminf$ $\mathcal{AMS}$ ⌜$\varliminf$⌝

▷ $\varlimsup$ $\mathcal{AMS}$ ⌜$\varlimsup$⌝

▷ $\varnothing$ $\mathcal{AMS}$ ⌜$\varnothing$⌝76–77. (Class 0: ordinary.)

▷ $\varOmega$ $\mathcal{AMS}$ ⌜Ω⌝ See \Omega.

▷ φ ⌜φ⌝ 18–19, 118–119. ϕ produces ϕ.

▷ $\varPhi$ $\mathcal{AMS}$ ⌜Φ⌝ See \Phi.

▷ ϖ ⌜ϖ⌝ π produces π.

▷ $\varPi$ $\mathcal{AMS}$ ⌜Π⌝ See \Pi.

▷ $\varprojlim$ $\mathcal{AMS}$ ⌜$\varprojlim$⌝

▷ $\varpropto$ $\mathcal{A}_{\mathcal{M}}\mathcal{S}$ $\lceil \propto \rfloor$ (Class 3: relation.)

▷ $\varPsi$ $\mathcal{A}_{\mathcal{M}}\mathcal{S}$ $\lceil \Psi \rfloor$ See \Psi.

▷ ϱ $\lceil \varrho \rfloor$ ρ produces ρ.

▷ ς $\lceil \varsigma \rfloor$ σ produces σ.

▷ $\varSigma$ $\mathcal{A}_{\mathcal{M}}\mathcal{S}$ $\lceil \Sigma \rfloor$ See \Sigma.

▷ $\varsubsetneq$ $\mathcal{A}_{\mathcal{M}}\mathcal{S}$ $\lceil \subsetneq \rfloor$ (Class 3: relation.)

▷ $\varsubsetneqq$ $\mathcal{A}_{\mathcal{M}}\mathcal{S}$ $\lceil \subsetneqq \rfloor$ (Class 3: relation.)

▷ $\varsupsetneq$ $\mathcal{A}_{\mathcal{M}}\mathcal{S}$ $\lceil \supsetneq \rfloor$ (Class 3: relation.)

▷ $\varsupsetneqq$ $\mathcal{A}_{\mathcal{M}}\mathcal{S}$ $\lceil \supsetneqq \rfloor$ (Class 3: relation.)

▷ ϑ $\lceil \vartheta \rfloor$ 70–71. θ produces θ.

▷ $\varTheta$ $\mathcal{A}_{\mathcal{M}}\mathcal{S}$ $\lceil \Theta \rfloor$ See \Theta.

▷ $\vartriangle$ $\mathcal{A}_{\mathcal{M}}\mathcal{S}$ $\lceil \triangle \rfloor$ (Class 3: relation.)

▷ $\vartriangleleft$ $\mathcal{A}_{\mathcal{M}}\mathcal{S}$ $\lceil \triangleleft \rfloor$ (Class 3: relation.)

▷ $\vartriangleright$ $\mathcal{A}_{\mathcal{M}}\mathcal{S}$ $\lceil \triangleright \rfloor$ (Class 3: relation.)

▷ $\varUpsilon$ $\mathcal{A}_{\mathcal{M}}\mathcal{S}$ $\lceil \Upsilon \rfloor$ See \Upsilon.

▷ $\varXi$ $\mathcal{A}_{\mathcal{M}}\mathcal{S}$ $\lceil \Xi \rfloor$ See \Xi.

▷ \vbox ∗ 66, 98–99, 104–105, 107, 110–113, 216, 224, 226, 228. ($\mathcal{A}_{\mathcal{M}}\mathcal{S}$-TeX alert: don't \redefine.) Allows material to be put in a "vertical box." A principal practical use of the command is to keep material (like a table) on a single page (page breaks are not permitted within a \vbox).

▷ **Vcent** See atom.

▷ $\vcenter$ ∗ 66, 100–101. A mathematics mode command rather like \vbox or \vtop, except that it centers the constructed box vertically. For example,

One   Three
Two   Four
        Five

is produced by typing

```
\medskip
\hbox{$\vcenter{\hsize 1in \noindent One\hfil\break Two}$
        $\vcenter{\hsize 1in \noindent Three\hfil\break
                Four\hfil\break Five}$}
\smallskip
```

▷ \vcorrection $\mathcal{A}_{\mathcal{M}}\mathcal{S}$ Adjusts vertical page position; e.g., \vcorrection{1in} moves the entire printed page down an inch on the paper. The command is defined by

```
\def\vcorrection#1{\advance\voffset#1\relax}
```

▷ $\vdash$ $\lceil \vdash \rfloor$ 126–127. (Class 3: relation.)

▷ $\vDash$ $\mathcal{A}_{\mathcal{M}}\mathcal{S}$ $\lceil \vDash \rfloor$ (Class 3: relation.)

▷ $\Vdash$ $\mathcal{A}_{\mathcal{M}}\mathcal{S}$ $\lceil \Vdash \rfloor$ (Class 3: relation.)

▷ $\vdots$ ⌈⋮⌉ 40–41, 100–103, 120–121, 128–129, 226.

▷ $\vec$ 166. $\vec a$ produces $\vec a$.

▷ $\Vec$ 𝒜ℳ𝒮 See \accentedsymbol.

▷ $\vee$ ⌈∨⌉ 82–83, 124–125. (Class 2: binary operation.) Alternate name: \lor.

▷ $\veebar$ 𝒜ℳ𝒮 ⌈∨̲⌉ (Class 2: binary operation.)

▷ **verbatim** The verbatim, or literal, reproduction of input text is achieved by switching off the special category codes assigned to characters like \, $, etc. Commands that do this are shown in Chapter 6. It is commonly asked if the contents of an entire file can be used as verbatim input; the answer is "yes." One way to do this is to use one of the standard verbatim commands, but to add to it the statement \catcode`\+=10, where it is assumed that + does not occur in the file that is to be used as input. (If it does, any other character that does not occur may be used instead.) The verbatim command being used will have chosen some temporary escape character, say @. Then, the line

```
@input+filename
```

within the scope of the verbatim command will do the trick.

▷ $\vert$ ⌈|⌉ 16–17, 26–27, 34–39, 62–63, 76, 94–95, 98–99, 102–103, 118–123. (Class 0: ordinary.) (𝒜ℳ𝒮-TEX alert: don't \redefine.) Alternate name for |.

▷ $\Vert$ ⌈‖⌉ 36–37, 44–45, 76, 94–95, 120–121. (Class 0: ordinary.) Alternate name for \|.

▷ \vfil * 66, 108–109, 216, 219, 224, 225, 228, 232. Fills a vertical region with white space.

▷ \vfill * 66, 228. Like \vfil, only stronger.

▷ \vfootnote Normal footnotes in TEX are tied to particular pieces of text. The \footnote command marks the text to which the footnote is being attached, then uses the same marker on the footnote. That is, of course, how one wants things. However, the command does not work from deep within complicated constructions (like formulas within formulas) or from within inserted material (like a \topinsert). The \vfootnote command accommodates such situations by providing a footnote mechanism not tied to any piece of text. The command is placed somewhere on the page where it is guessed that the relevant text will fall, and that text is independently marked with the same marker that is used with \vfootnote.

▷ $\vmatrix$ 𝒜ℳ𝒮 See matrices.

▷ $\Vmatrix$ 𝒜ℳ𝒮 See matrices.

▷ \voffset * 58, 74, 228. Sets the overall vertical position of the page.